Finding Time for the Old Stone Age

A History of Palaeolithic Archaeology and Quaternary Geology in Britain, 1860–1960

ANNE O'CONNOR

OXFORD

UNIVERSITY PRESS

OXFORD
UNIVERSITY PRESS

Great Clarendon Street, Oxford ox2 6DP

Oxford University Press is a department of the University of Oxford.
It furthers the University's objective of excellence in research, scholarship,
and education by publishing worldwide in

Oxford New York

Auckland Cape Town Dar es Salaam Hong Kong Karachi
Kuala Lumpur Madrid Melbourne Mexico City Nairobi
New Delhi Shanghai Taipei Toronto

With offices in

Argentina Austria Brazil Chile Czech Republic France Greece
Guatemala Hungary Italy Japan Poland Portugal Singapore
South Korea Switzerland Thailand Turkey Ukraine Vietnam

Oxford is a registered trade mark of Oxford University Press
in the UK and in certain other countries

Published in the United States
by Oxford University Press Inc., New York

British Library Cataloguing in Publication Data

Data available

Library of Congress Cataloging-in-Publication Data
O'Connor, Anne.
Finding time for the old stone age : a history of Palaeolithic
archaeology and quaternary geology in Britain, 1860–1960 / Anne O'Connor.
p. cm.
ISBN 978–0–19–921547–8
1. Paleolithic period—Great Britain. 2. Archaeological
geology—Great Britain—History. 3. Geology, Stratigraphic—Quaternary.
4. Tools, Prehistoric—Great Britain. 5. Antiquities,
Prehistoric—Great Britain. I. Title.
GN772.22.G7027 2007
936.1—dc22   2007012021

Typeset by SPI Publisher Services, Pondicherry, India
Printed in Great Britain
on acid-free paper by
Biddles Ltd., King's Lynn, Norfolk

ISBN 978–0–19–921547–8

1 3 5 7 9 10 8 6 4 2

For my family

Acknowledgements

This book has been helped on its way to publication by the kindness of numerous people over the past six years. I am particularly grateful to my brother, Ralph O'Connor, at Aberdeen University, for his encouragement. The committee and members of the History of Geology Group replenished my motivation through their heartfelt appreciation of the history of geology; as did Jim Secord, Anne Secord, and Martin Rudwick at the University of Cambridge. I am deeply indebted to my reviewers—Marianne Sommer, at the Swiss Federal Institute for Technology, Zurich; and David Oldroyd, at the University of New South Wales, Australia—for going over the manuscript so thoroughly and making so many valuable suggestions. The manuscript also gained clarity from a patient perusal by my father, John O'Connor; and I am grateful to my sister, Clémence O'Connor, and my good friend, Anne-Louise Russell, for checking my French translations. This book would not have appeared were it not for the open-minded attitude of my editor, Hilary O'Shea at Oxford University Press, when first faced with the unconventional miscellany in my manuscript; and I am much obliged to the staff at OUP for their support thereafter.

I have not forgotten the thoughtful advice given back in 2000 by Derek Roe, then at the University of Oxford, when these ideas were first taking shape; or the generosity of John McNabb, at Southampton University, in sharing his own research on my visit that same year. For friendly discussions about the history of archaeology, thanks are due to Peter Rowley-Conwy, Margarita Diaz-Andreu, and Richard Hingley, all at Durham University; Roger Jacobi at the British Museum; Raf de Bont at the Katholieke Universiteit Leuven; and C. Stephen Briggs. Phil Howard, at Durham University, kindly explained to me the complexities of mapping software; whilst Jeff Veitch, at the same institution, contributed his photographic skills. The archaeological core of this book comes from my Ph.D. thesis, which was funded by the AHRC.

I have experienced valuable assistance and congenial conversation amongst the archives explored in this work, but particular thanks are due to: Jeannine Alton, Stella Brecknell, Eliza Howlett, and Jim Kennedy at the Oxford University Museum of Natural History; Graham McKenna, Gill Nixon, and Joan Bird at the British Geological Survey Library; Susan Snell, Polly Tucker, and Andrew Currant at the Natural History Museum; Susanne Bangert at the Ashmolean Museum, University of Oxford; Claire Freeman and Jeremy Coote at the Pitt Rivers Museum, Oxford; Wendy Cawthorne at the Geological Society of London; and the staff in the Rare Books Room and the Department of Manuscripts, Cambridge University Library.

The author would like to thank the following for permission to reproduce published material: the University of Aberdeen (Figs. 4.4 and 4.5); *Anthropos* (Fig. 8.11); Figure 5.9 is reproduced by permission of The Geologists' Association from *Proceedings of The Geologists' Association*, A. O'Connor, 'Geology, Archaeology and "the raging vortex of the eolith controversy"', 114, pp. 255–262, fig. 3 © 2003 The Geologists' Association; and by permission of the Prehistoric Society. Figure 5.10 is reproduced by permission of Blackwell Publishing from the *Journal of the Royal Anthropological Institute*: S. H. Warren, 'The Experimental Investigation of Flint Fracture and its Application to Problems of Human Implements', 44, pp. 412–450, fig. 11.

Images and extracts from archives are reproduced by permission of University of Cambridge Museum of Archaeology & Anthropology (Fig. 10.3: W21/1/3); the Ashmolean Museum, University of Oxford; the BGS Library Archive (Figs. 2.5, 2.7, 8.2); the Bodleian Library, University of Oxford; the British Library; Derbyshire County Council: Buxton Museum and Art Gallery (wbd colln); the College Archives, Imperial College London; the Institute of Archaeology, University College London (Fig. 10.1); the Institute of Archaeology, University of Oxford (jacket picture and Fig. 5.2); Llyfrgell Genedlaethol Cymru / The National Library of Wales; the National Portrait Gallery, London (Fig. 5.4); the Pitt Rivers Museum, University of Oxford (Figs. 5.1, 5.3, 5.6, 6.5, 8.9); the Prehistoric Society (Figs. 5.9, 6.9, 6.10, 8.10, 9.2, 9.5, 9.6); Salisbury & South Wiltshire Museum; the Society of Antiquaries of London (Figs. 8.1, 8.6); the Syndics of Cambridge University Library (Figs. 1.1, 6.4, 7.2, 9.1, 9.3, 10.4); Edinburgh University Library, Special Collections Department (Fig. 2.1); the Oxford University Museum of Natural History; Figures 6.2 and 6.7 © Copyright the Trustees of The British Museum. Many thanks to James Dyer and Derek Roe for supplying the images for Figures 3.7 and 5.2 respectively.

Images and extracts from archives are reproduced courtesy of the Geological Society of London (Figs. 1.2, 1.4, 1.5, 2.3, 2.4, 5.5, 5.7, 6.3, 6.6, 8.3, 8.4, 8.5, 8.12, 10.2); Moray Council Museums Service (Fig. 1.3); and by courtesy of the University of Liverpool Library (Fig. 7.1). All material from the NHM archives is used here by permission of the Trustees of the Natural History Museum, London (including Figure 9.4, © The Natural History Museum, London). Archive material is used courtesy of the Sedgwick Museum of Earth Sciences, University of Cambridge.

Permission to quote from letters has kindly been given by Francis Burkitt; A. N. Greenwell; Jane Fawcett (for the papers of T. McKenny Hughes); Nina

Laborde (for the papers of E. Ray Lankester); the Executors of the Leakey Estate; Lord Lyell; Erik T. Paterson (for Fig. 10.3); Charles A. I. Wood; and the nieces of Donald Baden-Powell. A reasonable effort has been made to trace copyright holders; if, however, anyone has further information that may require a change to subsequent editions, please contact the author through Oxford University Press.

Contents

List of Figures

List of Tables

Introduction

Ancient tool-makers roamed Britain long ago, chipping stones into useful implements. Their discarded handiwork lay sealed in gravels, sands, clays, and cave earth for millennia. Though occasionally retrieved by later inhabitants, it was not until the middle of Queen Victoria's reign that the dwellers of southern England began to recover these prehistoric relics in large numbers from busy gravel pits and brickfields, their eyes sharpened by the recent acceptance of human antiquity. From the 1860s onward, the once-abandoned artefacts were seized once again, this time by wine merchants, diamond merchants, draughtsmen, clerks, officers of the Geological Survey, university professors, and museum curators. This book follows the efforts they made over the next hundred years, between c.1860 and c.1960, to arrange these tools in sequences through time and connect them to the great changes they thought had afflicted the landscape, sediments, climate, and animals of ancient Britain.

The human antiquity debates of the late 1850s and the geological sequences that were developed earlier in the nineteenth century have received detailed attention from historians of science. The tool-makers have been the subject of several authoritative books on the history of research on fossil hominids and human evolution (palaeoanthropology).[1] But the old stone tools of the Palaeolithic period and their treatment after the antiquity debates had subsided are more of a mystery; the classifications and chronological sequences that were offered over the following century are still obscure; the individuals who made these suggestions have been largely forgotten. It is often noted that they drew inspiration from geological zone-fossils in their early work, seeing in the series of distinctive, fossilized animals a model for the arrangement of equally distinctive stone tools. The influences that their Palaeolithic sequences later exerted on geological interpretations are, however, rarely mentioned. Nonetheless, the complicated, uneasy, and often remarkable relationships that developed between these researchers as they worked on the links between

[1] On the events leading to the acceptance of human antiquity, see Gruber 1965 and Riper 1993. On chronologies of earth and human history, see Grayson (1983), Rupke 1983, Rudwick (1986, 2005), and Lewis and Knell (2001). The history of palaeoanthropology has been examined by Oakley 1964a, Spencer 1984; 1990, Bowler 1986, Regal 2004, and Sommer 2005.

geological and Palaeolithic patterns, offer an intriguing subject for historical enquiry.

In 1859, when papers were read to learned societies in London on the great antiquity of humankind, stone tools were finally accepted to date back to Quaternary times. This period, which was so ancient in terms of human history, was very recent in terms of geology. It was only in the previous few decades that geologists had taken much interest in these 'recent' or 'superficial' sediments, and there was confusion about their origin and sequence. Many of those who became interested in the new problem of finding patterns through time in the stone tools had also been working on geological questions: the character of the Glacial epoch (or Ice Age), the age of extinct fauna, and the connection between river and glacial sediments. Indeed, a large number would have described themselves as 'geologists' rather than 'archaeologists'. The less loaded term 'Palaeolithic researcher' is used here for those with a particular interest in the stone tools.[2] The geologists—those who defended their opinions about the date and sequence of sediments or bones—are sometimes distinguished as glacial geologists, stratigraphical geologists or palaeontologists, when discussing their interpretations of these different branches of geology.

This study of British research is centred on English deposits because stone tools peter out in northern counties. Many of the connections between Palaeolithic and geological sequences took place in southern England, near the richest sources of stone tools. Nonetheless, glacial research in Scotland and in the caves of Wales is also explored and the influence of Continental research is examined. The earlier tools of the Palaeolithic period (those belonging to the 'Lower' and 'Middle' Palaeolithic of today's archaeologists) supply the main focus for this book. This emphasis follows a distinction common in the nineteenth century, when researchers tended to restrict the term 'Palaeolithic' to the 'Drift' or 'Mammoth' period, and assign later tools (of today's 'Upper' Palaeolithic) to the 'Cave' or 'Reindeer' period (see Chapter 4). It should also be noted that this book is concerned with the search for relative time, rather than the years of absolute time.

Four sources of geological time have been already been mentioned: glacial deposits, animal bones, river drifts, and Palaeolithic implements. All four need further introduction. Glacial geology is given considerable attention in these

---

[2] Riper 1993, 205–207 has drawn a distinction between 'geologists' who busied themselves with chronological questions, and 'geological archaeologists' who adopted a more cultural and ethnographic perspective of the British Palaeolithic. These expressions have not been adopted here because the researches of many geologists discussed in the course of this book were driven by a wider spectrum of concerns and conflicts. Some of the most violent arguments in twentieth-century British Palaeolithic research emerged from controversial *combinations* of chronological and cultural arguments, proposed by geologists and palaeolithic researchers alike.

chapters because it provided past researchers with two important time-markers. In the nineteenth century, the idea of a great submergence of the land beneath the sea—a submergence that divided pre-glacial from post-glacial times—often entered discussions about the relationship between the Glacial epoch, the sediments of the Stone Age, and the bones of ancient animals. If a few gaunt islands sticking out of a cold, desolate sea were all that might have been seen of Britain during the mid-glacial submergence, then this had important implications for its inhabitants. As Andrew Crombie Ramsay (1814–1891), famed for his work on glacial geology, remarked in 1860, 'When the country was down 2,300 feet of course there were no elephants on the little islands that now form our mountain tops'.[3] In the twentieth century, when these ideas of a great submergence had subsided, interglacials (lengthy warm phases) provided researchers with finer sub-divisions of the Glacial epoch. The glacial–interglacial sequence became a popular time-scale for dating stone tools and linking layers of sediment (or 'strata') from different sites into a broader stratigraphical picture. East Anglia, rich in glacial deposits, attracted those who sought to make such connections.

The patterns seen by palaeontologists in animal bones sometimes conflicted with the sequences claimed by their peers from stratigraphy (the order and relative position of the strata). In the 1870s, for example, William Boyd Dawkins (1837–1929), an outspoken expert in fossil mammals, became embroiled in argument with James Geikie (1839–1915), a glacial geologist. Geikie, who supported the controversial theory of interglacial phases, attacked Dawkins, who favoured the post-glacial age of deposits in caves and river drifts, 'on grounds partly geological, partly zoological & partly common-sense-ical' (Chapter 2).[4] Dawkins had come to his conclusions about the date of these deposits from a palaeontological perspective. Their disagreement peaked in 1881, when Geikie exploded that Dawkins was 'a nincompoop in physical geology'.[5] Similarly, in the twentieth century, Martin A. C. Hinton (1883–1961), who worked on voles and lemmings, and his friend Alfred Santer Kennard (1870–1948), who studied shells, maintained a monoglacialist standpoint that looked increasingly obsolete as the multiplying interglacial episodes gained respectability. In 1920, Kennard advised Miles Burkitt (1890–1971), the prehistorian: 'You give up the multiple Ice Age theories & stick to our views & you will be right'.[6]

Sequences of ancient river drifts, built up from spreads of gravel and the rich-brown loam (known as brickearth) that lined the sides of river valleys,

---

[3] Ramsay to Lyell, 27 March 1860: ULE: Gen 115/5037–5038.
[4] J. Geikie to Ramsay, 12 December 1871: ICL: KGA/Ramsay 8/412/3.
[5] J. Geikie to Peach, 17 February 1881: BGS: GSM1/321/68.
[6] Kennard to Burkitt, 13 April 1920: ULC: 7959, Box 3.

provided another way to date Palaeolithic tools. The Thames Valley became a famous source of stone tools and river drifts alike, and the task of arranging the deposits in order attracted the efforts of stratigraphical geologists, like Joseph Prestwich (1812–1896) and William Whitaker (1836–1925). They also tried to link these river drifts to the glacial chronology; but the relationship was unclear and provoked many arguments—partly because the glacial drifts lay much further north, in East Anglia. Dawkins, for example, argued in the 1860s that Prestwich and Whitaker had been mistaken in their connections between certain river drifts and the Glacial epoch. His opinions, founded on the character of the Thames Valley fauna, did not match their conclusions, founded on stratigraphy. Another geologist, Searles V. Wood Jr. (1830–1884), disagreed with Dawkins and confided to a friend that he had little confidence in the fieldwork skills of this palaeontological expert.[7] These nineteenth-century arguments about geological patterns are introduced in Chapters 1 and 2. The connections between river-drift and glacial sequences, and between the Thames Valley and East Anglia, would continue to trouble twentieth-century researchers.

Chapters 3 and 4 add the Palaeolithic to this broad picture of geological time. Certain aspects of Stone-Age research receive particular attention. My discussion of nineteenth-century classifications looks beyond the famous divisions and sequences developed by John Lubbock (1834–1913) and Gabriel de Mortillet (1821–1898) to observe the opinions of John Evans (1823–1908) and some little-known researchers working around London: Worthington George Smith (1835–1917), John Allen Brown (1831–1903), and Flaxman C. J. Spurrell. Their relationships with curio-collectors, forgers, and workmen are also examined. The debates over the Eolithic or Pre-Palaeolithic stones from Kent and East Anglia, scrutinized in Chapters 5 and 6, offer a different perspective on expectations of the Palaeolithic sequence. These stones (which today are recognized as natural products) were subjected to the same kinds of classification as true palaeoliths: 'tool-types' were identified and arranged in progressive sequences.

Another interesting development in the history of Stone-Age research occurred after tools from various different localities had been grouped together on the basis of their distinctive character into sets (as 'cultures' or 'industries'), arranged in order, and linked to periods of time. Despite the variety of Palaeolithic classifications and sequences suggested in the nineteenth century, it was not until the early twentieth century that British stone-tool industries received standard labels. Chapters 7 and 8 tell how they were

---

[7] Wood to Harmer, 13 January 1867: BGS: GSM1/542/52.

assigned more specific positions in relation to the chronologies introduced above: the sequences of glaciations, fauna, and river drifts.

This connection between the character of industries and their relative age had two far-reaching results. Both draw the geological themes mentioned above back into the story. First, the Palaeolithic sequence came to be used as another time-scale alongside glacial, faunal, and river-drift sequences, and industries then played a more prominent role in the struggles between geologists to defend their own correlations and chronologies. Second, Palaeolithic anomalies were recognised: assemblages of tools that could not be matched in both character *and* age to any of the slots on the accepted and expected sequence. Anomalies could be accommodated by modifying the old Palaeolithic sequence; but another solution was to reinterpret the age of the geological sediments in which they lay. Chapters 9 and 10 untangle the connections between changing expectations of the Palaeolithic sequence and the reshuffling of geological sequences, a particularly busy activity during the 1930s. The book draws to a close in the 1950s, when different questions about the Palaeolithic grew more popular after a spate of attacks on Palaeolithic classifications and chronologies: attacks which had also encouraged a split between geologists and Palaeolithic 'archaeologists'.

These, then, are the Palaeolithic and geological themes to be explored in the following chapters. But alongside these themes, the backgrounds and characters of the researchers themselves are also introduced. Their employment, personal networks of interaction, public forums for discussion, favoured arenas of publication, and styles of persuasion form an integral and interesting part of this tale. Some were employed by the Geological Survey of Great Britain; some worked in museums or universities; many others, however, did not work in an allied field. Prestwich was a wine-merchant. Hinton was a clerk in barristers' chambers. Kennard worked in the warehouse of a London firm. They and their peers became known in national and local societies, whose rooms and journals became battlegrounds for competing interpretations. These social dimensions of research moulded perceptions of, and arguments about, the ancient British past.

A variety of different sources has been used to recover these opinions: primary and secondary, published and unpublished. Primary published material includes major treatises, synthetic works, journal articles, reviews, newspapers, and accounts of meetings and exhibitions. A further level of detail is supplied by unpublished material: notebooks, photographs, maps, drawings, and private letters scattered in archives across the country.[8] The importance of published works, which contain the publicized opinions of the main

[8] For the archives used in this book, see the Manuscript Sources section in the Bibliography.

protagonists, is self-evident and needs little elaboration. But it is important, when using such sources, to be aware of the restrictions imposed on the style, format, and content of academic publications, which had to conform to traditions of scientific narrative to achieve plausibility. An example of these constraints is given in the following comment on the Reports of Field Meetings, published in the *Proceedings of the Geologists' Association* in 1954:

> The reports, while generally adequate on the geological side, tend to be somewhat inhuman; no-one, reading the staid and factual accounts would suspect that geologists could ever be guilty of 'letting their hair down'. As is truly stated, in the report of the Swiss meeting of 1947, on Sunday, 7 September, 'a lunch was provided at Guggisberg'. No mention is made of the highlight of the occasion, namely, that between courses, the party was regaled with a concert by Professor R. F. Rutsch on the concertina and Dr. K. Arbenz on the double bass, with songs by a bevy of the local beauties clad in their cantonal costumes, yet this was an outstanding feature of the meeting.[9]

The personal side of public debate that tended to be missing from published works was often revealed in private letters. James Geikie could describe the prestigious journal *Nature* as a one-sided 'kiss-my-arse of Macmillan & Co' with equanimity to a colleague on the Geological Survey, although he used what he described as more 'parliamentary language' in print.[10] Some researchers even broke out in occasional verse to describe and defend their opinions. But letters are amongst the most interesting and valuable sources of past interaction. Letters might give a friendly greeting, pass on an interesting slice of new information, or share opinion with more freedom and in more detail than was possible in published works. They could reinforce alliances, sustain antagonisms, and direct the strategies that might win over the opposition.

Letters also give an idea of the individuals behind academic debates: character can be glimpsed in style, tone, anecdotes, and handwriting. Personal idiosyncrasies survive better in letters than in print: Charles Lyell's robotic wish-lists of information required for his next edition; Hugh Falconer's script, curling like elephant tusks; Joseph Prestwich's massive, indecipherable scrawl; Benjamin Harrison's fountain-pen flourishes; James Geikie's ribald stories of pre-professorial years on the Survey; A. S. Kennard's rounded letters and lack of punctuation; Hazzledine Warren's upright hand and kindly courtesy; Henri Breuil's spiky, illegible hieroglyphs which drift to the right as he fills the page.[11]

---

[9] Himus 1954, 8.

[10] J. Geikie to Peach, 9 March 1881: BGS: GSM1/321/69; J. Geikie to Ramsay, 27 February 1881: ICL: KGA/Ramsay 8/412/15.

[11] All spellings, abbreviations and emphases reproduced here in quotations are original, unless otherwise stated; this avoids the disruptive 'sic,' and the repetitive 'original emphasis'. Emphasis expressed as underlining in the original is reproduced here using italics.

Secondary sources on the history of Palaeolithic research in Britain after
*c.*1860 are surprisingly sparse. The human antiquity debates are explored by
Bowdoin Van Riper, and the development of geological chronologies in
preceding decades is examined by Donald Grayson and, in more detail, by
Martin Rudwick.[12] The next phase, though, when Palaeolithic tools were
classified and connected to geological sequences, has attracted little notice
from historians of science. Historians of geology have devoted more time to
solid geology than to the superficial drifts of Palaeolithic times, and the earlier
nineteenth century has proved more popular than the later periods.[13] Historians of palaeoanthropology have examined theories about the tool-makers,
but are less interested in their tools.

The history of archaeology was, for a long while, dominated by the
pioneering books of Glyn Daniel; but his accounts of the history of Palaeolithic research are brief, and focus on the antiquity debates and later work in
nineteenth-century France.[14] The history of the French Palaeolithic has received some attention: Annette Laming-Emperaire explores the history of
prehistoric archaeology in France from medieval times to the end of the
nineteenth century; but the grand scope of her book leaves little space for a
detailed analysis of the last few decades. The life and work of Boucher de
Perthes (1788–1868) and his contribution to the human antiquity debates is
discussed by Claudine Cohen and Jean-Jacques Hublin, and by Noël Coye;
but, again, this is a mid-nineteenth-century story. James Sackett explores the
history of research on the French Upper Palaeolithic (or Reindeer period)
over a similar span of time to that covered here; but there is little overlap
between these cave-based researches in France and contemporary British
work on the older Palaeolithic of the river drifts and caves.[15]

Publications of practising scientists also touch on the history of geological
and Palaeolithic research. Recent geological texts, such as David Bridgland's
Quaternary of the Thames (1994), outline the geological history of classic
Quaternary areas, such as the Thames Valley and East Anglia, describe past
interpretations of major sites, and refer briefly to the attempts made in
the middle of the twentieth century to date deposits from the Palaeolithic

[12] Grayson 1983; Riper 1993; Rudwick 2005.
[13] The excellent account of the history of British geomorphology by Herries Davies 1969 ends
in 1878. Wilson 1985 offers an overview of the last 150 years of the Geological Survey. Oldroyd
1990 examines the social history of nineteenth-century geology with particular attention to the
work of the Geological Survey.
[14] Daniel 1975, 99–109 describes the researches of Édouard Lartet and Gabriel de Mortillet in
France in the 1860s and 1870s, but makes little mention of contemporary work on the British
Palaeolithic.
[15] Laming-Emperaire 1964; Cohen and Hublin 1989; Sackett 1981; 1991; Coye 2000.

sequence.[16] But little explanation is supplied about how these industries gained credibility as a chronological tool or about the relationships that developed between individuals working on these different time sequences; the names of past researchers are unaccompanied by information about their lives or how they persuaded others to adopt their beliefs. The same could be said of historical overviews produced by Palaeolithic archaeologists, such as Mark White and John McNabb. This is, of course, not surprising: archaeological publications and geological textbooks are produced to explain and describe current interpretations.[17]

It would, however, be unrealistic and unfair to draw a sharp boundary between science and the history of science. Many splendid works in the history of science have been written by scientific practitioners, such as George Stocking and Frank Spencer, and many historians of science have a background in science, such as Martin Rudwick. Nonetheless, there is still a strong feeling amongst many archaeologists writing on the history of their discipline that the main point of examining past research is to improve present scientific knowledge, and this brings a Whiggish tone to their publications. Historians of geology or palaeoanthropology might think that this horse has been flogged to death by now, but it still needs a good thrashing in archaeological circles.

The problems of histories of archaeology written by archaeologists have been analysed succinctly by Anders Gustafsson, who concludes his paper with the remark, 'it is satisfactory that the history of archaeology in the form of *archaeology* is bad as history. A good history of archaeology in the form of *history* would be unusable—at least directly—in contemporary archaeological practice'.[18] Past events are often arranged to fall into a progression of discoveries that seem to climb to the pinnacle of accurate modern opinions. Heroes tend to be identified and lauded for being the first to develop an advanced interpretation or approach on the basis of a superficial resemblance to a current theory or practice, often one that is close to the author's own stance in contemporary debate. Archives are used to answer the questions of today's researchers: the historical context of old notes, plans or letters might be acknowledged, but be valued most as a tool for distilling or translating useful scientific evidence from a biased or obscurely-expressed source.

Bruce Trigger has examined different trends in archaeological interpretation in his excellent overview, *A History of Archaeological Thought*, but there are surprisingly few detailed analyses of the development of particular

[16] Bridgland 1994, 20; Gibbard 1994, 162.
[17] McNabb 1996; White 2000.
[18] Gustafsson 1998, 292.

sub-disciplines—such as Palaeolithic archaeology—which draw on primary sources and avoid a Whiggish tone.[19] This gap is partly because the aims of the writers are not historical—either through choice, or through pressure to engage in current archaeological debate in today's specialized academic institutions where appointments, publications, and research funding (for university departments as well as for individuals) are governed by restrictive peer reviews. But another curious tone can be heard in historical publications by archaeologists, one that harmonizes with recent trends in archaeological theory.

In the last few decades, university-based archaeologists have been offered more motivation for exploring the history of archaeology. Since the 1980s there has been a growth in concern that interpretations of the archaeological past can, sometimes unwittingly, be twisted to reflect and justify present social and political circumstances. This concern gave historical studies some justification: as illustrations of the connection between the social context of research and interpretations of the past. However, the call for greater critical self-awareness, besides nurturing the historical researches of archaeologists, has also tended to restrict their gaze to the study of broad social and political biases in the works of their intellectual forebears—a stance which, ironically, justifies their own worries about archaeological methodology and the manipulation of the past. Meanwhile, the intellectual content of past interpretations has been left largely to archaeologists who specialize in today's equivalent of that past branch of research. They tend to produce histories that historians would describe as Whiggish, for reasons explained above. In summary, and to paraphrase Gustafsson, it is difficult to be a good archaeologist and also to do good history for historians.

Although it is reasonable to suggest that no one can step entirely outside the concerns of the present, and it is likely that many researchers of the past were indeed motivated by concepts of empire, colonialism, and other forms of social, political and economic control, and legitimation, it seems equally reasonable to suggest that more attention might be directed to other themes and explanatory notions less closely linked to current archaeological worries. In this book I have returned to the neglected middle ground that lies between these general studies of past social and political biases and the specific questions asked by the archaeologists and geologists of today. The approach has been inspired by publications on the history of geology written by historians of science; the content is presented as a history of science for any interested reader and is not aimed solely at practising scientists or historians of science.

---

[19] Trigger 1989.

No programmatic connections will be drawn between past interpretations and present research; this book will not be bringing the story up to the present day, but will end around 1960. There are two reasons why this date has been chosen. First, the mid twentieth century provides a convenient break: it was coloured by great changes in theory and methodology, and in the social and institutional practice of Palaeolithic research, changes that will be introduced in Chapter 10. The second reason lies in the practical problem of space: it would take at least another book to cover the shifting views of classifications and chronological sequences over the last half-century of research on the British Palaeolithic. The story of the preceding century of research, from *c.*1860 onwards, is, however, a fascinating one. But this will be a tale of conflicting beliefs, not of clear-sighted heroes fighting prejudiced villains. The question has not been 'who was right?' but 'why did they think that they were right?' and 'why did some interpretations enjoy greater popularity than others?' The answers require a journey into a world of forgotten researchers who developed and defended different interpretations of the sequence of events in Britain's ancient past, when tool-makers once wandered the country.

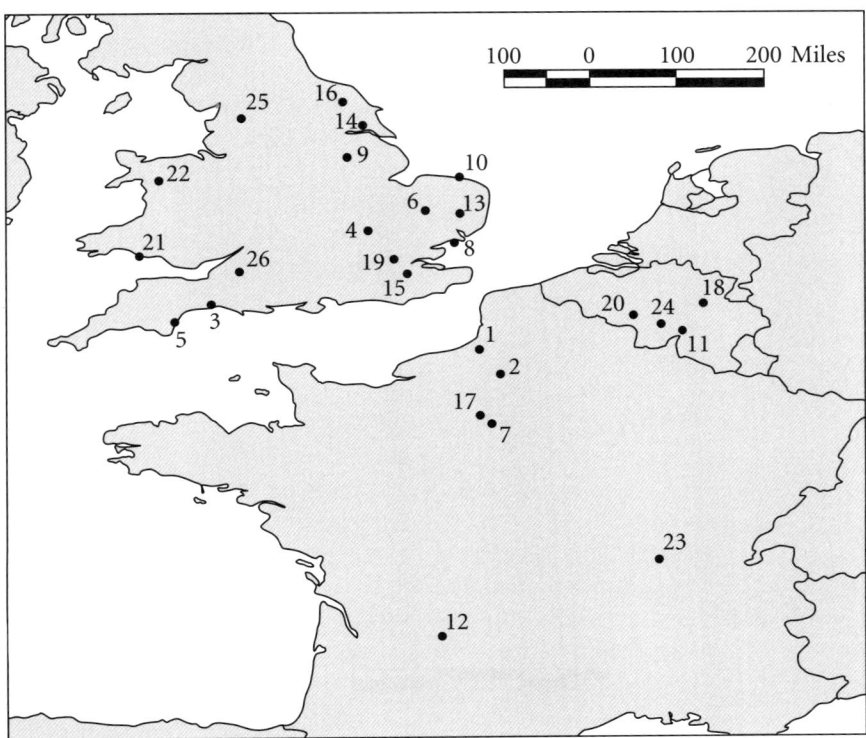

Map 1. Location of major Palaeolithic sites mentioned in the text: Britain, France and Belgium.

1 Abbeville, Somme Valley
2 Amiens, Somme Valley
3 Axminster, Devon
4 Bedford, Caddington and Dunstable, Bedfordshire
5 Brixham Cave and Kent's Cavern, Devon
6 Brandon, Suffolk
7 Chelles, Seine-et-Marne
8 Clacton-on-Sea, Essex
9 Creswell Crags, Derbyshire
10 Cromer, Norfolk
11 Dinant, Belgium
12 Dordogne, Aquitaine (La Madelaine, Le Moustier, Les Eyzies)

13 Hoxne, Suffolk
14 Hull
15 Ightham, Kent
16 Kirkdale Cave, Yorkshire
17 Levallois-Perret, Hauts-de-Seine
18 Liège
19 London (Baker's Hole, Hornchurch, Swanscombe)
20 Mesvin, Hainaut
21 Paviland Cave, Gower, Glamorganshire
22 Snowdonia, Wales
23 Solutré, Macon
24 Spy, Namur
25 Victoria Cave, Settle, Yorkshire
26 Wookey Hole, Somerset

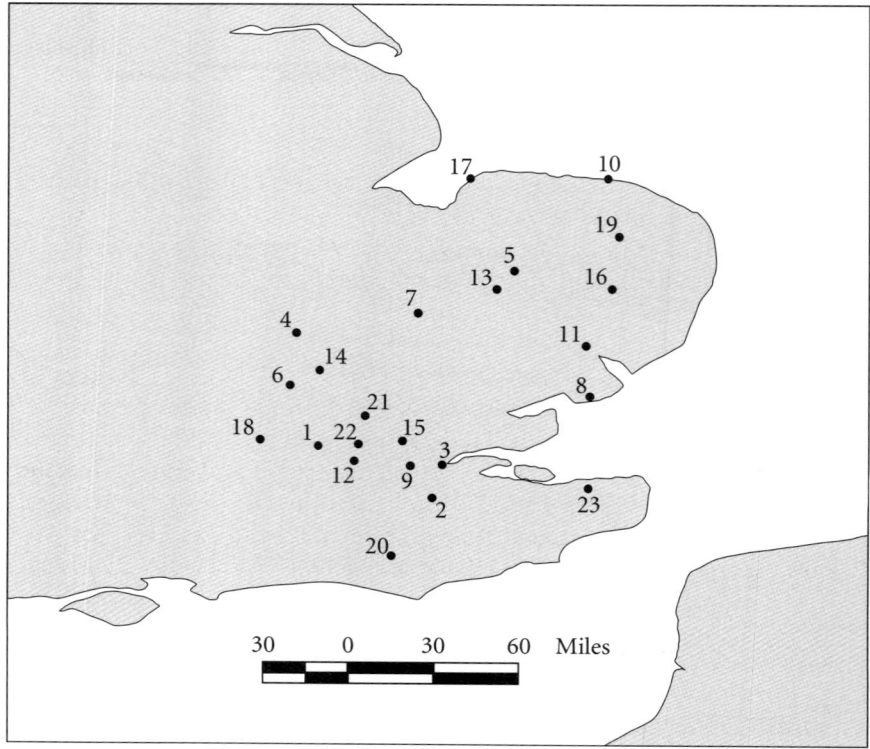

Map 2. Location of major Palaeolithic sites mentioned in the text: south-east England.

1 Acton and Ealing, London
2 Ash and Ightham, Kent
3 Baker's Hole, Gray's Thurrock, Northfleet and Swanscombe, London
4 Bedford
5 Brandon, Suffolk
6 Caddington and Dunstable, Bedfordshire
7 Cambridge
8 Clacton-on-Sea, Essex
9 Crayford, Dartford and Erith, London
10 Cromer, Norfolk
11 Foxhall Hall, Foxhall Road, Ipswich

12 Gray's Inn Lane, London
13 High Lodge, Suffolk
14 Hitchin, Hertfordshire
15 Hornchurch and Ilford, London
16 Hoxne, Suffolk
17 Hunstanton, Norfolk
18 Maidenhead
19 Norwich, Norfolk
20 Piltdown, Sussex
21 Ponder's End, Lea Valley, Middlesex
22 Stoke Newington, London
23 Sturry, Kent

Dramatis Personae

Many individuals contributed to research on the British Palaeolithic over the time-span covered by this book. Their number, and the occasional similarity of their names, might be confusing, so a brief description is given below for those who are to be most frequently mentioned in the following pages, with an indication of their role in the narrative.

Agassiz, Louis
: (1807–1873) Professor of Natural History in Switzerland, expert in fossil fishes and advocate of the land-ice theory.

Bagford, John
: (d. 1716) London bookseller and founder member of the Society of Antiquaries who, in 1715, pondered the age of an ancient stone implement.

Barrow, George
: (1853–1932) Survey geologist who objected to the Palaeolithic work carried out by his colleague, Dewey.

Boswell, Percy
: (1886–1960) Schoolteacher, later Professor of Geology, authority on the glacial deposits of East Anglia in the 1920s and 1930s.

Boule, Marcellin
: (1861–1942) French Professor of Palaeontology, Director of the Institute of Human Palaeontology, Paris, and eolith sceptic.

Breuil, Henri
: (1877–1961) French priest, Professor of Prehistoric Ethnography at the Institute of Human Palaeontology, Paris, and influential authority on the British Palaeolithic and Pre-Palaeolithic.

Brown, John Allen
: (1831–1903) Diamond merchant and geologist from Ealing who began working on the Palaeolithic drifts of London in the 1870s.

Buckland, William
: (1784–1856) Flamboyant geological lecturer at Oxford University who associated the drift with a deluge before adopting the land-ice theory.

Burkitt, Miles
: (1890–1971) Pupil of Breuil, later lecturer on prehistoric archaeology at Cambridge, who promoted Breuil's ideas about the Palaeolithic.

Chambers, Robert
: (1802–1871) Scottish author, folklorist, and editor who supported the land-ice theory.

Chandler, R. H.	Geologist and Palaeolithic researcher who worked with A. L. Leach on the Thames Valley drifts in the early twentieth century and observed the Swanscombe exposures through the 1920s and 1930s.
Christy, Henry	(1810–1865) Banker, collector, and excavator of the Dordogne caves with Édouard Lartet.
Clarke, W. G.	Reporter from Norwich and discoverer of a Pre-Palaeolithic type-fossil, the 'Norwich test specimen'.
Commont, Victor	(1866–1918) French science teacher, geologist, and palaeontologist who inspired British Palaeolithic classifications in the 1910s through his work in the Somme Valley.
Conyers, John	Apothecary and antiquarian of London in the late seventeenth and early eighteenth centuries, discoverer of the Gray's Inn Lane hand-axe.
Croll, James	(1821–1890) Scottish millwright, later Secretary on the Geological Survey, who turned to astronomy to explain glacial cycles.
Cuvier, Georges	(1769–1832) French palaeontologist who distinguished fossil from modern elephants.
Dawkins, William Boyd	(1837–1929) Palaeontologist, once of the Survey and later Professor of Geology at Manchester, who classified Quaternary mammals and criticized James Geikie's interglacial episodes.
Dewey, Henry	(1876–1965) Survey geologist and Palaeolithic researcher who worked at Swanscombe in the 1910s with Reginald Smith of the British Museum.
Dupont, Édouard	(1841–1911) Belgian geologist and cave-excavator who defined different stone ages by the presence of mammoth or reindeer.
Evans, John	(1823–1908) Manager of paper mills, numismatist, geologist, and expert in flint implements, who developed an early classification for the British Palaeolithic in the 1860s and later attacked the Kent eoliths.
Falconer, Hugh	(1808–1865) Surgeon, botanist, palaeontologist, catalyst for the human antiquity case of 1859, and expert in fossil elephant and rhinoceros who promoted a post-glacial Palaeolithic in the early 1860s.

Fisher, Osmond	(1817–1914) Reverend Fellow of Jesus College, Cambridge, who worked on the drifts and introduced the term 'Trail'.
Flint Jack	a.k.a. Edward Simpson, notorious forger of implements from the 1860s.
Frere, John	(1740–1807) Suffolk antiquary, discoverer of Palaeolithic tools at Hoxne in the late eighteenth century.
Garrod, Dorothy	(1892–1968) Palaeolithic archaeologist, pupil of Breuil, later Disney Professor of Archaeology at Cambridge who promoted parallel Palaeolithic cultures in the 1920s and advised different approaches to interpretation in the 1940s.
Geikie, Archibald	(1835–1924) Survey geologist, later Director-General of the Survey and Professor of Geology at Edinburgh, who described layers of peat in the Scottish boulder clays that were later identified as interglacial deposits.
Geikie, James	(1839–1915) Survey geologist, brother of Archibald, later Professor of Geology at Edinburgh, who admired Croll's work, defended interglacial episodes and criticized Dawkins's seasonal–migration theory.
Godwin, Harry	(1901–1985) Professor of Botany at Cambridge who used pollen to refine the date of Palaeolithic sites, dating the Clacton deposits in the early 1950s.
Godwin-Austen, R. A. C.	(1808–1884) Geologist who worked on the drifts of southern England and defended their pre-glacial date in the early 1860s.
Goodwin, A. John H.	(1900–1959) Archaeologist based at the University of Cape Town who worked on the Palaeolithic of South Africa from the 1920s and later criticized the industrial sequences of Europe.
Greenwell, William	(1820–1918) Durham clergyman, antiquary, collector, and archaeologist who had an interest in the Palaeolithic.
Harmer, Frederic	(1835–1923) Norfolk geologist from a Norwich woollen firm, later Lord Mayor of Norwich, who worked on the East Anglian drifts with Wood from the 1860s and encouraged Boswell's researches.

Harrison, Benjamin	(1837–1921) Diffident village shopkeeper of Ightham and discoverer of the Kent eoliths, which he promoted with the help of Prestwich.
Haward, Frederick N.	(1871–1953) Engineer and friend of Hazzledine Warren who studied the fracture of flint and opposed Reid Moir's pre-palaeoliths.
Hawkes, Christopher	(1905–1992) Archaeologist and member of the Swanscombe Research Committee who worked under Reginald Smith at the British Museum in the late 1920s and the 1930s.
Hicks, Henry	(1837–1899) Welsh doctor and geologist who worked in the caves of North Wales and promoted a pre-glacial Palaeolithic in the 1880s.
Hinton, Martin A. C.	(1883–1961) Clerk and expert in small fossil mammals, later Keeper of Zoology in the British Museum (Natural History) who worked with Kennard on the Thames Valley drifts in the early twentieth century.
Holmes, T. V.	(1843–1923) One-time Survey geologist whose descriptions of the Hornchurch section in the 1890s strengthened belief in the post-glacial date of the London drifts.
Hughes, T. McKenny	(1832–1917) Survey geologist, Professor of Geology at Cambridge from 1873, who defended the post-glacial date of the Palaeolithic.
Kennard, A. Santer	(1870–1948) Mollusc expert and monoglacialist who worked with Hinton in the early twentieth century on the geological history and Palaeolithic sequence of the Thames Valley.
King, William B. R.	(1889–1963) Professor of Geology at University College London who worked with Oakley in the 1930s on the geological stages and Palaeolithic sequence of the Thames Valley.
Lane Fox	See Pitt-Rivers.
Lankester, E. Ray	(1847–1929) Retired zoologist and prominent supporter of Moir's pre-palaeoliths in the 1910s and 1920s.
Lartet, Édouard	(1801–1871) French palaeontologist whose excavations in the Dordogne caves with Henry Christy in the early 1860s inspired British classifications of the Palaeolithic.

Leach, Arthur Leonard   (1869–1957) Schoolteacher and geologist who worked on the sequence of Thames deposits with Chandler in the early twentieth century.

Leakey, Louis   (1903–1972) Archaeologist and palaeontologist who worked on the Palaeolithic of the Thames Valley before marrying Mary Nicol and winning fame for hominid discoveries in East Africa.

Leakey, Mary   (1913–1996) Archaeologist who excavated at Clacton-on-Sea with Oakley in 1934 and later worked on the lengthy Palaeolithic sequence at Oldovai, East Africa. Criticized European industrial sequences.

Lowe, C. van Riet   (1894–1956) Professor of Archaeology at the University of Witwatersrand who worked on the Palaeolithic of South Africa and pointed out the differences to European patterns in the 1940s.

Lubbock, John   (1834–1913) MP and polymath, author of *Prehistoric Times* (1865).

Lyell, Charles   (1797–1875) Geologist and lawyer who wrote the influential synthesis: *The Geological Evidences of the Antiquity of Man* (1863).

Marr, John Edward   (1857–1933) Lecturer, later Professor of Geology at Cambridge who helped to ascertain the great geological age of Moir's pre-palaeoliths in 1911 and excavated at High Lodge in 1920.

Marston, Alvan T.   (1889–1971) Dentist from Clapham and discoverer of the Swanscombe skull.

Moir, James Reid   (1879–1944) Discoverer and defender of the East Anglian pre-palaeoliths who ran a tailoring business in Ipswich, worked with Boswell on the Palaeolithic, and excavated at Hoxne in the 1920s.

Morris, John   (1810–1866) Chemist, later Professor of Geology at Kings College, London, who worked on the drift fossils of the Thames Valley.

Mortillet, Gabriel de   (1821–1898) French political radical who worked at the Musée de Saint-Germain-en-Laye, founded the journal *Matériaux pour l'histoire positive et philosophique de l'homme* in 1864, and popularized the Chellean-Acheulian-Mousterian classification of Palaeolithic epochs.

Movius, Hallam L.   (1907–1987) Professor in the Department of An-
                         thropology at Harvard University who worked on
                         the Palaeolithic of East Asia and questioned the
                         industrial patterns of Europe in the 1940s and
                         1950s.

Oakley, Kenneth   (1911–1981) Geologist and Palaeolithic researcher
                         who worked on British geological and Palaeolithic
                         sequences from the 1930s with Mary Leakey, King,
                         Warren, and Hawkes.

Obermaier, Hugo   (1877–1946) Bavarian geologist and prehistorian
                         who studied with Penck, worked with Breuil, and
                         adapted the Palaeolithic sequence of de Mortillet
                         in the early twentieth century.

Paterson, Thomas T.   (1909–1994) Cambridge-based Palaeolithic archae-
                         ologist who worked on the industries of East Anglia
                         and India in the 1930s and 1940s, and adapted
                         Breuil's industrial classification to encompass more
                         variety.

Penck, Albrecht   (1858–1945) German glacialist, friend of James
                         Geikie, and author (with Brückner) of *Die Alpen
                         im Eiszeitalter* (1901–1909) in which he described
                         four glaciations in the Alps: Günz, Mindel, Riss,
                         and Würm.

Pengelly, William   (1812–1894) Schoolmaster and geologist of Tor-
                         quay, and careful excavator of Brixham Cave and
                         Kent's Cavern.

Perthes, Boucher de   (1788–1868) Customs officer from Abbeville,
                         North France, discoverer of implements in the
                         Somme and early proponent of human antiquity.

Peyrony, Denis   (1869–1954) French prehistorian and curator of
                         the National Museum in Les Eyzies who reported
                         contemporaneous Mousterian cultures at Le
                         Moustier and La Micoque in the early 1930s.

Phillips, John   (1800–1874) Professor of Geology at Oxford,
                         nephew of William Smith, and author of the
                         1855 *Manual of Geology*, a standard reference
                         work.

Pitt-Rivers, A. H.   (1827–1900) Formerly A. H. Lane Fox: archaeolo-
                         gist, anthropologist, and collector who developed
                         an evolutionary typology of weapons and found

	Palaeolithic tools at Acton and Ealing in the late 1860s.
Prestwich, Joseph	(1812–1896) Wine merchant who succeeded John Phillips as Professor of Geology at Oxford, authority on the superficial drifts, and early believer in their post-glacial date, and defender of the Kent eoliths.
Ramsay, Andrew C.	(1814–1891) Survey geologist who rose to become Director–General and was inspired by Welsh topography to support the land-ice theory.
Read, C. Hercules	(1857–1929) Keeper at the British Museum and Reginald Smith's superior, Read negotiated with the Geological Survey in 1912 and 1913 to assist Smith's Swanscombe excavations.
Reid, Clement	(1853–1916) Survey geologist and palaeobotanist who investigated the post-glacial date of the Palaeolithic at Hoxne and Hitchin in the 1890s, and supported Dewey's work at Swanscombe in 1912 and 1913.
Reinach, Salomon	(1858–1932) Curator of the Musée de Saint-Germain-en-Laye who attacked de Mortillet's Palaeolithic classification.
Rutot, Aimé	(1847–1933) Belgian engineer, geologist, later curator at the Royal Belgian Museum of Natural History, and proponent of eoliths, whose early Palaeolithic industries inspired British classifications.
Skertchly, Sydney B. J.	(1850–1926) Survey geologist who mapped the Fenland area, discovered Palaeolithic tools beneath the boulder clay at Brandon in 1876, and supported the interglacial Palaeolithic of James Geikie.
Smith, Reginald	(1874–1940) Assistant Keeper at the British Museum who worked at Swanscombe with Dewey in 1912 and 1913, helping to establish a standard Palaeolithic sequence for Britain.
Smith, William	(1769–1839) Canal engineer and mineral surveyor who popularized the use of fossils to connect stratigraphical sequences in the late 1810s.
Smith, Worthington G.	(1835–1917) Draughtsman, fungus expert, and prehistorian who worked on the Palaeolithic 'living floors' of north–east London, collected many

	tools from the 1870s onward, and refitted flakes to reconstruct ancient tool-working techniques.
Sollas, William J.	(1849–1936) Professor of Geology at Oxford from 1897 and author of the synthesis *Ancient Hunters* (1911), who irritated Moir and Lankester with his vacillating views on their pre-palaeoliths from East Anglia.
Spurrell, Flaxman	Antiquarian, natural historian, and prehistorian of Kent, whose reconstructions of Palaeolithic tool-working techniques from debris on 'living floors' inspired Worthington Smith.
Strahan, Aubrey	(1852–1928) Dewey's superior on the Geological Survey who controlled permission for his work with Smith on the Palaeolithic.
Sturge, W. Allen	(1850–1919) Retired physician, collector, first President of the Prehistoric Society of East Anglia and supporter of pre-palaeoliths.
Teall, Jethro J. H.	(1849–1924) Director of the Survey for England and Wales who, with Strahan, controlled Dewey's role in the British Museum collaboration.
Tiddeman, Richard H.	(1842–1917) Survey geologist, excavator of Victoria Cave, Settle, who defended a pre-glacial Palaeolithic in the 1870s and was sympathetic to the land-ice theories of James Geikie.
Treacher, Llewellyn	(1859–1943) London geologist and market gardener who worked on the Thames Valley sequence.
Trimmer, Joshua	(1795–1857) Published early papers on the drifts of south–east England, and observed the shelly drift high up Moel Tryfaen in Wales.
Warren, Samuel Hazzledine	(1872–1958) Geologist from Essex, opponent of eoliths and pre-palaeoliths who studied flint fracture, named the 'Ponder's End' cold stage, and worked on the Clactonian industry.
Whitaker, William	(1836–1925) Survey geologist and popular member of the Geologists' Association who built on Prestwich's work on the drifts around London and identified three old river terraces in the Thames Valley.
Wood, Searles V. Jr.	(1830–1884) Suffolk solicitor who began to survey the East Anglian drifts in the mid 1860s with

Frederic Harmer, dividing them into a Lower, Middle, and Upper Glacial series.

Woodward, Henry

(1832–1921) Editor of the *Geological Magazine*, Keeper of the Department of Geology at the British Museum (Natural History), and uncle of Horace B. Woodward.

Woodward, Horace B.

(1848–1914) Survey geologist who worked on the London drifts and published the 1909 memoir on *The Geology of the London District*.

Wyatt, James

(1816–1878) Bedford worthy who discovered Palaeolithic implements at Biddenham Pit in 1861.

Zeuner, Frederick E.

(1905–1963) German geologist and palaeontologist who moved to England before the Second World War, lectured in geochronology at the London Institute of Archaeology, and, like Boswell and Oakley, used industries to help his geological correlations in the 1930s.

1

Before the Stone Age Existed

What is the difference between temptation and geological time? The one
is a wile of the devil and the other is a devil of a while.

(Dawkins to Hughes, 17 March 1870: SMC: TMH)

In the early eighteenth century, John Conyers, an apothecary and antiquary
of London, discovered the body of an elephant as he was digging for gravel at
Gray's Inn Lane. Nearby lay a flint implement (Fig. 1.1). Today we might well
call his elephant a 'mammoth' and refer the implement to the 'Palaeolithic'
period; in 1715, however, Conyers's beast was dated to the reign of Claudius,
the Roman Emperor. This was the belief of John Bagford, an old friend of
Conyers, a bookseller and one of the founder members of a tavern-based
antiquarian club that was soon to become the Society of Antiquaries of
London. At the time, a Roman elephant attacked by an Ancient Briton seemed
a likely scenario to account for the curious occurrence of this animal in
London, far from its hot and distant homeland. It would be a century and a
half later when our ancestors were acknowledged as the contemporaries of
such enormous animals: they would then be pictured in a newly-discovered
geological world, more ancient than the time of the Romans or even the
British natives described by Caesar.[1]

For Bagford and his contemporaries, the time allotted to humans, and even
to the Earth itself, was not long. Their knowledge about the distant past was
gathered from folklore or historical texts, and the Bible supplied a particularly
important source of chronological information. Back in the seventeenth
century, James Ussher (1581–1656) had famously calculated the age of the
Earth and the Creation to date to 4004 BC. But Ussher did not, as is often
believed, reach this date by counting back through the generations of the
Bible; indeed, he could not. As John Fuller has observed, there is no fixed
point from which to start counting: a vague gap divides the last of the Hebrew
books from the year AD 1. In fact, it had been a common belief even in
medieval times that four thousand years had elapsed between Adam and

---

[1] Hearne 1770, lxiv. On Bagford's place in the history of the Society of Antiquaries, see Evans
1956 and Grayson 1983, 7–8.

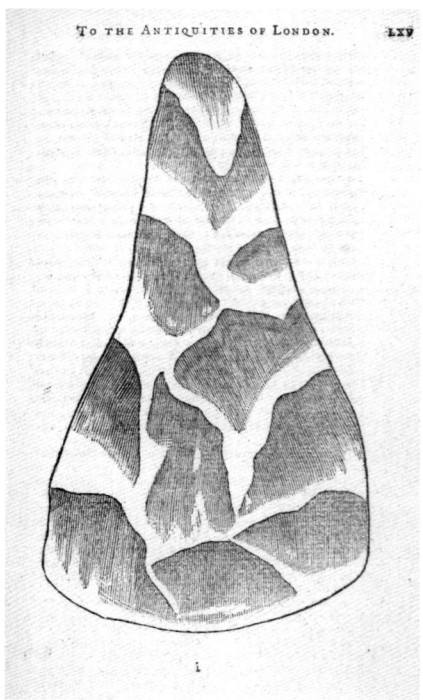

TO THE ANTIQUITIES OF LONDON.   LXV

Figure 1.1. This sketch of the Gray's Inn Lane implement illustrated the account of John Bagford, which was published in the preface to Thomas Hearne's *Joannis Lelandi Antiquarii de Rebus Britannicis Collectanea* (Hearne 1770, lxv).

Christ; the additional four years had been added on in the sixteenth century to correct a mistake in the Early Christian calendar regarding the exact date of Christ's birth.[2]

In the centuries that followed Conyers's discovery at Gray's Inn Lane, a detailed time-scale would be developed for the ancient stone tools of Britain, but much of this work took place outside the great libraries. Information about the distant past was gathered not from books, but from the stratigraphic records of ancient rocks and the relics of past lives that they contained. The once uncertain middle ground between geological time and historical texts—the period that became known as the Quaternary—was gradually laid out by a varied army of researchers. Some used the sequences of sediments left by ancient rivers, seas, icebergs, or glaciers to divide this period into stages; others based their conclusions on the succession of animals that had once lived in Britain, arriving and departing as the climate and landscape changed. Even the fragile remains of shells played their part in

---

[2] Barr 1985, 578–579; Rudwick 1986, 300–302; Fuller 2001. My thanks to John Fuller for kindly sharing with me his notes on Ussher and the year 4004 BC.

the construction of time-scales. By the latter half of the nineteenth century, researchers would be able to assign Palaeolithic tools to particular portions of the geological past. But they would disagree violently about who had painted the most reliable picture of that past, see different merits in their own sequences, and draw different patterns to their peers. Argument would continue into the twentieth century.

This chapter introduces the first few decades of Quaternary research, preparing the way for later decades explored in subsequent chapters. It sweeps swiftly over events in the earlier half of the nineteenth century that have been documented in detail by Donald Grayson and Bowdoin Van Riper, charting the entrance of ancient Palaeolithic tool-makers into geological realms, and observing the early development of chronological guides that would be used to date stone tools.[3]

1.1 STONE TOOLS, ANCIENT BEASTS, AND THE FLOOD

The name of John Frere (1740–1807) is closely associated with early discoveries of British Palaeolithic implements. Frere was a respectable antiquarian from Suffolk who, in the last years of the eighteenth century, found stone tools in a brick pit at the village of Hoxne on the Norfolk border. These implements lay alongside the 'extraordinary bones' of unknown animals, a curious circumstance that Frere reported to the Society of Antiquaries of London. This Society had expanded in size and influence since the days when Bagford and his friends frequented the London taverns, and had gained its Royal Charter in 1751. Frere speculated in his letter that these implements belonged to 'a very remote period indeed; even beyond that of the present world'.[4]

There was little response to his statement. Although Frere drew his conclusions nearly a century after Bagford, it was still difficult to relate such finds to time-markers founded on stratigraphy and fossils. Detailed time-sequences did not yet exist.[5] It was only in the latter decades of the eighteenth century that the new discipline of geology had started to emerge: the Geological Society of London would not be formed until 1807, the year of Frere's death. The matter was forgotten for now, but the site of Hoxne would be

---

[3] Grayson 1983; Riper 1993.
[4] Frere 1800, 204, 205. Evans 1956, 203, footnote 2 has noted that the antiquary William Stukeley also reported a find of flint implements in 1764. Stukeley thought they might be antediluvian.
[5] Gruber 1965, 378; Grayson 1983, 55–59.

remembered by future generations as they tried to refine the relationship between tools, sediments, and bones.

In the early nineteenth century, the tiny group of researchers working on the Quaternary period began to ponder two questions of particular import-ance for the development of time-markers. First, they queried the origin and age of the stiff, bluish, rock-studded clay and the spreads of gravel known as 'diluvium' (because of their association with a supposed great flood), and the far-travelled 'erratic' rocks associated with this period. Second, they tried to decipher the relationship between these distinctive deposits and the fossil bones of extinct animals.

Some of these British researchers had been inspired by the work of Georges Cuvier (1769–1832), the famous French palaeontologist. Cuvier recon-structed the earth's history from fossil bones. He compared the bones of fossil elephants to those belonging to elephants of his present world and saw that the two sets were very different. The fossil beasts, he decided, had lived in Europe long *before* the time of the Romans—their bones had, after all, been found in lands far distant from the Roman Empire. Cuvier believed that the recently-extinct quadrupeds, whose remains now lay in caves and the gravel beds of ancient rivers, were antediluvian: they pre-dated the last watery inundation of the Earth. Since he knew of no convincing evidence of human bones amongst such antique species, Cuvier concluded that humans had not existed at this period.[6]

In Britain, William Buckland (1784–1856) was promoting his own influen-tial opinions on the matter. Buckland was the first incumbent of the Oxford University Readership in Geology, established in 1819, where he won fame for his flamboyant style of lecturing. Like many of his contemporaries, Buckland believed that the diluvial clays and gravels had been left by the last in a series of catastrophic floods that had moulded the surface of the Earth into its present form. He described the relics of this flood in his book *Reliquiae Diluvianae* (1823).

Buckland attacked a view held by many of his contemporaries—that the bones of hyenas, elephant, and rhino found lying beneath the diluvial clay in the bone-caves had been washed to their final resting places on floodwaters from some more southerly country—and argued instead that they repre-sented a group of animals that had once roamed Europe. Inspired by his researches at Kirkdale Cave in Yorkshire, Buckland recounted, in enthusiastic detail, how the bones he had recovered bore witness to generations of animals that had lived and died over thousands of years. He explained that these fossilized relics had been gathered together in caves by a variety of natural

---

[6] Cuvier 1821, 198–204; Buckland 1823, 41, 96; Rudwick 1997, 90, 91–94, 232–234.

processes, accumulating gradually over a great period of time before the diluvial clay of the floodwaters had covered them. Buckland agreed with Cuvier that this last universal inundation, which seemed to have extinguished the antediluvian species, had probably occurred some five or six thousand years ago. Like Cuvier, Buckland also believed that humans had occupied Europe in postdiluvian times, *after* the land had emerged from the flood-waters.[7]

Although the bones and handiwork of humans had been reported from antediluvian deposits, such discoveries were generally treated with caution. Cave deposits might have been disturbed in the long course of ages after they were laid down, and the ease with which remnants of different periods could become mixed together within a single layer of cave earth was well known. Buckland had no hesitation in assigning a human skeleton from Goat's Hole, Paviland, on the coast of South Wales, to the time of the Roman invasion.[8] A few years later, he gave a similar interpretation to the flint tools from Kent's Cavern, a cave in a small hill near Torquay harbour in Devon. These tools had been found with the bones of extinct animals by Father John McEnery, a local Catholic priest who had started excavating the Cavern in 1825.[9]

Buckland's conclusions are sometimes portrayed as an attempt to square geological observations with theological beliefs. Buckland certainly suggested in the preface to *Reliquiae Diluvianae* that the geological evidence of a univer-sal deluge supported the Biblical testimony of a Mosaic Flood.[10] Nicolaas Rupke, however, has drawn attention to the awkward position occupied by Buckland, who was dealing with a sensitive area of the past, where geology met Biblical history; but who was also a representative of the new and fragile discipline of geology at Oxford. Buckland could not afford to alienate the clerical establishment who controlled most university appointments and pat-ronage. It is not surprising that he worded his geological communications to retain harmony between the new science and theology.[11]

Another point, often forgotten when considering the then controversial question of antediluvian humans, is that the *lack* of human bones or tools in the diluvium had cast *doubt* on the Flood connection. The antediluvian generations between Adam and Noah would have been caught up in the Biblical floodwaters, but no convincing human remains had yet been recovered from

[7] Buckland 1823, 44–45, 51, 97, 231. Cuvier (quoted in Rudwick 1997, 93) argued in 1806 that elephants had inhabited the countries where their bones now lay.

[8] Buckland 1823, 87–92.

[9] Buckland's geological achievements are summarized in his obituary by Portlock 1857. Comprehensive accounts of his cave researches and interpretations are given by Boylan 1967, Rupke 1983, Grayson 1983, and Sommer 2004a.

[10] Buckland 1823, iii.   [11] Rupke 1983, 56–59, 200–209.

European deposits. When Adam Sedgwick (the distinguished Woodwardian Professor of Geology at Cambridge) recanted his belief in a single great Biblical Flood in 1831, he considered the lack of humans and their handiwork to have hindered the link between a Flood and the diluvial deposits. Initially, Buckland had solved the dilemma by suggesting that the earliest humans would be found not in Europe, but in Asia: the cradle of the human race. By 1836, however, he had another solution: the geological deluge must have *predated* the Biblical Flood. It must have occurred before the creation of man. This reasoning, which has been examined in detail by Donald Grayson, pushed the diluvial marker back in time and consigned the monstrous extinct animals to a far more remote period, long before *any* humans might have existed.[12]

Finally, it is important to note that Buckland's interpretations were tempered with caution. His prudence was praised by Charles Lyell (1797–1875), who was renowned in geological circles for a self-conscious avoidance of speculation and earnest attempts to establish geology as a rigorous and reputable science. Lyell had attended Buckland's lectures as an undergraduate at Oxford. His interest in geology had continued after he was called to the bar in 1822. Lyell argued in the second of his three influential volumes on *The Principles of Geology* (1832) that the bones now lying in caverns or fissures might easily have become mingled together during their lengthy accumulation.[13]

Lyell and Buckland both visited the caves around Liège in Belgium where Dr Phillipe-Charles Schmerling had recovered bones of humans and extinct quadrupeds, bones that he thought might be coeval. They both departed unconvinced. In the late 1820s, Marcel de Serres, Paul Tournal, and Jules de Christol had recorded human bones lying with those of extinct animals in the caves of southern France. These observations received a similar reception.[14] Lyell asked:

Must we infer that man and these extinct quadrupeds were contemporaneous inhabitants of the south of France at some former epoch? We should unquestionably have arrived at this conclusion if the bones had been found in an undisturbed *stratified* deposit of subaquaeous origin [...] But we must hesitate before we draw analogous inferences from evidences so equivocal as that afforded by the mud, stalagmites and breccias of caves, where the signs of *successive* deposition are wanting.[15]

---

[12] Buckland 1823, 170; Sedgwick 1831, 313–314; Buckland 1836 (i), 95, footnote. The changing ideas about the connection between fossils and floods are explored by Grayson 1983, 43–86, Rupke 1983, 82–94, and Sommer 2004a, 68–69, 71–72.

[13] Lyell 1830–33 (ii), 219–220, 226–227. For a brief summary of Lyell's life, see Wilson 1998.

[14] Lyell 1830–33 (ii), 224–225; Buckland 1836 (i), 105–106, 598; Lyell 1863a, 68. The discoveries and conclusions of Tournal, de Christol and de Serres are discussed in Lyon 1970.

[15] Lyell 1830–33 (ii), 225–226.

In the 1840s, stone tools were discovered with the relics of extinct animals in more compelling geological circumstances: lying in undisturbed deposits of sand and gravel left by the ancient River Somme. But the account published by their discover, Jacques Boucher de Crevecoeur de Perthes, a customs officer from Abbeville in North France, did not inspire confidence. His references to geological catastrophes left a flavour that tasted old-fashioned after the work of Lyell and others on gradual processes of geological change. His illustrations were small, poor in quality, and tainted by the presence of dubious stones alongside implements. Sceptics wondered if natural processes, rather than the hands of ancient humans, had formed all of his so-called 'tools'. The British geologist Gideon Mantell was not alone in criticizing de Perthes's 'antiquarian labours' for a lack of geological knowledge and recognizing the natural origin of some of his 'so-called antediluvian works of art'.[16]

Thus far, this story might sound relatively familiar, having followed the well-trod track that has led other historians to its well-known conclusion: the acceptance of human antiquity. But before continuing down this path, more information is needed about the different time-markers that would later be used to date those tools; more background to the sequences that would cause so much argument in the future. This requires a quick glance in three directions: at the patterns seen in the Glacial epoch, the bones and shells, and the sediments left by ancient rivers. The problem of proving human antiquity was not necessarily of primary concern for those who worked on these patterns, but their records and interpretations would be vital for ordering and interpreting the tools when they were cast back in time.

1.2 DILUVIUM, DRIFT, ICEBERGS, AND GLACIERS

For a long time, the Quaternary period had been neglected by geologists. In the earlier decades of the nineteenth century the older rocks (or 'solid' geology) attracted much more attention than these younger ('superficial', 'diluvial', 'drift', or 'recent') deposits that lay near the surface of the earth in thin, patchy layers. William Whitaker, the Survey geologist who devoted much of his life to the Tertiary and Quaternary strata, recalled decades later how 'The older geologists treated all Drift as of one kind, and did not trouble themselves about such perplexing matters as its divisions and their classification. They

---

[16] Mantell 1850–51, 238; Prestwich 1860a, 303, 308, footnote. On the role of Boucher de Perthes in the establishment of human antiquity, see Laming-Emperaire 1964, 155–172, Grayson 1983, 126–129, Cohen and Hublin 1989, and Coye 2000.

accepted this varying set of deposits rather as a necessary evil, obscuring the beds below'.[17]

The Director–General of the Geological Survey, Roderick Murchison (1792–1871), felt sorry for Whitaker, who had been sent to map the eastern counties where there was 'nothing but soft squashy materials, and no good hard rocks to hammer', and had to be reassured that Whitaker really *did* like to work there. The Survey had emerged in 1835 on a wave of enthusiasm for the work of Murchison's predecessor, Henry De la Beche (1796–1855), in Devon. Nonetheless, it would not be until the 1860s that officers of the Survey would be directed to turn their attention to the superficial deposits.[18]

Joshua Trimmer (1795–1857), another geologist who liked recent geology, complained in 1847 that superficial deposits had received the least attention from geologists. He condemned the hasty generalizations that had bestowed such a variety of names on recent sediments, but had not been matched by any effort to work out their sequence.[19] Admittedly, it would not be a simple task to order and arrange them. Lying so near the surface, they had often suffered erosion and disturbance; but Trimmer was one of a growing number of geologists who became intrigued by this problem. He had gained his geological knowledge superintending copper mines and working slate quarries for his father in North Wales, but his interest in recent geology had been stimulated when managing his father's farm in Middlesex, his childhood home.[20]

Of all the recent formations scrutinized during the earlier decades of the nineteenth century, the distinctive and controversial 'diluvium' attracted most attention. The fiercest arguments, however, were focused on its origin rather than its arrangement or sub-division. By the 1830s, few geologists linked these great spreads of boulder clays, gravels, and erratic blocks perched high up in the hills to the Biblical Flood. But watery agencies were still invoked to explain their origin: waves, currents, and submergences beneath a glacial sea were all called upon to account for their appearance, distribution, and the peculiar occurrence of Quaternary sea shells at high altitudes. But the 'diluvium' would become better known as 'drift', a term associated with Lyell's theory that icebergs floating on a frigid sea above a sunken land had dropped their stony load on the sea floor.

Lyell popularized his iceberg theory in *Principles of Geology*, but it was not until 1840 that he proposed the term 'drift' as an alternative to 'diluvium'

[17] Whitaker 1889 (i), 353.

[18] Murchison to Ramsay, quoted in George 2004, 53. On the origins of the Geological Survey in the early nineteenth century, see Wilson 1985, 3–12. For an overview of research on solid geology and the classifications of geological periods, see Berry 1987, 64–102.

[19] Trimmer 1847, 448–449.   [20] Portlock 1858, xxxiii.

(following a suggestion that had been made by Murchison, amongst others). He believed that 'drift' gave a better idea of permanent, rather than transient, submergence, and observed that the drift was analogous to the 'boulder formation' of Danish and Swedish geologists (so-called because it contained so many erratics: large rocks, scattered far from their origin). Lyell believed that the erratic rocks had fallen from icebergs as they drifted above the present-day hills, while finer sediments passed down through the waters to leave swathes of gravels and boulder clays below.[21]

Lyell's iceberg theory, though popular, did not stand unopposed. Supporters of the rival 'land-ice' theory employed glaciers rather than icebergs as agents of deposition, and envisaged a land scoured by the grinding action of vast glacial ice-sheets (although most of them continued to support submergence as well, and few abandoned icebergs altogether). In the late 1830s, Louis Agassiz (1807–1873), Professor of Natural History at the University of Neuchâtel, Switzerland, and an authority on living and fossil fishes, became a notorious proponent of land-ice activity. The case made by Agassiz for a gigantic ice-sheet won a small number of distinguished supporters in Britain. They included Robert Jameson in Scotland, Buckland in England, and even, briefly, Lyell himself.[22]

Buckland had first contacted Agassiz in the early 1830s on the subject of fossil fish. Inspired by Agassiz's work on the Alpine glaciers and by a field-trip they both took to the Highlands of Scotland in 1840 (following a Glasgow meeting of the British Association at which Agassiz had expounded his ideas), Buckland converted to the land-ice theory. When they returned from Scotland to London, Buckland and Agassiz read memoirs on this controversial subject to the Geological Society. Buckland drew on his observations in Scotland and north England to defend the theory, arguing that the features he had previously attributed to diluvial floods were much better explained by land-ice. The polished and striated rocks had, he thought, been carved by ancient glaciers; the erratic boulders had been carried along by these icy sheets; the spreads of unstratified gravels and boulders were glacial moraine. Watery agencies were not rejected entirely: Buckland thought that some erratics might have been transported by floating ice; whilst Agassiz distinguished between jumbled

---

[21] Lyell 1840a, 176. See also Buckland 1840a, 234 and Lyell 1830–33 (i), 299; (iii), 149–150. The decline of diluvialism and the rise of Lyell's drift and icebergs are described by Herries Davies 1969, 248–256. Imbrie and Imbrie 1979 summarize the changing opinions of the Glacial epoch from Victorian times to the twentieth century, and examine astronomical theories in detail.

[22] Lyell 1840b, 337–338; Argyll 1874, xxxviii–xli. On the roles of Agassiz, Jameson and Buckland in introducing land-ice theories to Britain, see Herries Davies 1969, 266–289; on Buckland, see Boylan 1981.

deposits left by glaciers, and the stratified gravels and mud deposited in water after the Glacial epoch. But both authors gave land-ice a prominent role, and their papers provoked a lengthy, and largely negative, discussion.[23]

Buckland presented another paper the following year on evidence for glacial action in Snowdonia, North Wales. Once again he did not claim an exclusive role for glaciers, but asserted that they had played their part alongside bodies of water and drifting icebergs. Buckland believed that the sea must have deposited the drifts that lay on the flanks of a hill known as Moel Tryfaen, 1,392 feet above sea level—drifts that contained sea shells. Trimmer had described these Moel Tryfaen drifts a decade before, when he was working slate quarries for his father in the area.[24] But it would take a few decades more before a modified version of the land-ice theory prevailed over the notion of icebergs in the minds of many British geologists.

Andrew Crombie Ramsay (Fig. 1.2) would emerge as another notable defender of the land-ice theory. He had joined the Geological Survey in 1841, when it was a small and relatively new institution and its officers still worked in the field wearing the smart but impractical uniforms of the Board of Ordnance: dark-blue, well-buttoned frock-coats. Ramsay entered their ranks armed with the standard geological hammer, thin leather map-case, prismatic compass, and full pockets of the field surveyor. Thus equipped, he set about mapping the rocks of Wales, declaring enthusiastically: 'there cannot be a better school for a practical geologist than the Ordnance Geological Survey'.[25]

Ramsay was also welcomed into the Geological Society of London, a major metropolitan focus for research. In 1841, the same year that he joined the Survey, Ramsay reported in overawed tones to his brother the experience of dining at the Crown and Anchor Inn with the Geological Club: an exclusive group of influential geologists from the Geological Society. He was introduced to Buckland, who recalled Ramsay's earlier work on the geology of the Scottish island of Arran, and offered 'two of his digits to shake'.[26] Ramsay began to attend the Society regularly but he did not, at first, support Buckland's ideas

---

[23] Agassiz 1840, 329; 1885 (i), 232–233; 248–249; Buckland 1840b, 332–333, 337, 348. The violence of the arguments that followed these papers is captured in Woodward 1907, 138–142 and Thackray 2003, 95–102.

[24] Trimmer 1831, 332; Buckland 1841, 583–584; Portlock 1858, xxxv.

[25] Ramsay to Murchison, 11 (November?) 1841: GSL: 838R1/1. Officers were able to choose their own civilian clothes after 1845, when the Survey was under the Office of Woods and Forests. The equipment, tasks, and conditions of Survey fieldwork are outlined by Geikie 1924, 44–49; for a more detailed account, see Oldroyd and McKenna (2005).

[26] A. Ramsay to W. Ramsay, 25 March 1841: NLW: 11590 D/4. On the Geological Club, see Thackray 2003, ix–x.

Figure 1.2. Andrew Crombie
Ramsay (1814–1891), glacial
geologist (GSL: P44/04/03).

about the Glacial epoch. After one meeting in 1845, he noted in his diary:
'Jolly night at the Geological. Buckland's glaciers smashed'.[27]

But as Ramsay mapped the solid geology of North Wales, he became
impressed by the evidence for glacial action. In 1851, he entered the rooms
of the Geological Society in a different spirit and agreed with Buckland's case
for the ancient extension of glaciers in the Welsh mountains. A 'wild dream
on glaciers and the glacial period' was how Mantell described Ramsay's
account of the glacial sequence in North Wales.[28] Ramsay divided an earlier
great glacier period, which had left erratics and carved grooves in the mountainsides, from a later deposition of drift over a land submerged under water
2,300 feet deep. He went on to describe how a *second* glacial period had
followed this submergence, when glaciers formed once again on the higher
parts of the emerging land.[29]

[27] Ramsay 1845, diary entry quoted in Geikie 1895, 64.
[28] Mantell 1851, diary entry quoted in Thackray 2003, 173.
[29] Ramsay 1852, 373–374, 376.

Ramsay had found the mountain scenery of Wales persuasive, but his views were stifled in the metropolis of London. Up in Scotland, however, land-ice theories would flourish. Robert Chambers (1802–1871), author, folklorist, and editor of the popular weekly paper *Chambers's Edinburgh Journal*, attacked the iceberg theory with ferocity through the 1850s. Chambers had won notoriety through his suspected authorship of the anonymous *Vestiges of the Natural History of Creation* (1844), a book that associated the diluvium and erratic blocks with a submergence and icebergs.[30] After a trip to Scandinavia in 1849, he exchanged icebergs for widespread land–glaciation.

Chambers announced to the Royal Society of Edinburgh in 1852 that great sheets of ice, far more vast than Buckland's glaciers of the Highlands, had caused the erosion and debris of the Glacial epoch. He described a series of submergences and elevations of the land; but his ideas, in their general outline, confirmed Ramsay's conclusions: there had been two pulses of glacial activity in Scotland and north England—an early, widespread glaciation which had left the boulder clay (but which had not necessarily been associated with submergence), and a more recent extension of local valley–glaciers. Chambers also distinguished between the boulder clay, produced by glaciers, and the drift (or 'till', as it was known in Scotland) that had been deposited on the sea floor.[31] Adolf van Morlot (1820–1867), geologist, archaeologist, and curator of the Archaeological Museum at Berne, had observed a similar sequence of events in Switzerland. He drew analogies to Chambers's work on the glacial deposits of Scotland. Morlot described two glacial periods, and believed that the mammoth had lived in Switzerland during the milder conditions of an intervening diluvial period.[32]

Ramsay and Chambers were not alone in their attempts to carve Britain's confusing mass of glacial deposits into smaller slices. In the 1840s, Trimmer identified distinct upper and lower strata within the drift of Norfolk as he surveyed the area for the Royal Agricultural Society—who were more interested in his contributions to the practical problem of drainage. He described a Lower Drift (a till or boulder clay), which seemed to be a marine formation; an Upper Drift; and, above the Upper Drift, 'a deposit from turbid waters returning to a state of tranquillity': the 'warp of the drift'. Trimmer also observed that the boulder clay had not reached south of the Thames Valley,

---

[30] Anon [Chambers] 1844, 137–140. The excitement aroused by this book, which ranged from geology to theology and from astronomy to anthropology in its account of the origins and future of the earth, is brought to life by Secord (2000). On the earlier reception of the land-ice theory in Scotland, see Finnegan (2004).

[31] Chambers 1853, 279–280. See also Chambers and Chambers 1853, 76, a source that Roger Jacobi drew to my attention, and Chambers 1855, 97–98.

[32] Morlot 1855, 20–21, 24–25; Huxley 1869, xxxi.

a distribution that would confound later attempts to link the glacial drifts of East Anglia to the river drifts around London.[33]

More detail was recorded within the glacial drifts over the 1840s and 1850s; more layers were identified, named, and connected to different phases of glaciation. The land-ice theory also gathered support, but submergences were not restricted to those who favoured the iceberg theory. Ramsay and Chambers both incorporated submergences alongside their descriptions of land-ice, and would continue to do so in the 1860s. This is an important point, because the idea of a mid-glacial submergence between two phases of glaciation became a popular division of geological time. John Phillips (1800–1874), who gained the Chair of Geology at Oxford in 1856, was one of many researchers who saw in the drift the reflection of turbulent waters rushing over a submerged land. In his *Manual of Geology* (1855) Phillips used the term 'preglacial' to refer to the epoch before this great mid-glacial submergence; his 'postglacial' deposits had been left after the re-emergence of the land above the sea.[34] A glacial chronology was starting to emerge.

1.3 SHELLS, BONES, AND HUGH FALCONER

Quaternary sequences were also built on the fossilized bones of animals and shells of molluscs. Phillips's uncle, William Smith (1769–1839), had popularized the use of fossils as time-markers in the late 1810s when he published *Strata Identified by Organized Fossils*—with its handsomely-coloured prints of the fossils Smith thought characteristic of different strata—and his *Stratigraphical System of Organized Fossils*.[35] Unfortunately, Smith's system only reached up to the time of the London Clay, a Tertiary deposit, and arguments over the following period were fierce. Palaeontologists had as many battles as glacial geologists in the middle of the nineteenth century. They disagreed about species identification and the reliability of different animals as indicators of time and climate.

Lyell had suggested, in the third volume of his *Principles of Geology* (1833), that the Tertiary formations should be divided into four age groups—the Eocene, Miocene, Older Pliocene, and Newer Pliocene—based on the shells they contained. These age groups were identified by comparing the proportions of living species to extinct species of shells: the Eocene, for example, had

---

[33] Trimmer 1847, 461–465; 1851a, 25.
[34] Phillips 1855, 411, 421, 430; Murchison 1868 (ii), 587.
[35] Smith 1816–1819; 1817. On Smith's concept of strata identified by fossils, see Knell 2000, 20–27.

the lowest proportion of living species. Lyell later added a fifth division—
the Post-Pliocene—for deposits with fossil shells that were all identical to
living species. A change of name was proposed in 1846 by Edward Forbes
(1815–1854), Professor of Botany at King's College London and Palaeontologist to the Geological Survey, who suggested that Lyell's term 'Pleistocene' was
a clearer label for the beds of the Glacial epoch than his 'Newer Pliocene.'
Forbes thus shifted the basis of the definition from fauna to climate (as Berry
has pointed out). This Pleistocene or Newer Pliocene period is referred to as
the 'Quaternary' in this book.[36]

Lyell's divisions became extremely popular. But some palaeontologists
working on fossil mammals believed they could offer a finer chronology for
recent deposits. Changes in the proportion of shell species could provide a
reliable time-scale for the long course of ages over which most sediments had
accumulated but, as the numbers of extinct species dropped through time,
Lyell's shelly time-markers became too coarse for classification: eventually *all*
the shells were of living species. His outspoken opponent, Dr Hugh Falconer
(1808–1865), proposed an alternative. Falconer (Fig. 1.3) was a rising young
palaeontologist, surgeon, and sometime superintendent of the East India
Company's botanical gardens. It was partly on his authority that the China
tea plant and the quinine-yielding Cinchonas were introduced to India, and
questions were asked about the overexploitation of Indian teak forests; but
Falconer would also influence British perceptions of geological time.[37]

Falconer preferred to use mammals, rather than molluscs, to date the
more recent geological deposits. He based his own chronology on fossil
pachyderms, identifying different species of hippopotamus, rhinoceros, and
elephant, and grouping them together in clusters that changed across space, as
well as through time, because their prehistoric ranges would have shifted as
the climate altered. Falconer argued that these mammals had evolved and
been extinguished in shorter pulses than Lyell's molluscs; they revealed a finer
grain in the texture of time.[38]

This interest in fossil pachyderms had been inspired in India. Falconer had
built his knowledge and his name in the Siwalik Hills, a little-known palaeontological paradise near the Botanic Gardens at Suharunpoor where he

---

[36] Lyell 1830–33 (iii), xiii, xvii, 61; Lyell 1841 (i), 210, 212; Forbes 1846, 403. Lyell's sub-
divisions of the Tertiary are described and explained by Berry 1987, 103–114.

[37] On Falconer's mammoth researches and the palaeontological arguments of the time, see
Cohen 1994, 132–139, 141–142. Falconer's memoirs and notes were collected after Falconer's
death and published by Charles Murchison 1868 in two volumes, which included much
material that had not previously been printed in full. Murchison (1868 (i), xxiii–liii) provided
a biography of Falconer; see also Anon (1867a, xvi–xvii) and Moore 2004.

[38] Falconer 1857, 359.

Figure 1.3. Hugh Falconer (1808–1865), palaeontologist, botanist, and surgeon (FMF: HF, 381).

worked. With no convenient museum to turn to, he robbed the living fauna of their skeletons and built his own comparative collection. With his friend, Captain Proby T. Cautley, Falconer identified a number of distinct species of elephant in addition to the mammoth. All were magnificently illustrated in their joint publication on the *Fauna Antiqua Sivalensis*.[39] When Falconer retired in broken health from the Indian Service and returned to England in 1855, he exchanged his Indian pachyderms for European problems.

Cuvier's early work on the mammoth and mastodon had been followed by decades of disagreement among palaeontologists about the number of fossil elephant species. Some, like Richard Owen, still believed that all Tertiary elephants of the northern hemisphere could be incorporated within Cuvier's single species: the 'mammoth'; others divided his mammoth into as many as ten species. Falconer took a line between the two.[40] The Geological Society of

[39] Falconer and Cautley 1845.
[40] On Cuvier's opinions about the mammoth, see Owen 1843, 210–214; 1846, 232–234, 243 and Rudwick 1997, 16–24; 2005, 361–362. The variety of views on the mammoth was recorded by Falconer and Cautley 1846, 2–3 and Falconer 1857, 309, and has been discussed more recently by Cohen 1994, 130–131.

London heard two accounts of fossil mastodon and elephant from Falconer in 1857. He used their enormous teeth and jaws to identify different species, describing crowns, plates, tubercles, ridges and tusks in enthusiastic detail. For Quaternary times, Falconer's most important conclusion was that the mammoth (*Elephas primigenius*, Blum.) was distinct from the straight-tusked elephant (*E. antiquus*, Falc.); he also added several rhinoceros species to the recent geological deposits of England and described an earlier elephant species: *E. meridionalis.*[41]

Having identified such variety, Falconer turned to the patterning of these species in time and space—a topic that had, in his view, received little attention since Buckland's pioneering researches. Most palaeontologists still believed that the bones in all British caves dated to a single period of geological time. They tended to consign them to the earlier end of the scale provided by the glacial chronology. Phillips, for example, followed Buckland and Cuvier in placing the elephant, hippopotamus, hyena, and other extinct animals of the bone-caves in the pre-glacial era. Lyell, in the fifth edition of his *Manual of Elementary Geology* (1855), suspected that the bone-rich brick-earths of the Thames Valley (with their mammoth, rhinoceros, and hippopotamus) pre-dated the time of the mid-glacial submergence. Trimmer described two elephantine groups: a pre-submergence cluster in the bone-caves and Crag, and a post-submergence group in the river–valley gravels of the Thames and other areas.[42] Falconer's opinions about the connection between his pachyderms and the Glacial epoch will, however, be left until Chapter 2, because they would become closely connected to later arguments about the age of stone tools. Another time-sequence must be introduced before those tools can enter the story.

1.4 RIVER DRIFTS AND JOSEPH PRESTWICH

The river drifts of the Thames Valley had yielded one implement to Conyers in the early eighteenth century. In the 1860s they would reveal many more, and patterns from the gravels, sands, and brickearths in the river valleys of southern England would be used alongside sequences from glacial drifts and

[41] Falconer 1857; 1858; Murchison 1868 (i), xliii–xliv; (ii), 104–144, 158–211.
[42] Trimmer 1851a, 25; Phillips 1855, 411; Lyell 1855, 153–154. On Falconer's views about the stagnation of palaeontological researches and the importance of the mid-glacial submergence as a time-marker, see Murchison 1868 (ii), 588 and Prestwich 1873, 472.

Figure 1.4. Joseph Prestwich (1812–1896),
stratigraphical expert (GSL: P53/26).

fossils to position these tools in time. By then, the river drifts were regarded as
the particular province of Joseph Prestwich, a London wine-merchant.

Prestwich, pictured in Figure 1.4, was devoted to geology. When commercial work for his family business took him to France or Belgium, he expanded
his knowledge of superficial strata and his circle of academic colleagues. When
in Britain, Prestwich made diligent use of evenings and weekends to observe
and record local geology. His immense energy is evident from the mass of
localities, collections, and individuals described in his field-notebooks, and he
was eager to connect his own interpretations to the findings of others.[43]

In the late 1850s, Prestwich became close friends with Falconer, who
described him as 'a quiet observer, of matchless sagacity and indomitable
perseverance', and gave a glowing assessment of his geological skills: 'in the
Quaternary sands and gravels he was unrivalled'. In 1855, when Falconer left
India, he was intending 'to apply to the [Pliocene and Quaternary] deposits,
palaeontologically, the same kind of analysis which Mr. Prestwich was applying to them stratigraphically, but without concert with or even personal
knowledge of him at the time'.[44] For the next decade (the last of Falconer's
life) they would make a formidable team. Prestwich would later marry

---

[43] On Prestwich, see Hicks 1897, Evans 1897a, Prestwich 1899, and Thackray 2004.
Prestwich's field notebooks are held by the Library of the Geological Society of London
(GSL: 794).

[44] Murchison 1868 (ii), 584, 585.

Falconer's niece, Grace McCall, although Falconer did not live to see their happy union.

Already known for his work on the London Tertiaries, Prestwich attacked the drifts with vigour over the 1850s. Once so neglected, these drifts were at last attracting more attention. 'We are evidently on the eve of a revival of the study of what used to be called "diluvial" deposits', declared Forbes in his 1854 Anniversary Address to the Geological Society.[45] Prestwich walked for miles across country, visiting sections exposed in cuttings and gravel pits; examining collections of geological specimens, bones, fossils, and shells; recording accounts of local gravel extractions; and meeting up with other geologists in the field. He explained his plan to Lyell in 1849:

In my observations on the Drift period I have taken Essex as my base, for I have there found the characters of the different deposits by far the best defined. From this as a centre I have worked over the district to the north as far as the coast of Norfolk, to the west to Devizes, eastward to the Channel, & am now proceeding over the ground southward, for I feel that the phenomena, altho' presenting great variety & infinite modifications must be viewed in connection over large areas.[46]

Of all the drifts in southern England, those of the Thames Valley were notorious for their complexity. In 1855, after spending some time in the field with Prestwich, Lyell observed to another young colleague: 'We are making progress in classifying the gravels of the Thames but it is a laborious work & very curious in the details'.[47] These gravels would become even more controversial after the acceptance of human antiquity, when they produced increasing numbers of implements to rival Conyers's early discovery at Gray's Inn Lane. In the 1850s, though, Prestwich was concerned with its geological puzzles. When and how was the valley formed? What was the relationship between the valley and the gravels lying along its sides? Had these gravels and brickearths all been deposited at once or could they be divided into several stages? How had they been deposited: by the river, the sea, icebergs, or glaciers? How were they related to other drifts, to the Glacial epoch, to Quaternary fauna?

Back in 1838, Prestwich's friend John Morris (1810–1866), a pharmaceutical chemist from Kensington and later Professor of Geology at University College London, had suggested that the mammal bones in the Thames Valley brickearths had been accumulated under freshwater, fluviatile (river) conditions *after* the diluvial gravel had been laid down. Trimmer had also

[45] Forbes 1854, xliv.
[46] Prestwich to Lyell, 1 August 1849: ULE: Gen 115/4879–4880.
[47] Lyell to Lubbock, 16 June 1855: BLL: Add.Ms.49638/16.

concluded that the bone-rich deposits of the Thames Valley post-dated his Boulder Clay (which lay further north in Norfolk and Suffolk).[48] Prestwich accepted this post-glacial date. Others disagreed. Robert Alfred Cloyne Godwin-Austen (1808–1884), the drift-geologist to whom Forbes bequeathed his geological papers, was convinced that the old land-surface of southern England, over which mammoth had once roamed, *pre*-dated the great mid-glacial submergence. Lyell also dated the extinct mammals of the Thames Valley drifts and the caves to a time before the submergence.[49] The outcome of this disagreement, which would colour ideas about the date of the British Palaeolithic will, again, be examined in Chapter 2.

Behind the question of human antiquity, so often pushed to the front of the stage, lay an intricate background of geological time, woven from tangled sequences of river drifts, bones, and glacial drifts. Before 1859, geologists were all looking in different directions and reaching conflicting conclusions as they each sought answers to their own questions. When stone tools were found to date back to these times, although they slid into the vision of researchers like Prestwich, Falconer, and Ramsay, and even entered their arguments, they did not glide straight to the centre of their attention. It is important to clarify this point, which has been made earlier. Geological sequences provided the means to order the British Palaeolithic—but the beliefs, motivations and controversies that lay *behind* those sequences provide the key to understanding why they ordered the Palaeolithic in the way they did, and why that task provoked such controversy.

1.5 THE ESTABLISHMENT OF HUMAN ANTIQUITY IN 1859

In 1858 a new cave was discovered during quarrying operations above the small fishing town of Brixham in Devon. Falconer wrote to the Geological Society, hoping to win their interest and financial backing for an excavation, arguing that this cave might well contain the bones on which he could build a detailed chronology of extinct mammals. That question, he reminded the Society, had received little attention since Buckland's publication of *Reliquiae Diluvianae* in 1823:

---

[48] Morris 1838, 545; Trimmer 1851a, 25–26. On Morris, see H. Woodward 1878.
[49] Godwin-Austen 1850; 1851, 132–133; 1855; Lyell 1853, 75–76, 737–740; Prestwich 1856, 132–133. On the life of Godwin-Austen, see H. B. Woodward 1885. Godwin-Austen was known as 'Austen' before 1854, but all his publications are here grouped under 'Godwin-Austen.'

I am strongly of the conviction that, with our present advanced knowledge, the thorough investigation of a well-filled virgin cave in England would materially aid in clearing up the mystery, either of the contemporaneity of the Pliocene Mammalian Fauna with the commencement of the Postpliocene Fauna, or of the conditions and association under which the former was replaced by the latter.[50]

The Windmill Hill Cavern, also known as Brixham Cave, was excavated between 1858 and 1859. This excavation was not begun in order to prove human antiquity—even if it has been portrayed as such in retrospect—but it did lead, nonetheless, to a sympathetic reanalysis of the tools that de Perthes had been finding in the Somme Valley since the 1840s. This outcome was ironic in view of the uncertainty attached to cave stratigraphy, upon which, as Lyell had observed, a reliable case for such associations could not be built. But Brixham differed in two important ways from earlier, rejected cases: in the method of excavation, and the people who directed it. The role of this Cave in the human antiquity question has been explored thoroughly by Bowdoin Van Riper.[51] Only a brief summary is offered here.

Excavations were conducted by a committee of the Geological Society, assisted by a local sub-committee. They were funded largely by the more ancient and exalted Royal Society. Some of the committee members will be familiar: Falconer was the Chairman and Secretary; Prestwich, the Treasurer with responsibility for directing the excavations. Ramsay was another member: he had been elected President of the Geological Society by the time of his visit to the Cave in September 1858. Lyell also brought his illustrious name to the group. Another important, but less lofty member of the committee was William Pengelly (1812–1894), a Fellow of the Geological Society who lived nearby. It was Pengelly who had first drawn attention to the discovery of this Cave in his neighbourhood.[52]

Pengelly had left school at the age of twelve to join the crew of his father's fishing boat, and he later became a schoolmaster in Torquay. He had dug in Kent's Cavern long before discovering Brixham Cave, and would come to observe that he was the only member of the Geological Society's committee with extensive practical experience in cave excavation. Unsurprisingly, it was Pengelly who had charge of the day-to-day running of the new site. He employed an unusually careful excavation technique, inspired by the system

---

[50] Falconer to the Secretary of the Geological Society, 10 May 1858, quoted in Prestwich 1873, 474.

[51] For a rebuttal of the myth that Brixham Cave was a test case for human antiquity, see Gruber 1965, 385. Riper 1993 gives a thorough analysis of the role played by this cave in the question of human antiquity.

[52] Murchison 1868 (ii), 493; Prestwich 1873, 475–476.

he had used at Kent's Cavern back in 1846: peeling away the layers of sediment horizontally, one by one, to avoid mixing material from different levels.[53]

Excavation progressed, and led to the famous discovery of 'Flint Knives' alongside the bones of extinct animals, all neatly sealed beneath stalagmite. Despite the care taken in excavation, the association of tools with these bones was not sufficient to lead to the acceptance of human antiquity because cave deposits had a dubious, possibly mixed origin. But, unlike previous reports of similar associations, the Brixham excavations had been conducted under the eyes of some of the most important geologists of the day. As Riper has observed, the significance of the Brixham discoveries was that they sparked a genuine personal interest among a group of individuals who had the intellectual weight to convince their peers of a greater antiquity for humankind.[54]

In September 1858, Falconer wrote to Grace, his niece, about a planned trip abroad: 'I am loath to the last degree to leave England at present, as I have so much unfinished work on hand: but I have already had a reminder that the least check or cold will lay me up for the winter'.[55] Whilst away, he fulfilled an old promise made to de Perthes when they first met in 1856: to visit his collections at Abbeville. With the recent discoveries at Brixham on the surface of his mind, Falconer lingered thoughtfully over the French finds. The tools from Brixham had come from the sediments of a frequented cave that might well have been disturbed in the past, but these French implements lay stratified and sealed by river drifts in a far more reliable geological context. Intrigued, Falconer wrote to Prestwich to suggest that he might want to see them too:

Abbeville is an out-of-the-way place, very little visited, and the French *savants*, who meet him in Paris, laugh at Monsieur de Perthes and his researches. But after devoting the greater part of a day to his vast collection, I am perfectly satisfied that there is a great deal of fair presumptive evidence in favour of many of his speculations regarding the remote antiquity of these industrial objects, and their association with animals now extinct.[56]

Prestwich, who had also been impressed by the association of the Brixham tools with animals now extinct, set out for France at the end of April 1859. There he was joined, only a few hours after his arrival, by John Evans, the numismatist

---

[53] On Pengelly, see the memoir by Hester Pengelly 1897. William Pengelly recorded his remarks on the excavation experience of committee members in a letter to Prestwich, dated 21 July 1865 (GSL: 8, p.3). On the techniques of excavation employed at Brixham, see Murchison 1868 (ii), 493–494, Pengelly 1873, 790, Prestwich 1873, 476, 482, and Riper 1993, 86–89.

[54] Riper 1993, 96. See also Lyell 1859, 93 and Prestwich 1860a, 280.

[55] Falconer to McCall, 25 September 1858: FMF: HF, 106.

[56] Falconer to Prestwich, 1 November 1858: FMF: HF, 322.

Figure 1.5. John Evans
(1823–1908), authority on
stone tools (GSL: P56/Evans).

and geologist (Fig. 1.5). Prestwich had first met Evans in a railway carriage in
1851. Then, they had been travelling as rival geological experts on their way to
the same water-rights case. Now, when they met at the Abbeville railway station,
they were both on the same mission: to see de Perthes and answer their questions
about the associations and antiquity of stone tools.[57]

Evans ran a family business, John Dickinson & Co., manufacturing paper at
Nash Mills, Hemel Hempstead in Hertfordshire. He already had a sound
reputation as a numismatist, and would extend his authority from coins to
stone tools.[58] The morning after their arrival in France, de Perthes collected
Prestwich and Evans, taking them first to the gravel pits where he had found
his implements, and then to his house where they saw the collection. Evans
wrote to Frances Phelps, his future wife, to share his thoughts on visiting this
curious house:

a complete Museum from top to bottom full of paintings, old carvings, pottery &c
and with a wonderful collection of flint axes & implements found among the beds of
gravel, and evidently deposited at the same time with them—in fact the remains of a
race of men who existed at the time when the deluge or whatever was the origin of

---

[57] On the visit by Prestwich and Evans to de Perthes, see Prestwich 1899, 121–124 and Evans
1943, 100–104. On the life of John Evans, see Evans 1943 and Foote 2004.
[58] Evans 1843, 96; Sollas 1909, lviii.

these gravel beds took place. One of the most remarkable features of the case is that nearly all if not quite all of the animals whose bones are found in the same beds as the axes are extinct.[59]

Evans and Prestwich were convinced that humans had, indeed, lived at the time of Drift and extinct mammals. They had been particularly impressed by the extraction, in front of their eyes, of a flint axe from the undisturbed gravels.[60]

On their return to Britain, they presented a carefully argued case for human antiquity to two prominent learned societies. Their reports focused on discoveries in the Somme Valley, but an old British site made a strategic reappearance. Evans had inspired Prestwich to revisit Hoxne, where Frere had recovered his hand-axes over half a century before, and this British site was presented as an analogous case to the French sites.[61] Brixham Cave, once so provocative, was passed over as eyes were drawn to France. To Pengelly's irritation, it would be nearly fifteen years after work at Brixham had ended before the final report was published in full: a delay that was due, in part, to the ill-health and early death of Falconer in January 1865.

On 26 May 1859 Prestwich delivered their conclusions to the Royal Society. In the role of the geological authority, he answered every possible objection or alternative hypothesis in turn: the flints were the products of human hands, not natural processes; they were ancient, not modern; they had not been introduced subsequently into the geological beds; they were associated with the bones of extinct animals; and they were post-glacial in date. Evans, as the stone-tool expert, contributed a few words on the implements. Prestwich had forgotten to bring the prepared Appendix with him, so Evans had to recall his thoughts on the spot: he reinforced Prestwich's point that these had undoubtedly been made by humans, and explained that they differed in shape and manufacture from the polished stone axes of the relatively recent Celtic period.[62]

The stone tools of River-Drift times attracted widespread interest. This news soon seeped into other societies. Reports infiltrated geological, antiquarian, and anthropological institutions; and research on the British Palaeolithic sequence would retain this scattered institutional identity for many decades more. On 2 June 1859 Evans gave another account of the tools to the Society of Antiquaries of London, conservative descendent of Bagford's old tavern-based club. This paper was summarised at length in *The Athenaeum* on 11 June, the

---

[59] Evans to Phelps, 1 May 1859: AMO: 1927.6033/1927.4006.
[60] Prestwich 1860b, 52.   [61] Prestwich 1860a, 304–308; 1860b, 57.
[62] Prestwich 1860a, 294, 308–309; 1860b, 59. Prestwich's forgetfulness was recorded by Evans 1943, 103.

organ of the exclusive London club for learned gentlemen. Falconer delivered reports to the Geological Society in June on further associations of stone tools and extinct mammals, this time from the Grotta di Maccagnone in Sicily.[63]

Lyell, who had visited the Somme Valley gravel pits soon after the departure of Prestwich and Evans, upheld the antiquity of stone tools in his 1859 Presidential Address to the Geological Section of the British Association for the Advancement of Science. The British Association had been founded in 1831 to encourage communication between members of the provincial learned societies and the prominent national societies clustered in London. Itinerant yearly meetings were held every year in a different town: popular events that attracted many interested bystanders.[64] Papers in favour of human antiquity were also heard by members of French learned societies, such as la *Société géologique de France* and *l'Académie des sciences*. Prestwich's conclusions were summarized in the Paris *Comptes rendus hebdomadaires des séances de l'académie des sciences* as well as in popular British journals like the *Annals and Magazine of Natural History*.[65]

In Britain, the opinions of Prestwich and other advocates were accepted in many quarters of learned society. There were, however, some objections. Thomas Wright (1810–1877), an antiquary and one of the founders of the British Archaeological Association, wrote to *The Athenaeum* in 1859, criticising Evans and suggesting that the flints from Abbeville had been formed naturally.[66] For some years afterwards, letters condemning the anti-Scriptural nature of these supposedly pre-Adamite finds appeared in the popular press: Evans collected such clippings in his scrapbook on 'Flint Implements from the Drift'.[67] Even a decade and a half after the delivery of the papers on human antiquity, a clerical gentleman from Durham felt provoked to write to Evans: 'The tide seems to have ebbed again. I thought some time ago that in a few years everyone would admit the antiquity of man, but it seems we are a good way from that yet'.[68]

There was even some squabbling among the defenders of human antiquity. Falconer and Prestwich, infuriated by a lack of acknowledgement in Lyell's book on *The Geological Evidences of the Antiquity of Man* (1863), shot a series of angry

[63] Anon 1859a, 781–782; Falconer 1860a.

[64] Lyell 1859, 94. On the origins and role of the British Association of the Advancement of Science, see Morrell and Thackray 1981.

[65] Prestwich 1859; Anon 1859b.

[66] Wright 1859. Wright's criticisms are examined by Riper 1993, 120, 122–123.

[67] John Evans collected a large number of newspaper clippings attacking human antiquity in his Scrapbook. These include articles by Scipion Gras 1862 in *The Parthenon*; Charles Watkins 1864 in *The Northampton Herald*; Nicholas Whitley 1874 in *The Globe*; and William Robinson 1874 in *The Cambridge Independent Press* (Evans, undated, 'Implements from the Drift': AMO: JE/D/1/10, pp 50, 70, 82, 84).

[68] Greenwell to Evans, 5 May 1874: AMO: JE/B/1/13.

letters into *The Athenaeum*, which embarrassed Evans.[69] Nonetheless, Lyell's book reinforced the message that the peoples who had left Palaeolithic implements in river drifts were coeval with the extinct animals that had once roamed the world. Human antiquity had, more or less, been established. The extent of that antiquity was, however, still to be decided; the connection to the varied and conflicting geological sequences outlined above was still unclear. With so many different visions of the Quaternary—enduring a Glacial epoch, traversed by ancient rivers, and inhabited by successive swarms of creatures that now included tool-makers—this would be a controversial task.

[69] Falconer 1863; Prestwich 1863; Lyell 1863b; Evans to Lyell, 15 April 1863: ULE: Gen 110/1203–1204. Bynum 1984, 162–169 addresses the allegations of plagiarism by Falconer and Prestwich. Grayson 1985, 111–118 examines the alterations made by Lyell to subsequent editions as a result of these criticisms.

2

Arguments over the Ice Age

And now there came both mist and snow,
And it grew wondrous cold:
And ice, mast-high, came floating by,
As green as emerald.

(The Rime of the Ancient Mariner,
S. T. Coleridge; Fry 1999, 31)

The acceptance of human antiquity in the mid-nineteenth century fed a desire
to know more about the age of these chipped stone tools from the drift. In
1863, Canon William Greenwell (1820–1918), the antiquary, archaeologist,
and collector from Durham, declared: 'The great question which has yet to be
settled is this—at what period was the drift in which the flints are found
deposited? And side by side with this was another important query—down to
what time did these now extinct animals occupy any part of our continent?'[1]
This chapter seeks to untangle the web of time that was spun around the stone
implements of Britain over the last four decades of the nineteenth century.

Greenwell's great question was a popular one, and 'what period' was often
answered by connecting the implementiferous drifts to the Glacial epoch. The
mid-glacial submergence, entertained by geologists like Ramsay and Phillips,
provided a convenient division between pre-glacial and post-glacial times. On
each side of this great division, detailed patterns were being drawn in stratigraphy and bones. As decisions were made about the pre-glacial or post-glacial
date of sediments from river drifts and caves, rich in tools and bones, the
glacial chronology was, meanwhile, being revised and subdivided too. During
the latter decades of the nineteenth century there was great activity and little
agreement about the order of events in these distant times. Researchers
immersed in different material, gathered from different geographical areas,
and asking different questions would not find it easy—or even desirable—to
mesh their findings into a single coherent sequence. Attempts to date the
stone tools of Britain entered a contentious arena.

---

[1] Greenwell 1863, 6.

2.1 A CONFUSION OF QUATERNARY CHRONOLOGIES

The chronological indicators scrutinised by these researchers—river drifts, glacial drifts, and bones—offered few clear answers to Greenwell's question. The sands, gravels, clays, and brickearths of Quaternary times were so scattered, patchy, and variable that even Prestwich found it difficult to understand their sequence. In 1859 he admitted to Lyell that he was at a loss when it came to arranging the drifts of southern England in their correct order:

As for the exact order of succession they [the southern drifts] are so complicated that as often as I imagined I had detected it as often have I been thrown out again. When I think about it, some 3 or 400 sections & facts flit before me some tempting me one way & some another until I feel fairly bewildered. In the great coast sections the matter is clear enough but when we come inland the confusion is great.[2]

The diagram (Fig. 2.1) that accompanied this letter illustrated Prestwich's despair of finding a simple linear sequence in these sediments. He used a circular format instead of the usual vertical list to avoid making any definite decisions about their relationship, observing: 'I do not attempt any order but give them round robin fashion'.

Falconer was equally bewildered by the sequence and date of extinct mammals. One problem, he argued, was that the views of his colleagues had changed very little since Buckland's time. As explained in Chapter 1, the animals of most interest to Falconer—fossil elephants, rhinoceros, and hippopotamus—were still clumped in a single geological period, which was either consigned to antediluvian times (now 'pre-submergence' or 'pre-glacial' times) or left suspended in uncertainty.[3]

Another problem was the lack of information about their patterning through time. As Falconer travelled across Europe, visiting large collections gathered in previous decades, he found few records of the stratigraphic position once occupied by elephant bones. Most collectors had assumed that all extinct elephants could be labelled 'mammoth'; Falconer, however, saw more variety in their collections.[4] If questioned about the date of their 'mammoth' bones, these collectors would probably have replied that their date was

---

[2] Prestwich to Lyell, 6 July 1859: ULE: Gen 115/4931–4936.

[3] Buckland 1823, 97; 1826, 5–6; Murchison 1868 (ii), 587–588.

[4] Murchison 1868 (ii), 78, 141. Rhinoceros species, like elephants, had been clumped together and designated *Rhinoceros leptorhinus* (Falconer 1857, 352). Lubbock's 1869, 288 explanation of the rhinoceros case began thus: 'It is necessary therefore to bear in mind that the *R. leptorhinus* of Owen is not the *R. leptorhinus* of Cuvier, but that it is the *R. hemitoechus* of Falconer, while M. Lartet maintains that it is identical with the *R. Merkii* of Kaup. On the other hand [...].' For this reason—confusion of terminology—rhinoceros research has not been

Figure 2.1. Prestwich enclosed this diagram of the drifts of southern England with his letter to Lyell, written in 1859 (ULE: Gen 115/4936).

given as much prominence as it perhaps deserves in this chapter. Elephants, instead, have been used to illustrate the problems faced by nineteenth-century palaeontologists. In the 1860s, the species known as *Rh. leptorhinus* (or Merck's rhinoceros) was usually grouped with the warm pre-glacial fauna, whilst *Rh. tichorhinus* (the woolly rhinoceros) was seen as a contemporary of the mammoth.

obvious: they were associated with the glacial drift. Falconer was not satisfied with such a simple assertion. He believed that this icy image of the mammoth had been impressed on their minds by the preserved mammoth, found 'frozen in, flesh and bone, at the mouth of the Lena' in Northern Siberia during the first few years of the nineteenth century.[5] Falconer could identify a variety of species in bone collections; to discover their date would be more difficult.

The troubles experienced by Falconer, Prestwich, and others in their search for order in these sediments and bones provoked Ramsay, the glacial geologist, to write in dejection to Lyell early in 1860:

in spite of what Falconer has done, the relation of the bones in Caves to the Superficial deposits of late Tertiary date is I think still unsettled. How many ups & downs there may have been, & how much obliteration of strata there may have been before the final grand subsidence that attended the deposition of the great boulder beds is another point on which we are in the dark. There is a horrid gap I fear.[6]

The efforts to fill this 'horrid gap' would continue, nonetheless. As the 1860s dawned, Falconer carried out his own investigations in bone-caves to collect more information. He had already tackled Brixham Cave (with its famous results for Stone-Age research) and the patterns he announced from the Welsh caves, a few months after Ramsay wrote his letter, would dispel some of this despair. Prestwich still searched for a sequence in the southern drifts; Ramsay, in the glacial drifts further north. Over the next few decades, a swarm of investigators, asking different questions about estimations of geological time and giving different answers about the date of stone tools, converged on two areas of England: the Thames Valley and East Anglia. (Map 2 gives the location of the major sites mentioned in the text.)

The Thames Valley drifts, which had given up one of their stone implements to Conyers in the early eighteenth century, later attracted geologists like Trimmer and Prestwich in their search for the origin and age of the sediments, and palaeontologists looking for bones. These river drifts also attracted commercial exploitation. London was pockmarked with pits, dug to extract gravel and brickearth for Britain's expanding roads, towns, and cities. Geologists, who found such activities convenient to their purpose, gathered around these windows into the ancient history of the River Thames. Collectors, too, visited the busy gravel pits and brick pits, hoping to pick up bones—and, from the 1860s, increasing numbers of stone tools.

East Anglia was another centre for commercial extraction—Frere's implements had been found twelve feet deep in a Hoxne brick pit—but geologists

---

[5] Murchison 1868 (ii), 78–79; Cohen 1994, 135–136; Rudwick 2005, 559–561.
[6] Ramsay to Lyell, 27 March 1860: ULE: Gen 115/5037–5038.

also valued this area for its glacial drift. For Prestwich, the most interesting feature of Hoxne was the relationship between its implements and the boulder clay. Boulder clays were distinctive glacial time-markers. They were thought to have been deposited during (or, some now claimed, before) the great mid-glacial submergence of the land: anything lying above them would be dated to post-glacial times. Boulder clays, sometimes known as the northern drift, extended up from East Anglia through the northern counties and into Scotland, but none seemed to have reached as far south as the Thames Valley. Implements, conversely, were concentrated in southern England and were rarely found far north. The boulder clay of East Anglia, as the most southerly example, would play a pivotal part in dating stone tools. One of its names was the 'Chalky Boulder Clay'. William Whitaker of the Geological Survey described it as 'a bluish-grey clay, mostly of a dark shade, crowded with boulders'; boulders that were dominated by chalk blocks.[7]

A valuable remnant of the time *before* the Chalky Boulder Clay could also be found in East Anglia: the Cromer Forest Bed, generally seen as the reflection of a warm, pre-glacial period. Bones and fragments of vegetation could be found in some of these peaty-looking deposits, which lay on the coasts of Norfolk and Suffolk above the Pliocene Crags. East Anglia and the Thames Valley would both be examined closely over the following decades by researchers searching for pattern and order in the recent geological past.

2.2 FALCONER, PRESTWICH, AND THE POST-GLACIAL RIVER DRIFTS

Ancient stone implements from the gravel pits of the Somme Valley, curious and plentiful, were coveted in Britain. Geologists and collectors crossed the Channel to France after the announcements of 1859, to buy specimens and view the famous sites of St Acheul and Montières, near Amiens and Abbeville. But it was not long before a trickle of tools from the drifts of England joined Conyers's solitary Gray's Inn Lane implement from London and Frere's few tools from Hoxne.[8] Whitaker later observed that the river drifts attracted far more interest once ancient stone implements had been found in them.[9]

---

[7] Prestwich 1860a, 307–308; Whitaker 1875, 53.

[8] Notable early reports of Palaeolithic discoveries in England were made by Prestwich 1861a, Wyatt 1862a, 1862b, 1864, Evans 1864a, Flower 1867 and Lane Fox 1869, 1872, amongst others.

[9] Whitaker 1889 (i), 338.

Prestwich, whose secluded geological interest in the river drifts pre-dated this rush of excitement, had already decided that the sediments and, by association, the tools that they contained dated to post-glacial times. In his 1859 report to the Royal Society on the new evidence for human antiquity he argued that the river drifts from the Somme Valley and from similar sites in Britain (such as Ilford and Brentford) post-dated the glacial boulder clay. Prestwich supported this post-glacial date with a description of Hoxne, Frere's old site in Suffolk, where boulder clay lay *beneath* deposits that had once yielded implements. This was not welcome news for the palaeontologists, who believed that cave fauna and river-drift fauna were pre-glacial, or for the geologists—Godwin-Austen and Lyell for example—who had supported their pre-glacial date in the past.[10] Prestwich knew that his post-glacial conclusion would not be popular. Although he declared this conclusion clearly at the end of his paper, his words are sometimes interpreted as conservatism or restraint on the subject of human antiquity:

> The evidence, in fact, as it at present stands, does not seem to me to necessitate the carrying of man back in past time, so much as the bringing forward of the extinct mammalia towards our own time; my previous opinion, founded upon an independent study of the superficial drift or Pleistocene deposits, having likewise been certainly in favour of the latter view.[11]

Falconer was one of the palaeontologists who believed that some river drifts were older than Prestwich thought. The range of species that they contained led him to disagree with his friend. Falconer had split Quaternary mammals into two groups: one pre-glacial, the other post-glacial. His pre-glacial group stretched from the time of the Pliocene Crags up through the Cromer Forest Bed to the mid-glacial submergence and included *Hippopotamus major* and the straight-tusked elephant (*Elephas antiquus*). Unlike those who relied on the shell chronology, Falconer saw no abrupt break within this pre-glacial period; he placed the major faunal change in post-glacial times, when woolly rhinoceros (*Rhinoceros tichorhinus*) and mammoth (*E. primigenius*) occupied a land which he thought had recently emerged from the sea.[12]

Although Falconer was satisfied that the low-level *gravel* beds of the Thames Valley were post-glacial, he believed that the bones from the *brickearth* gave these particular deposits a pre-glacial date. Remains of straight-tusked elephant and hippopotamus, familiar from the pre-glacial Cromer Forest Bed, could be found in the low-level brickearths at sites like Grays

---

[10] Godwin-Austen 1851, 132–133; 1855; Lyell 1853, 75–76, 737–740; Prestwich 1860a, 302, 308–309. On palaeontological beliefs in a pre-glacial date for the cave and river-drift fauna, see Prestwich 1864, 248 and Murchison 1868 (ii), 209, 582–583.
[11] Prestwich 1860a, 309.   [12] Falconer 1858, 83–84; Murchison 1868 (ii), 206–210.

Thurrock in Essex. The animals that Falconer considered post-glacial—woolly rhinoceros and mammoth—were absent. Many of these low-level brickearths also contained the shell *Cyrena fluminalis* (Fig. 2.2), which was then regarded as a purely pre-glacial species.[13]

Falconer's opinion about the pre-glacial date of these brickearths changed when he worked on the great haul of bones from the Welsh caves gathered by his friend, Lieutenant-Colonel Edward Robert Wood (1819–1876) of Stout Hall. Wood had spent a decade collecting these bones from the caves of the Gower peninsula in Glamorganshire. His collection offered Falconer a chance to refine his theories about the sequence of extinct mammals, and in 1858 Falconer began to examine the distribution of these bones. They included species from both his groups: pre-glacial and post-glacial. He found some puzzling patterns. In Bacon Hole and Minchin Hole, for example, the bones of his supposedly pre-glacial rhinoceros and straight-tusked elephant had been found *above* a deposit of sea-sand, which contained living species of shells, which hinted that they post-dated the mid-glacial submergence. In most caves the pre-glacial and post-glacial elephant species were found in discrete groups, as Falconer had expected; but the cave of Spritsail-Tor contained a mixture of the two.[14] Such anomalies did not match his definitions.

In September 1859, Falconer urged Prestwich to come to Wales and give an opinion on the stratigraphy: 'I am so grieved to think that you are coming on here so late. [...] It is of the utmost importance that we should have seen the caves together. [...] Some of the points in the problem are very complex, and we might have examined them together with much advantage'.[15] When Prestwich finally arrived a few weeks later, he examined the physical geology of the nearby area and discovered two deposits that connected Falconer's bones to the mid-glacial submergence (marking the period between pre-glacial and post-glacial times). The first was a boulder clay; the second, a raised beach (one of the ancient shorelines that were thought to have been left above sea level as the land emerged). Prestwich linked the raised beach to the marine sands and shells that lay beneath the bones in the caves, since they lay at a similar level, and concluded that both deposits were later than the boulder clay.[16] He saw no evidence for any subsequent submergence that might have disturbed the bones.

---

13 Falconer 1858, 83; Murchison 1868 (ii), 209, 582–583, 586.

14 On Falconer's theories about the Gower caves and their influence on his faunal sequence, see Murchison 1868 (ii), 498–499, 501–503, 529–530, 586, 589.

15 Falconer to Prestwich, 9 September 1859: FMF: HF, 324.

16 Prestwich to Falconer, 17 May 1860, quoted in Prestwich 1899, 149–151.

Falconer delivered his controversial conclusions about the post-glacial age of British cave fauna to the Geological Society of London on 30 May and 13 June 1860. He informed his audience that this surprising mixture of species all dated to post-glacial times—even those he had once considered pre-glacial: '*Elephas antiquus* with *Rhinoceros hemitoechus*, and *E. primigenius* with *R. tichorhinus*, though respectively characteristic of the earlier and later portions of one period, were probably contemporary animals'.[17] These conclusions strengthened Prestwich's case for the post-glacial date of the Thames Valley river drifts, where the same range of species had been found. They were not welcomed by the Society, and the paper only appeared as an abstract in their journal. After the first reading in May, Falconer warned Prestwich about the atmosphere:

You know what a fierce onslaught was made on me by Lyell and [Godwin-] Austen. I thought the latter was going to eat me up. The whole subject will be up again at the next meeting: when the main brunt of the attack will fall on you. There is no wavering in the aspect of the mammalian evidence, it is coming out stronger than ever.[18]

After contributing to the Gower paper, Prestwich made two further observations that would finally establish the post-glacial date of the river drifts and cave deposits. First, he re-dated a species of small freshwater shell, *Cyrena fluminalis* (Fig. 2.2), previously regarded as a marker of pre-glacial times. *Cyrena fluminalis* shells were known from the Norwich Crag, a formation far older than the boulder clay of Norfolk. Their presence in the Thames Valley brickearths had added weight to a pre-glacial date. When these shells were found with flint implements at Abbeville, the question of their age gained more urgency. In the summer of 1860, Prestwich was pleased to find *Cyrena fluminalis* lying *above* the boulder clay in a gravel pit near Hull, proof that the species had lived on in Britain into post-glacial times and another blow to the pre-glacial date of the Thames Valley river drifts.[19]

Prestwich struck his second blow when he announced to the Royal Society in 1862 that Falconer had identified hippopotamus and straight-tusked elephant (both members of his old pre-glacial group) lying alongside mammoth and woolly rhinoceros in the low-level valley-gravels of Bedford, in the Ouse Valley. These valley-gravels cut into, and therefore post-dated, the boulder clay: a stratigraphical relationship that confirmed Falconer's belief in the post-glacial date of these mammals and Prestwich's views of the sediments.[20]

---

[17] Falconer 1860b, 491. See also Murchison 1868 (i), xliv; (ii), 531, 536, 586–587 and Prestwich 1861b, 446.

[18] Falconer to Prestwich, 2 June 1860: FMF: HF, 340.

[19] Prestwich 1861b; Murchison 1868 (ii), 583, 590.   [20] Prestwich 1864, 283–285.

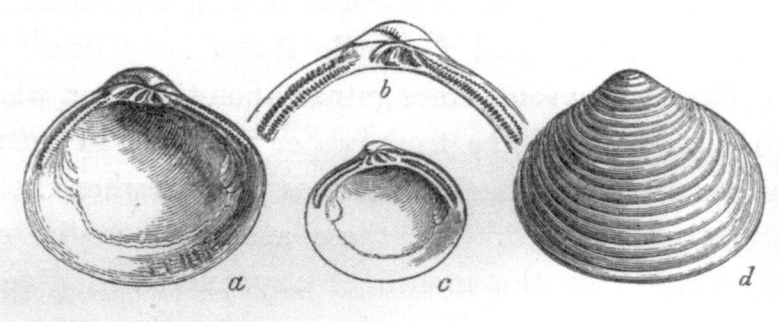

Figure 2.2. *Cyrena fluminalis*, the fossil shell that proved such an important time-marker (Lyell 1863a, 124, Fig 17).

The gravels near Bedford also produced flint implements. James Wyatt (1816–1878), a Bedford worthy and editor of the local paper, had been hopefully haunting the nearby gravel pits since his visit to the Somme Valley in 1860. He discovered his first English implement on 8 April 1861 at the Biddenham pit above the River Ouse, alongside the bones of extinct mammals. This rare find attracted visits by Evans, Prestwich, and Lyell, who wrote to Wyatt: 'Of all the sections given by Mr. Prestwich in his last Paper yours is the best, as the Flint implements occurred at the bottom of the gravel, and the relation of the northern drift is so clear'.[21]

In the same paper that tackled the contentious connections to the Glacial epoch, Prestwich reported the latest results of his steady work on the drift sequence in and around the river valleys. He divided the post-glacial valley-gravel into a high-level series, lying around 50 ft to 100 ft (or even 150 ft) above the current river, and a younger low-level series. The high-level valley-gravels had, Prestwich believed, been left on the valley sides as the torrential ancient river, charged with floodwaters and blocks of ice, carved its way down through the valley floor. He warned that the two spreads were not always distinct, and that the division between the two was further complicated in southern England by their common origin in a still higher spread of hill-gravels.[22] Though simple and indistinct, this sequence of gravel spreads would supply the basis for future elaborations and would offer a finer date than 'post-glacial' for the tools that lay within these sediments.

The Hoxne section, the Gower caves, Prestwich's re-dated *Cyrena fluminalis* and the Bedford sections all helped to promote the post-glacial age of the river drifts and cave deposits. Godwin-Austen still could not accept that the bones

[21] Wyatt 1862b, 154.   [22] Prestwich 1864, 250–251, 255–256, 264–265, 301.

in the southern river valleys were post-glacial, and argued in 1863 that these had been washed out from the boulder clay. In essence, as Falconer put it to Evans, Godwin-Austen concluded that 'all your Geological and Palaeontological, and time inferences are Buncombe'.[23] Lyell, however, agreed that the deposits of the Thames Valley were post-glacial. He explained in *Geological Evidences of the Antiquity of Man* (1863) that they post-dated the time of the great submergence of Central England, when the marine drift (boulder clay) had been deposited beneath the glacial sea. This opinion reached a large audience: Lyell's book, the first major synthesis of the new evidence, ran to three editions in its first year.[24] The post-glacial date of tool-bearing deposits rapidly gained credibility.

2.3 DAWKINS AND THE STRATIGRAPHICAL GEOLOGISTS

As Falconer approached the last years of his life, another palaeontologist entered the Quaternary arena: William Boyd Dawkins. Dawkins had excavated in the 'Hyaena-den' of Wookey Hole, Somerset, as an undergraduate at Oxford in the late 1850s. He gained more knowledge of the superficial drifts in the Geological Survey, which he joined not long after graduating, and began work on a lengthy monograph on the recent fossil mammalia for the Palaeontographical Society. Dawkins also wrote several papers about the relationship between the bones and sediments of the Thames Valley and the Glacial epoch.[25] Perceptions had changed since the work of Falconer and Prestwich in the early 1860s, but there was still tension between palaeontological and stratigraphical interpretations.

In 1867, Dawkins gained notoriety for his suggestion that the lower brick-earths of the Thames Valley were not post-glacial, but glacial or pre-glacial in date. This opinion was delivered to the Geological Society, but his arguments reached a wider readership through a summary in *The Philosophical Magazine*.[26] His conclusions provoked criticism from Whitaker and Searles V. Wood Jr., the East Anglian geologist. Both saw a different pattern in the stratigraphy. Pre-glacial river drifts were no longer popular, but the glacial

---

[23] Falconer to Evans, 19 October 1863: FMF: HF, 363U; Godwin-Austen 1863, 68.
[24] Lyell 1863a, 163–174. See Grayson 1985 on 'The First Three Editions of Charles Lyell's *The Geological Evidences of The Antiquity of Man*'.
[25] On the Wookey Hole excavations 1859–1862, see Dawkins 1862; 1863. For more on Dawkins, including his work outside palaeontology, see Tweedale 1991; 2004.
[26] Anon 1867b; Dawkins 1867.

Figure 2.3. William Boyd Dawkins
(1837–1929), palaeontologist (GSL: P47/
28/03).

chronology was still central to perceptions of the recent geological past. This is
evident from the vigour of their reaction to Dawkins.

Whitaker, the only son of a wholesale perfumerer, brush manufacturer, and
wine merchant, had joined the Survey in 1857. A few years later, in 1863,
Roderick Murchison had decided that the Survey's maps of solid geology
ought to be joined by maps of the drift. London was one of the earliest areas
to be re-surveyed for drift geology on the one-inch scale. Several officers were
assigned to the London district, including Dawkins, but it was Whitaker who
contributed most to the new maps and explanatory memoirs. The officers
surveyed the area methodically, from west to east. The maps they produced
built on the earlier work of Prestwich, and had significance for agriculture,
horticulture, sewers, water supply, building, and extraction as well as the
British Palaeolithic.[27]

Whitaker's black, leather-bound Survey notebooks reveal how geologists set
about mapping and correlating these complicated deposits. Mapping was
laborious work. Whitaker was alert to any openings in the ground that
might explain the relations between different layers of drift. He walked
many miles to collect records from hundreds of road-cuttings, railway-
cuttings, gravel pits, sewer trenches, and building foundations. Special note
was made of any incline or disturbance that could bridge the gaps between
pits and build up a picture of the beds below. Though he noted the presence

---

[27] Bristow in Whitaker 1875, introductory note; Whitaker 1889 (i), 9; Geikie 1898, 7–8, 11,
19; Flett 1937, 112.

of implements as well as bones and shells, he did not give them particular attention, whilst connections to glaciations were rarely mentioned—Whitaker believed the London drifts to be post-glacial in date.[28]

Like the other officers, Whitaker carried Ordnance maps into the field. He covered the engravings of house plots and field boundaries with ink lines, watercolour washes, and descriptive notes to illustrate the geology beneath. When surveying was completed, officers would retire to the office to work up their portion of the final map. They had to fit their observations to the standard Survey colours and classifications: red-orange for old river-gravel, orange for gravel and sand, and yellow for brickearth and loam. These colours simplified a complicated story: the many different stretches of gravel recorded on Whitaker's working copy had to be mapped as one mass on the Survey Map. More detail could be given in the memoirs that accompanied the maps, but the difficulties of classification were evident. Whitaker cautioned that the colours on the Survey's maps did not always indicate a definite opinion, and were sometimes merely suggestions based on the balance of probabilities.[29]

Whitaker was not self-sufficient when mapping the drift: he relied on the memories of foremen, the records of engineers, and the observations of geologists to supplement his own records. In his notebooks, he would speculate about how his ideas related to their findings—and he often disagreed with them. Whitaker's biographer, Bill George, has observed that although he was known as an exceptionally kindly man by most of his peers, his superiors on the Geological Survey saw a more dogmatic and intractable side to his character; Whitaker was unafraid of speaking his mind.[30]

As Whitaker worked for the Survey on the river drifts of London, Wood was mapping the Crag and glacial drifts of East Anglia. Wood was a Suffolk solicitor, the son of the well-known authority on Pliocene molluscs, S. V. Wood, Sr. He had started work in the mid 1860s, shortly before the Survey arrived in the area with their official instructions to survey the drift. A friend, Frederic Harmer (1835–1923), gave him some assistance; the two men had met in 1864 when geologising on the Norfolk coast at Mundesley beach. When Wood delivered his classification of the glacial drift of East England to the Geological Society in 1864 he received scant attention from their Council, who suspected that his work was derived from Trimmer's research. Undaunted, Wood printed his map privately. A few years later, he deposited his enormous bound manuscript and his maps that described these little-known deposits in the Geological Society.

---

[28] Whitaker 1866–1874, Notebooks 14–17: BGS: NGRC 994–997; Dewey 1926, 233.
[29] Whitaker 1875, 49; 1889 (i), 300–301, 389.   [30] George 2004, 51, 61–63.

Wood's esteem grew with his pile of publications, studded with sections and forbidding in their detail. The old terms 'glacial drift' and 'river drift' began to fade as geologists like Wood assigned to these deposits a new variety of more specific names. The tripartite division of the glacial drift into Lower Glacial, Middle Glacial Sands, and Upper Glacial Deposits, developed by Wood, became a standard framework for geologists working on the East Anglian drifts (see Table 2.1).[31] It was also Wood who had named Dawkins's disputed beds in the Thames Valley: 'the Lower Brick-earths.' In 1867, when Dawkins read his paper to the Geological Society 'On the Age of the Lower Brick-earths of the Thames Valley', Wood was in the audience. He reported the whole affair in a letter to Harmer: 'We had a great fight on Wednesday over Mr. Dawkins' paper on the Thames Brick-earths', he declared with satisfaction.[32]

Prestwich's arguments had gained support over the last few years, and Wood was one of a majority who now believed in the post-glacial date of the Thames Valley river drifts.[33] Dawkins agreed that *most* of the implementiferous river

Table 2.1. A simplified summary of the sequence of Quaternary deposits in East Anglia and the Thames Valley that was widely used from the 1860s until the end of the nineteenth century. Prestwich (1864, 273) grouped the lower brickearths with the low-level river gravels, but Dawkins (1867, 109) suggested a link to the Chalky Boulder Clay.

Geological divisions	Geological deposits
Post-glacial	Low-level river gravels High-level river gravels
Glacial deposits	Upper Glacial series (Chalky Boulder Clay) Middle Glacial series (Sands and Gravels) Lower Glacial series (Lower Boulder Clay; Contorted Drift; Cromer Till)
Pre-glacial	Cromer Forest Bed series
Pliocene	Crags

[31] On Wood's life, see Bonney 1885 and H. Woodward 1885. The suspicions of the Geological Society Council about Wood's work were recounted by Woodward 1907, 236. Wood's manuscript and maps in the Geological Society Library are dated January 1867 (GSL: 71/13E). On Wood's tripartite division of the glacial beds, see Wood 1870a and Harmer 1902, 451.

[32] Wood to Harmer, 11 January 1867: BGS: GSM1/542/51.

[33] On Wood's support for the post-glacial age of the Thames Valley gravels, see Wood 1866, 62, 105; 1867, 404–416. The year before Dawkins read his paper, Wood 1866, 59–61 considered the lower brickearths of Ilford and Erith to be early post-glacial in date, and to pre-date the deposition of the main spread of valley-gravel in the Thames, with the Grays

Table 2.2. A summary of the faunal divisions used by Dawkins (1867, 103–109) to argue his case for the pre-glacial date of the Thames Valley lower brickearths.

Geological divisions	Geological deposits	Selected mammals
Post-glacial	River gravels and caves (mixture of cold and warm fauna)	Reindeer (post-glacial cold); *Elephas primigenius*; *E. antiquus*; *Hippopotamus major* (pre-glacial warm)
Glacial	The Trail of Fisher; Chalky Boulder Clay	
Pre-glacial	Lower brickearths of the Thames Valley (cold fauna absent)	*E. primigenius*; *E. antiquus*; *H. major* (all pre- and post-glacial). *E. priscus*; *Rhinoceros megarhinus* (all pre-glacial)
	Cromer Forest Bed series	*E. primigenius*; *E. antiquus*; *H. major*; *E. priscus*; *Rh. megarhinus*

drifts and cave deposits post-dated the boulder clay, but argued that the bones in some of the lower brickearths (Crayford, Erith, Ilford, and Grays Thurrock) belonged to pre-glacial animals. In an argument that recalled Falconer's earlier work, Dawkins supported his case by pointing to the Pliocene species (familiar from the pre-glacial Cromer Forest Bed) and warmth-loving pachyderms in these brickearths (such as straight-tusked elephant and hippopotamus). He also emphasised the strange lack of mammals associated with arctic conditions (such as reindeer, lemming, or musk-sheep), usually abundant in post-glacial deposits. His divisions of the mammals are given in Table 2.2.

Dawkins concluded that the lower brickearths dated to pre-glacial times, before a refrigeration of climate encouraged arctic species to enter Britain and mingle with its earlier inhabitants. He suggested that the distinctive deposits known as 'Trail', which lay above the brickearths, were contemporary with the boulder clay of East Anglia. Trail was a name given by Osmond Fisher (1817–1914), the Cambridge geologist, to a deposit that he thought had been formed by frost, rain and other gradual, subaerial agents of denudation.[34]

---

brickearth being deposited much later in time, after the gravel sheet. The year after Dawkins had read his paper, Wood 1868 decided that all these brickearths dated to this late time, after the main sheet of valley-gravel had been laid down in the Thames Valley.

[34] Dawkins 1867, 106–109. On Dawkins's belief in the post-glacial date of the majority of the river drifts, see Dawkins 1862, 119–120; 1863, 273–274. On Fisher's Trail, see Fisher 1866. Fisher based his theories about the 'Trail' on Trimmer's work. Trimmer believed that this superficial soil had been deposited late in the Glacial epoch (after his Upper Drift), and termed it 'the Warp of the Drift, or the Erratic Warp' (Trimmer 1851b, 32).

This paper provoked strong reactions in the Geological Society. Wood disagreed with Dawkins's pre-glacial date for the lower brickearths and sniffed dismissively at 'the Geology of Palaeontologists' in his letter to Harmer. He was delighted to pass on Whitaker's response: that 'if the palaeontologists cannot make their palaeontology square with the Geology so much the worse for the former!' (Whitaker was referring to the relationship between the deposits at Grays and the boulder clay.) Lyell was also present at the meeting, but did not give his support to either side and, to Wood's disgust, 'made a rambling speech meaning nothing except that the case was very obscure (which it is not) & that the more people looked at it the more difficult it was!!! & carefully keeping himself from being committed to any opinion'.[35]

Whitaker and Wood did not hold exactly the same opinions about the Thames Valley river drifts, but they were united in their irritation with Dawkins for refusing to accommodate his palaeontological sequence to the supposedly reliable framework established by field geologists. They believed that he had simply misunderstood the stratigraphical evidence. Although Dawkins was employed as an officer of the Geological Survey in the 1860s, Wood thought him a poor fieldworker, and confided to Harmer: 'I regard such a production as Mr. Dawkins as evidence of his incompetency to map the district on which he is engaged, however competent he may be to describe the palaeontology'.[36] He later revealed to Thomas McKenny Hughes (1832–1917), another Survey officer, that 'Dawkins did his work about here by taking my map & making use of as much of it only *as squared with his preconceived theories*, & mostly inventing the rest'.[37] There was personal rivalry behind this opinion: Dawkins had invaded Wood's private patch. But Wood made a perceptive point about the difficulties under which Survey officers laboured when, two years later, he remarked to Fisher:

Dawkins is not acquainted with the *entire* structure of the Thames valley; indeed none of the Geo[l] Survey men are; for they are like Tailors working at a coat; one takes the skirt, another the sleeve, another the collar, &c; & they do not, so Hughes told me, even pay much attention, often none, to what their colleagues close by are doing; so that they do not grasp the whole of a subject.[38]

---

[35] Wood to Harmer, 11 January 1867: BGS: GSM1/542/51.
[36] Wood to Harmer, 13 January 1867: BGS: GSM1/542/52.
[37] Wood to Hughes, February 3 1873: ULC: 9557/2/G/26. The Survey benefited from Prestwich's private work on the superficial deposits around London; he lent them his maps when they started work on the area in the 1860s (Prestwich 1899, 173–174). Lack of respect for Dawkins's fieldwork skills contributed to his failure in 1873 to secure the Woodwardian Chair of Geology at Cambridge (O'Connor 2005).
[38] Wood to Fisher, 11 May 1869: ULC: 7652/V/P/56.

Although Hughes had fallen out with Murchison, his Director–General, around this time, there was some truth in Wood's remark.[39] The job of the Survey men was to produce a general geological summary of the districts they were assigned, using standard classifications and colours. They were not necessarily interested in grander theories.

The dispute between Dawkins, Whitaker, and Wood was never resolved. Dawkins continued to defend an early date for the lower brickearths of the Thames Valley, though he became more wary of the term 'pre-glacial.' His opponents continued to riposte that he had not taken the stratigraphical evidence into account. In 1869 Dawkins left the Survey to take up curatorial and lecturing duties in Manchester, where he would become Professor of Geology. Whitaker, who remained in the Survey, was still bothered by the conflicting views of Dawkins and other palaeontologists. In his authoritative memoir on *The Geology of London* (1889), he wrote:

It is curious to note how purely palaeontological reasoning has led to error in assigning to our River Drift its place in the geologic series. [...] It is strange that whilst geologists are generally ready to give their full value to palaeontological reasons, yet palaeontologists sometimes show a tendency to dogmatise on geological matters, and to pass over stratigraphical considerations, or those which depend upon the character and position of the beds, as things of small importance, though the field-geologist may have worked long and hard at them. The two lines of evidence should be combined, though perhaps only stratigraphical evidence can decisively settle questions of the relative age of nearly related deposits.[40]

It is not surprising that researchers working on different lines of evidence defended different sequences, but it is difficult to classify their conclusions into groups that can be posted into different specialist pigeonholes. Palaeontologists argued with stratigraphical geologists, but they also argued with each other. Dawkins could find non-palaeontologists who supported his views: in 1867 he reported to Fisher, 'That paper of mine is endorsed by [Godwin-] Austen, Sir C. Lyell, and Ramsay'.[41] Equally, Wood and Whitaker might have agreed that the Thames Valley deposits were all post-glacial; but they differed about the relations between these deposits. Then there were other considerations: Wood was defending geographical as well as intellectual territory against Dawkins; young researchers had their names to make; older peers, their statuses to defend. Lyell had a good reason for rambling in the public discussion that

---

[39] On the dispute between Hughes and Murchison, see Oldroyd and McKenna 2005, 206 and O'Connor 2005, 445, 459.

[40] Whitaker 1889 (i), 334–335.

[41] Dawkins to Fisher, 10 July 1867: ULC: 7652/II/NN/18.

followed Dawkins's paper, sitting on the fence where he could tuck his reputation safely away from flammable arguments.

2.4 CARVING UP THE ICE AGE

The arguments about geological time considered thus far have circled around a single pivot: the supposed mid-glacial submergence that divided the pre-glacial from the post-glacial period. Some glacial geologists, however, began to suspect that the Glacial epoch had been interrupted by periodic warm phases: 'interglacials'. Their findings laid a finer mesh of alternating glacial and interglacial periods over the old mid-glacial submergence: a glacial sequence that could connect bones, river drifts, and tools to smaller periods of time. The existence of lengthy warm periods within the Glacial epoch also had important implications for the movements of ancient animals, including tool-makers. The interglacial theory led James Geikie, its most prominent supporter, to challenge the simplistic 'post-glacial' date of river drifts and cave deposits, and to question the age and distribution of Dawkins's mammals. Interglacials were not popular in the nineteenth century, but the arguments that surrounded them are illuminating.

If there was a standard vision in Britain of the Glacial epoch in the latter half of the nineteenth century, it might have looked something like this: in early glacial times, glaciers formed and converged into a massive ice-sheet that cloaked much of Britain. Boulder clay was produced beneath this great weight of gritty ice as it carved its way across the land. The animals of warm Pliocene times died out. Then the land sank slowly beneath the sea. During the great mid-glacial submergence, shelly drifts were deposited on submerged mountainsides in Wales, and layers of clay (the northern drift, or boulder clays of southern geologists) formed under the dreary, iceberg-laden sea. Icebergs broke away from the feet of northern ice-sheets and sailed above the submerged Midlands and over eastern England. The land south of the Thames seemed to have remained above the waves—or perhaps the glacial drifts were later scoured away in that area.

As the country emerged again, old beaches became raised high above the sea. Glaciers gathered once again in the mountains of Scotland, Wales, and Cumbria in this second glacial period, churning up the drifts of earlier times but never reaching the extremes of the first glacial ice-sheets. Further south, the Thames flowed a hundred feet above its present level, a tributary of the Rhine; the British Channel did not exist; much of the North Sea was dry land. As the climate in southern England became warmer, early tool-makers crossed over from the European mainland, accompanied by mammoth and woolly rhino. They made their way across the now-elevated drifts of East Anglia but did not penetrate far into colder northern regions. They rested by the sides of lakes and wandered down the ever-deepening valleys, where the lapping waters gradually covered discarded tools with the gravel, sand, and loam of the river drifts. Violent sweepings of debris from the surrounding area also dragged gravel and tools into these drifts.

This vision of a first and a second glacial period separated by a submergence was not universal. Some saw more variety in the deposits that had been left by the first great ice-sheet: additional layers that suggested that this period had not been uniformly cold. Three geologists working in Scotland shared a view of a finer glacial sequence: Archibald Geikie (1835–1924), James Croll (1821–1890), and Archibald's younger brother, James Geikie. All three were employed on the Scottish branch of the Geological Survey.

Archibald Geikie was the eldest, more ambitious, and less popular of the Geikie brothers. He spent his boyhood geologising around his home in Edinburgh, formed an early acquaintance with Robert Chambers and Hugh Miller, and had a fortunate and encouraging meeting with Ramsay in 1853. Ramsay was then the Local Director of the Survey and Professor of Geology at the School of Mines; Geikie joined the Scottish branch of the Survey in 1855. Not at all discouraged by Chambers's warning of the vast quantities of food he would be expected to consume as a Survey officer, Geikie looked on his early years in the field as a delightful time in his life. His brother, James, would come to share this feeling.[42]

The superficial drifts intrigued Archibald Geikie. He was paid to map the older solid geology, but admitted, 'it was not possible to shut one's eyes to those later deposits, and to the extraordinary problems presented by them'.[43] Geikie converted from icebergs to the land-ice theory: by 1861 he was convinced that the boulder clay had been ground and pounded into existence beneath a frozen cloak of slow-moving ice, the boulders it had grasped carving grooves into the bare rock-surfaces. In 1862 he reported to the Geological Society of Glasgow his observations of clearly stratified layers lying *within* the familiar jumble of contorted boulder clay. Some of these stratified deposits even contained the peaty remains of ancient plants, which suggested that the land had not remained frozen throughout the Glacial epoch, but had witnessed intervals when streams ran down the valleys, plants grew by the valley sides and reindeer and mammoth browsed on their leaves. Archibald Geikie seemed to have discovered a pre-submergence land-surface.[44]

---

[42] Geikie 1924, 17, 31, 52. See also Oldroyd 2004a.
[43] Geikie 1924, 94.   [44] Geikie 1863, 54–68, 88–89, 92–93, 96.

Geikie's account of subsequent events returned to the visions of Chambers and Ramsay. He described how the land then gradually sank beneath the sea and marine shells became incorporated within the boulder clays. When the land emerged, a 'stratified drift' was deposited: the wreck of the old boulder clay, distinguished from earlier drifts by its layers that had been carefully laid down by the waves. Then came the time of new, local glaciers, ploughing afresh through the accumulated debris of the lowlands after they had emerged from the sea.[45]

Theories about the origin of glacial deposits have attracted more interest among historians than theories about their order. Despite this interest, it is sometimes implied that the mid-glacial submergence was confined to adherents of the iceberg theory. This watery geological episode was certainly central to their view that icebergs had played a major role in depositing glacial drift, but submergence also appeared in the visions of the Glacial epoch held by Chambers, Ramsay, and the Geikie brothers, and was not opposed to their opinions about land-ice.[46]

To return to the stratified layers observed by Archibald Geikie in the pre-submergence drifts: these would be seen in a different light after James Croll applied his knowledge of physics and astronomy to glacial problems. Croll was the son of a stonemason from Perthshire in Scotland. He had trained as a millwright but an inflammation of the elbow forced him to stop work—the first of many bouts of ill-health that would plague his life. Croll's interest in physical science developed when he worked in Anderson's College and Museum, Glasgow, which held a good scientific library. In the spring of 1864, Croll turned to astronomical theories to discover the cause of Glacial epochs. Over the following two decades, he explained how cyclical changes in the orbit of the earth could generate cyclical changes in climate.[47]

Croll published the first account of his ideas in 1864. Ramsay, riveted by the paper, immediately wrote to Croll to express his interest.[48] Archibald Geikie,

---

[45] Geikie 1863, 98–106, 154–155.

[46] Oldroyd 1999, 186–187, 192 recognises the reliance placed by James Geikie on submergence. Geologists who might be cast into a 'land-ice' or 'iceberg' camp tended to give more prominence to their favoured agency rather than abandoning the other agency altogether. Lyell 1863a, 247, 301–305 was willing to give a prominent role to land-ice in his *Antiquity of Man* and move away from his earlier iceberg-dominated scenario, but his ideas about the erosive power of land-ice were more rigid: 'I do not go so far, you will see, with Ramsay in regard to erosion of lakes but I go some way in regard to the small ones' (Lyell to A. Geikie, 14 May 1863: ULE: Gen 1425/354). The previous year, Ramsay (1862) had given a controversial account of the erosive power of glaciers to the Geological Society.

[47] Croll's biography has been written by Irons (1896). On Croll's astronomical theories of climate change, see Croll 1864, 1875, Imbrie and Imbrie 1979, Oldroyd 1996, 152–154; 2006.

[48] Ramsay to Croll, 19 August 1864, quoted in Irons 1896, 113–114.

now Director of the Scottish Survey, was also impressed by Croll's work and offered him the post of Secretary in their Edinburgh offices in 1867. Sir William Thomson (Lord Kelvin) supported his engagement by the Survey.[49] When Croll failed the requisite Civil Service examination Geikie managed to engineer his entry without the need for a second attempt. He asked Murchison to raise the matter with the Lords of the Committee of Council on Education.[50] Murchison obliged:

I would ask their Lordships to allow me to enrol Mr. James Croll as an Assistant Surveyor, albeit he has failed in his examination solely through nervousness; which so incapacitated him, that, although he is an eminent mathematician, and has earned a wide reputation as an author by the application of Astronomical Science to Geology, was pronounced to be deficient in arithmetic![51]

After Croll's work on astronomical theories of glaciation, sediments could be interpreted differently. Geikie's stratified deposits within the Scottish boulder clay, for example, might be seen as the reflection of one of Croll's lengthy warm interglacial periods. Croll also took a closer look at palaeontological patterns. He applied his ideas about interglacials to the problem of 'mixed fauna': the presence in the same deposit of both cold- and warm-loving species. Cold species (such as mammoth, woolly rhino, and reindeer) could be found alongside warm species (such as *Cyrena fluminalis*, hippopotamus, and straight-tusked elephant) at sites like Grays Thurrock, Hoxne, and Bedford. This occurrence had puzzled Falconer, inspired Lyell and Dawkins to describe migrations of hippopotami across Europe, and moved Prestwich to suggest that these aquatic animals, now known from the tropics, had adapted to post-glacial cold by growing warm coats of seal-like fur—a suggestion that had horrified Falconer.[52]

In 1868 Croll published his own explanation for the mixture of 'cold' and 'warm' species. He mentioned the theories of Lyell and Dawkins, who had explained the blend by invoking oscillations of climate and seasonal migration. Lyell, for example, had suggested that herds of hippopotami might have swum up rivers in the summer to bask in the Thames and then retreated south, away from the advancing snow and ice of winter.[53] Croll, however, took a different stance. Drawing on his astronomical theory, he argued that 'the

---

[49] Thomson to A. Geikie, August 1867: ULE: Gen 525.
[50] A. Geikie to Murchison, 11 August 1867: GSL: 838 G2/20; Murchison, 2 August 1867: BGS: GSM1/8, p. 511. Croll's problems with the Civil Service examination were also mentioned by Irons 1896, 34–35 and Geikie 1924, 115.
[51] Murchison to the Secretary of Science and Art, 2 August 1867: BGS: GSM1/8, pp. 505–506.
[52] Falconer to Prestwich, 'Tuesday' 1862: FMF: HF, 348. See also Prestwich 1864, 285.
[53] Croll 1868, 381; Lyell 1863a, 156, 180–181; Dawkins 1867, 103–106. In the tenth edition of his *Principles of Geology*, Lyell 1867, 193 provided an alternative scenario that allowed the hippopotamus a more lengthy stay, during changes in climate that he suggested might be of

Figure 2.4. John Lubbock
(1834–1913), polymath
(GSL: P45/25/04).

intense cold of the glacial epoch was not continuous, but was broken up by one or more warm periods, during which the ice, to a considerable extent at least, disappeared for a long period of time'. He added that most of these ancient land-surfaces would have been swept away by the following glaciation, when the bones of cold species would become mingled with interglacial species.[54] In his explanation for the mixed fauna Croll employed a cycle of glacial and interglacial periods spanning centuries, rather than a yearly cycle of summer and winter seasons.

The seasonal migration theory and Prestwich's furry hippopotami were attacked again the following year by John Lubbock (Fig. 2.4). In the second edition of his *Prehistoric Times* (1869) Lubbock wrote that the presence of hippopotamus indicated a lengthy interval of mild climate. The first edition of this book, published in 1865, had gained him a popular reputation as a

astronomical origin. Lyell had been in close correspondence with Croll before he incorporated this theory into this book (see in particular the letters from Croll to Lyell, 1864–1866: ULE: Gen 109/582–633).

[54] Croll 1868, 384. Croll 1875, 236–291 elaborated on his ideas of a succession of warm periods within the so-called 'Glacial epoch' and the meaning of the mixed fauna in his book, *Climate and Time*.

synthesiser of archaeological, anthropological, and geological information; among his fellow researchers, though, he was regarded as somewhat of a dilettante and a man of impulsive interests. Lubbock became equally famous for his politics, his work on ants, and his establishment of Bank Holidays. He had been born, silver-spooned, into an influential family of bankers and his entrance into learned society was helped by an early acquaintance with Charles Darwin: Lubbock's neighbour at Down in Kent.[55]

Though Lubbock had been impressed by the calculations of Croll (and also by the work of Morlot on a Swiss interglacial period) he made no reference to Croll's interpretation of mixed fauna in *Prehistoric Times*. Nonetheless, he reached a similar conclusion: 'I am disposed then, on the whole, to consider that the quaternary fauna consists of two distinct groups, belonging to different periods and to two different conditions of climate, one warmer than the present, the other colder'.[56]

2.5 JAMES GEIKIE AND THE INTERGLACIAL PROBLEM

Neither Croll nor Lubbock became associated strongly with the interglacial interpretation of mixed fauna. It was James Geikie, Archibald's brother, who attacked seasonal migrations most vehemently and gathered volumes of geological evidence to support Croll's interglacial episodes. James Geikie had been apprenticed to a printer when he left school but was unhappy there and broke his apprenticeship in 1858. He was delighted to join his brother on the Scottish branch of the Survey in 1861 and enjoyed the harsh, rewarding life of a surveyor far more than the stuffy rooms and long hours he had known as an apprentice.[57] The merry, outspoken tone of his earlier letters is quite different to those written in later years, after he succeeded his brother as Murchison Professor of Geology at the University of Edinburgh. James Geikie was a close friend of Ramsay who declared to his wife Louisa in 1866: 'James Geikie is a man. Frank, courteous, sagacious, full of tact, conscientious, confident yet modest, in fact quite the cheese, the mealy root, the real Mackay'.[58]

When James Geikie joined the Survey he was set to the new task of mapping the superficial deposits of Scotland: first in Fife and the Lothians, later in Ayrshire. Enthralled by glacial geology, his investigations occupied much of

---

[55] Hutchinson 1914, 22–24.   [56] Lubbock 1869, 301.
[57] James Geikie's life has been recounted by Newbigin and Flett 1917, Woodward 1913 and Herries Davies 2004.
[58] Ramsay to L. Ramsay, 19 October 1866: NLW: 9636C/32.

Figure 2.5. James Geikie (1839–1915), glacial geologist. Source: BGS Library Archive: GSM1/639 (P575803).

his holidays as well as his official time. Geikie was one of many officers who contributed to the great expansion in records of the glacial drifts after they were added to the Survey staple of solid rocks in the 1860s.[59] But the interests of his fellow officers were often restricted. Wood's comparison between Survey officers and many tailors working on a single coat has already been mentioned. James Geikie was one of few to distil broader patterns from the wealth of new information.

Geikie saw much support in the Scottish drifts for the lengthy interglacials predicted by Croll. It is likely that he met Croll during the late 1860s, in the winter months when surveyors were frozen out of the field and gathered in Edinburgh to carry out their office work. Geikie later described Croll as 'the most philosophical physico-geologist we have had since Hutton'.[60] Indeed, Geikie would later be described as a disciple of Croll because of his fascination with the interglacials predicted by Croll's astronomical theories.[61]

It was not long before Geikie cast his glance down from Scotland to the glacial drifts of England. He was working at an auspicious time for glacial research, able to squeeze fellow Survey geologists, ripe with new information, about the superficial drift of their allotted regions. In October 1871, he wrote to his colleague and friend Benjamin Peach (1842–1926), who had been

[59] Flett 1937, 112.   [60] J. Geikie to Peach, 17 February 1881: BGS: GSM1/321/68.
[61] Howorth 1894, 496.

mapping the Scottish glacial drifts for the Survey, about 'some new theories with regard to the drift which I am thinking of perpetrating'.[62]

The following month, Geikie updated Peach with his progress. He described how the period of Scottish boulder clays 'had been broken up by a number of inter glacial periods of milder conditions', and that 'these great climatal changes must have left their impress upon the English drifts even more strongly than they have in the case of the Scotch deposits'.[63] The Scottish drifts had been ploughed through afresh with each glacial episode; the English drifts, lying further from the centre of glaciation, must have retained a clearer imprint of early glacial pulses and interglacial episodes.[64] As his gaze travelled still further south, the stone tools and bones in the drifts of southern river valleys also became drawn into Geikie's controversial glacial sequence.

Geikie held two particularly disruptive views: he promoted the existence of several lengthy interglacial periods before the mid-glacial submergence, and he attacked the post-glacial age of river drifts and cave deposits. In November 1871, he put his case to Peach: 'the Cave Deposits cannot be all of post glacial age—but on the contrary must belong to pre-glacial & inter-glacial times as well' and added: 'But I go a step further & insist that there is not the slightest ghost of proof that all the old river gravels of England are postglacial'. Geikie concluded:

I have sent to the Geol. Magaz. the first of a series of papers on the subject—for which I expect to draw some of the wrath of *the Dawkins* & others. But I am not afraid of my position. I have spent a long time looking at the joints of my armour & grinding my lethal weapons to a fine point. Unless the objections are polite they may look out for squalls.[65]

Geikie also warned McKenny Hughes, then working for the Survey on the English glacial drifts, that he intended to 'hazard some heresies about the English drifts' in the last of this series of papers.[66]

Geikie's defiant reference to the wrath of '*the Dawkins*' was connected to a paper that Dawkins had given in Edinburgh that year on 'The Relation of the Quaternary Mammalia to the Glacial Period'. This paper was summarised in the *Popular Science Review*, but Geikie had been in the large audience who heard the whole paper at the Edinburgh meeting of the British Association for the Advancement of Science. In 1871 Dawkins was less sure than he had been

---

[62] J. Geikie to Peach, 12 October 1871: BGS: GSM1/321/31.
[63] J. Geikie to Peach, November 1871: BGS: GSM1/321/32.
[64] Geikie 1872, 220–221. Oldroyd 1999 recognises the difficulties of working out glacial sequences in glaciated regions in his examination of the problems faced by geologists working in the Lakeland mountains during the nineteenth and early-twentieth centuries.
[65] J. Geikie to Peach, November 1871: BGS: GSM1/321/32.
[66] J. Geikie to Hughes, 8 November 1871: SMC: TMH.

in 1867 about the pre-glacial date of the Thames Valley lower brickearths, troubled by the discovery of a cold species at Crayford: the Musk-sheep. But Dawkins still thought that there had been a gradual drop in temperature after the warm Pliocene, and that animals now familiar from northern or arctic lands had started to appear in Britain as it grew colder. Geikie would have listened in annoyance. In place of Dawkins's steady increase in cold, he saw an alternation of glacial and interglacial periods.

Dawkins had also announced his solution to the mixture of cold (or northern) and warm (or southern) fauna in the post-glacial deposits of the Thames Valley (see Table 2.2). He followed the dramatic picture, painted by Lyell in *Geological Evidences of the Antiquity of Man*, of herds of hippopotami migrating northwards for the summer season. While these animals wandered over the land-bridge from the Continent and around the Thames Valley, Dawkins believed that the area north of the Thames had remained under ice.

Geikie disagreed. He was equally certain that several lengthy interglacial episodes had warmed the northern as well as the southern regions of Britain. Like Croll, Geikie placed centuries, not seasons, between interglacial periods (when southern species had overrun Britain) and glacial periods (when northern animals had dominated the country). Although Geikie responded to Dawkins in the discussion that followed this Edinburgh paper, he felt driven to elaborate his views in print. This started a conflict that would become more acrimonious through the years.[67]

James Geikie offered his own series of papers 'On Changes of Climate during the Glacial Epoch' to the popular monthly *Geological Magazine*. He had delivered short notes to the *Geological Magazine* before: his 'Note on the Discovery of Bos Primigenius in the Lower Boulder-clay of Scotland' (1868) was one of the finds that led Croll to wonder if warm fauna might have lived during lengthy warm interglacial periods.[68] But this new series of papers, published between December 1871 and June 1872, gave the first clear outline of Geikie's opinions on the Glacial epoch.

The *Geological Magazine* was not a casual choice. There were notable differences between forums for geological discussion in the late nineteenth century and Geikie was aware that the editor, Henry Woodward (1832–1921), often accepted controversial articles and accounts of local geology that might be rejected by the Geological Society.[69] There was no great difference in their price: a monthly issue containing one of Geikie's articles cost one shilling and sixpence; the quarterly journal offered by the Geological Society cost four shillings. But the Geological Society was regarded by some geologists as an

---

[67] Lyell 1863a, 156, 180–181; Dawkins 1869, 212–215; 1871a, 96; 1871b, 392–397.
[68] Croll 1868, 384, footnote.   [69] Sheets-Pyenson 1982, 184.

exclusive, high-brow clique, and it was also expensive to join: in 1880 one would be expected to pay an admission fee of six guineas and an annual contribution of two guineas.

The *Geological Magazine* had been founded in 1864 to promote discussion and widen geological participation: an unashamed alternative to the Geological Society. The example had been set by the Geologists' Association, established in Central London in 1858 to encourage communication and geological enjoyment, away from the disconcerting eyes of professors and advanced scientific discussions associated with the Geological Society.[70] The atmosphere of the Geologists' Association attracted researchers with few pretensions and less money, many of whom were interested in Quaternary geology and stone tools. In 1880, it cost only ten shillings to be admitted to the Geologists' Association and another ten shillings each year for membership. While on the subject of relative cost, it is interesting to see that the price of an implement from Biddenham Pit, Bedford, where Wyatt had made his discoveries in 1860, had remained stable at ten shillings; other purveyors of pit-finds were less accommodating.[71]

Consequently, it was in the welcoming and often controversial pages of the *Geological Magazine* that Geikie argued his way through the Glacial epoch. In the course of seven papers published over seven months he travelled from the beds in the Scottish Till, which he thought represented mild interglacial periods preceding the great submergence, to their equivalents in Europe and America. In March 1872 he finally seized on the glacial drifts of England, and in April he made a courteous assault on Dawkins's interpretation of mixed fauna in the southern river drifts and cave deposits.[72] Instead of seasonal migrations in post-glacial times, Geikie advocated lengthy climatic shifts between glacial and interglacial periods, with later mixture between the bones of animals that had lived in these quite distinct periods. He ended his paper with the statement that the post-glacial age of the older river-gravels of England was 'almost a physical impossibility'.[73]

Readers of the *Geological Magazine* must have looked forward with interest to the May number to see how Geikie would defend this contentious comment. In his final two papers, Geikie argued that glacial drifts had been deposited before *and after* the older valley-gravels: the implementiferous deposits at Hoxne might lie above boulder clay; but this glacial deposit did not necessarily belong to the *end* of the Glacial epoch.[74] Finally, he expressed his debt to Croll's astronomical theory.

[70] Smith *et al.* 1858.   [71] Smith to Evans, probably 1884: AMO: JE/B/2/80.
[72] Geikie 1871; 1872.   [73] Geikie 1872, 170.   [74] Geikie 1872, 218.

Geikie's geological case for interglacial episodes was closely related to Croll's thematic physics: a relationship that has been analysed by Christopher Hamlin.[75] Geikie was open about Croll's influence on his geological interpretations: 'the publication of this ingenious theory tended to revolutionize all our previous conceptions on the subject.' He acknowledged Croll's work on the mixed fauna of the valley gravels and saw his own role, thus inspired, as an attempt 'to correlate the river-gravels with undoubted interglacial deposits.'[76] These correlations and glacial divisions are given in Table 2.3, below. Two years later Geikie reflected that his papers in the *Geological Magazine*:

were the first attempt to prove, by correlating glacial deposits, that the Palaeolithic gravels of Southern England could not be of post-glacial age, but ought to be referred to inter-glacial and pre-glacial times. They also bring forward, for the first time, reason to show that a wide land-surface existed in the British area after the disappearance of the ice-sheets, and before the period of great submergence had commenced.[77]

Table 2.3. A simplified version of James Geikie's divisions of the Glacial epoch (after Geikie 1872, 110–111, 262–263).

Divisions of the Glacial epoch	Deposits	Physical conditions
Post-glacial period	Raised beaches and buried forests. Neolithic, Bronze and Iron periods	Continental conditions followed by partial submergence
Last Glacial period	River gravels; arctic fauna	Floating ice. Local glaciers appear as the land rises
Last Interglacial period	Cave deposits; river gravels; Palaeolithic tools	Southern fauna replace arctic fauna. Period ends with last great submergence of land north of the Thames, up to 2000 feet (deposition of Moel Tryfaen shells). Cold climate returns
Great cycle of Glacial and Interglacial periods	Cave deposits; river gravels; Palaeolithic tools. Till and boulder clay; southern and arctic fauna	Alternation of glacial conditions (great land-ice glaciers) with intermediate warm periods
Pre-Glacial period	Norwich Crag	Cold approaches

[75] Hamlin 1982, 566.   [76] Geikie 1872, 265.   [77] Geikie 1874, xi.

Geikie's first attempt won little support. He did, however, have the backing of a fellow Scottish glacial geologist: his friend Ramsay. Geikie had written to Ramsay in December 1871, shortly after the first paper of his series was published in the *Geological Magazine*. This letter clarified the task he saw before him, his reasoning, conviction and perception of the opposition:

I am glad to find that you are 'in accord with the leading ideas' contained in my short paper in the Magaz. [...] About a year ago I began to hunt up all the evidence in regard to the English Cave deposits & so-called 'postglacial gravels', and have come to the conclusion on grounds partly geological, partly zoological & partly common-sense-ical that you & [Godwin-] Austen are much more than justified in questioning the post-glacial age of all the Cave-deposits. I go a step further & question the post-glacial age of *all* the river gravels. If there is anything we can be certain of in Scotland it is this that *no great oscillation of climate has taken place since the deposition of our latest drift beds*—but there has been a gradual amelioration of climate from Arctic to temperate.

It is a long story and I won't think of humbugging you with a screed at present. Dawkins papers on the subject are full of the wildest absurdities. It is impossible that the hippopot. &c. can have occupied Britain since glacial times. I believe (whether I shall be able to prove is another matter) that hippopot. &c. are interglacial & pre-glacial: that, in short, the old high-level gravels of England are relics of interglacial times & perhaps preglacial times also. I am well aware that this involves the question of the age of the river valleys &c.—but I can only say that in my cogitations I have not left that out of sight—but am as conscientious & firm a believer in subaerial denudation as any of your disciples.

I have already written out nearly all I have got to say in a little book which I mean to publish by and by, and my reason for writing to the Magazine has simply been to have the matter discussed before finally kicking my lucubration before the foot-lights.

I note your suggestion about the mapping out of the English drifts. What appears to bother south-country geologists is that awful mixture of marine Boulder-clays which you have. Our succession, as you say, is much more easily made out.[78]

Despite the success of the post-glacial case made by Prestwich, Falconer, Lyell, and others in the early 1860s, a few doubters remained. James Geikie was one of the loudest. He continued to expand and modify his early versions of the glacial sequence and his 'little book' appeared in 1874: *The Great Ice Age and its Relation to the Antiquity of Man*. At twenty-four shillings, this was an expensive volume and was not little, but it sold well. Geikie dedicated the first edition to Ramsay, who was touched. Ramsay wrote to thank his friend, ending his letter: 'I must close, and will take the book home with me to read the dedication aloud in an impressive way after dinner'.[79]

[78] J. Geikie to Ramsay, 12 December 1871: ICL: KGA/Ramsay 8/412/3.
[79] Ramsay to J. Geikie, 20 January 1874, quoted in Newbigin and Flett 1917, 60.

Subsequent editions of *The Great Ice Age* reflect Geikie's changing ideas about the Glacial epoch. The English glacial deposits gained in detail; the interglacials grew in number. Some of the drifts that he had originally attributed to icebergs and marine action were ascribed to land glaciers in the second edition (1877). A number of terms were used to describe the drifts, but 'till', 'boulder clay', and 'ground moraine' all referred to deposits left by land-ice. Geikie retained several submergences but reduced their extent in his third edition (1894).[80] All three editions were steadfast in promoting inter-glacials and attacking the post-glacial age of English river drifts and cave deposits.

2.6 THE POST-GLACIAL PALAEOLITHIC UNDER ATTACK

Geikie announced in the introduction to the first edition of *The Great Ice Age*: 'until we clearly understand what the succession of changes during the Ice Age was, it is premature to speculate upon the geological age of those deposits which hold the earliest traces of man in Europe'.[81] The division between a pre-glacial and a post-glacial period had given a crude age to tools and bones from those deposits. Geikie promoted a finer succession of changes through the Glacial epoch, pushing tools back from their previous post-glacial position into the fluctuating cycle of glacial and interglacial periods. Although the glacial–interglacial sequence offered a more precise date for stone tools, Geikie's attack on the post-glacial theory was not welcomed, and sometimes not even understood, by many of his peers.

More than a decade had passed since Prestwich and Evans had read their papers on human antiquity. Prestwich and many of his colleagues believed these river drifts and their contents to be 'post-glacial' because they post-dated the familiar glacial marker provided by the Chalky Boulder Clay of East

---

[80] Geikie 1877a, xiv–xv, 247. On the origin of the boulder clay of East Anglia, compare the icebergs in the first edition of *The Great Ice Age* to the retraction in the second edition (Geikie 1874, 368; 1877a, 346). On submergences, compare the second edition to the third edition (Geikie 1877a, 192; 1894, 323). Dugald Bell 1827–1898, the Scottish geologist and clerk, fierce believer in land-ice and enemy to those who countenanced a marine submergence, saw Geikie's retention of submergences of five-hundred feet in his third edition of *The Great Ice Age* as a turnabout from a declaration made the previous year by Geikie 1893, 173, footnote that he no longer believed in the great submergence. According to Bell, one Glaswegian glacial geologist had been so excited when he read Geikie's 1893 abandonment of the submergence that he 'suddenly took off his smoking cap & gave three vigorous cheers' (Bell to J. Geikie, 23 June 1893: BGS: GSM1/321).

[81] Geikie 1874, viii.

Anglia. Geikie and his followers saw many more glacial periods reflected in the different boulder clays of Britain. For them, this relation to the Chalky Boulder Clay was not enough to justify a 'post-glacial' label for the river drifts: glacial drifts *later* than the Chalky Boulder Clay existed in Yorkshire and elsewhere in the north of England.[82]

The Chalky Boulder Clay was one of Geikie's earlier glacial deposits. He explained, 'There is an *older* boulder-clay than that "chalky till" [the Chalky Boulder Clay] and there are two separate boulder-clays which are *younger*, as Mr. S. V. Wood has demonstrated' (see Table 2.4).[83] Horace B. Woodward (1848–1914), a Survey geologist (and nephew of Henry Woodward) who had worked on the drift maps of London, tried to clarify Geikie's perspective: 'The Thames-Valley Deposits are no doubt later than the Chalky Boulder Clay, and are considered Post-Glacial in the sense of being newer than the latest glacial deposits in the district; but they are Inter-glacial if we include in the Glacial Epoch the later evidences of ice-action in North Britain'.[84] But clear explanations of the problem did not end the disagreement. Arguments over the post-glacial or interglacial date of the river drifts swung back and forth until the end of the decade, with few additions and little change.

This quibble over terminology had bubbled up from deeper differences in aims and perceptions. These geologists were arguing over the same deposits

Table 2.4. Wood's sequence of four English boulder clays and their relation to the emergence of the land (after Wood and Rome 1868 and Wood 1870b). The boulder clays are highlighted in bold type.

East Anglia	Yorkshire, Lincolnshire
	Hessle Boulder Clay (Older Post-glacial)
Emergence of land	
	Purple Boulder Clay (Upper Glacial)
Upper Glacial series (**Chalky Boulder Clay** of the Hoxne section)	
Middle Glacial series (Sands and Gravel)	
Lower Glacial series (**Lower Boulder Clay**; Cromer Till)	
Cromer Forest Bed	
Upper Crag	

---

[82] Geikie 1872, 220; 1874, 474–475; 1877a, xv; Skertchly 1877, 142; Miller and Skertchly 1878, 532–535.
[83] Geikie 1877b, 141. See also Geikie 1874, 426; 1877a, 489.
[84] Woodward 1886, 129.

but did not hold the same expectations. Geikie once observed that disagreements about the origin of the glacial drifts could often be linked to experiences of the geology in different geographical regions. His peers also remarked on the difference between the conclusions reached by geologists working in southern England and East Anglia and those reached by geologists working in northern England and Wales, where the glacial topography was very different.[85] Their arguments about the origin of the boulder clay, for example, were notorious. They even provoked an anonymous poem in the *Geological Magazine* of 1867, entitled 'Lines on a Scratched Boulder', its comic tone and anonymity permitting a freedom of expression denied to more pedestrian narrative forms:[86]

> Tell me, Geologists, I pray!
> What you mean by Boulder-Clay!
> Does it consist of beds contorted,
> Or layers of sand and clay assorted?
> Is it unstratified or stratified?
> Can each or either view be ratified?
> Did it on floating Icebergs travel?
> Or slide down Glaciers mixed with Gravel?
> Is it this latter reconstructed
> By Icebergs thawing when obstructed?
> Is it Moraine Clay or Marine?
> Or is it neither, but between?
> Was the whole country capped with ice,
> Which churned the rocks up in a trice?
> Did wave, mysterious of translation
> Perform the work of denudation?
> Can you declare, with voice emphatic,
> Its stones and not your views erratic?
> Could but the Boulder Clay or Till
> Find words' twould call you bolder still.[87]

The rejection of Geikie's interglacial arguments was also associated with differences in the scale of interpretation. For Geikie the southern drifts were only a small fraction of a great glacial sequence that he had constructed from glacial deposits across the whole of Britain. For his opponents the southern drifts filled their vision: the Chalky Boulder Clay of East Anglia was their

[85] The divergent opinions of distant researchers was mentioned by Geikie 1874, 355–356, Tiddeman 1878, 169, Dawkins 1878, 159–160, Hicks 1886a, 5–6, Whitaker 1889 (i), 387, and Smith (1894, 8).

[86] Rudwick 1992, 39–40, 56–57; Sommer 2004a, 60, 72–74; R. O'Connor 2007 forthcoming: *The Earth on Show*, Chapter 2.

[87] Anon 1867c.

benchmark that divided post-glacial from pre-glacial times; the river drifts post-dated this marker and were therefore post-glacial. Theories of later boulder clays that lay far to the north fell outside their sphere of relevance. In any case, it was notoriously difficult to correlate boulder clays from different regions because each took the character of the local bedrock; each region had its distinctive variety of boulder clay.

Within a region, such similarities made it difficult to build even a local sequence. And even if Geikie was right, the boulder clays of his northerly glacial periods had not physically touched southern England so it was questionable how much influence his supposed changes of climate had exerted on the animals and tool-makers that were restricted to those southern regions. The questions asked of the Glacial epoch might have sounded similar, but they were pitched differently. No consensus could be reached about the connection between the Glacial epoch and the river drifts until there was agreement about the scale of interpretation, not just the number of glaciations.

Some conflicts arose from experience of different geographical regions, others from differences in the scale of interpretation, but they were also stimulated by the use of different sources of time. Geikie saw the patterns of sediments, bones, and tools in terms of his glacial chronology; Dawkins filtered them through his faunal chronology; even stone tools were gaining potency as time-markers. Wood's definition of time did not rest on boulder clays or climatic cycles, but on tools and bones. (See Table 2.5 for Wood's connections between his own views and those held by Dawkins.) In his attack on the Chalky Boulder Clay as the divider of pre-glacial from post-glacial time, Geikie had referred in his arguments to two boulder clays that Wood had shown to be younger than the Chalky Boulder Clay. Wood agreed that some boulder clays, such as the Hessle, did indeed post-date the Thames Valley deposits but he would still not use Geikie's term 'interglacial' for the implement-bearing strata. He called these late boulder clays 'Post-glacial, in the sense that they are posterior to the introduction of Man and of the Great Pachydermata'.[88]

Geikie's glacial sequence was misunderstood on many different levels and rejected by most of his peers, guided by their own preconceptions and schooled in the simpler 'post-glacial' and 'pre-glacial' labels of times past. Geikie would find it difficult to stop them from referring to the Thames Valley

---

[88] Wood to Lyell, 23 June 1872: ULE: Gen 117/6273–6274. On post-glacial boulder clays, see also Wood 1867, 417. Wood's designation of the Hessle as a 'post-glacial' boulder clay also spared him from having to alter his Lower, Middle and Upper Glacial terminology (Baden-Powell to Warren, 20 February 1955: UMO: DB-P: K149).

Table 2.5. Wood's attempt to reconcile Dawkins's faunal divisions to his own glacial periods. Based on a table enclosed in a letter from Wood to Lyell, dated 3 July 1872 (ULE: Gen 117/6292–6293).

Wood's glacial periods	Dawkins's faunal divisions according to Wood
Fourth period: 'Newer post-Glacial' Maximum cold, spread of Reindeer	Third (or arctic) division
Third period: 'Older post-Glacial' Glacial deposits south of the Thames removed by the sea, accumulation of Thames brickearths; later deposition of marine boulder clay in the north	Second division. Linked to the Pliocene and the late Pleistocene.
Second period: 'The Glacial' Depression up to 2000 ft, synchronous with the gathering of an ice cap. Accumulation of glacial deposits	No mammalian remains
First period. Cromer Forest Bed	First division. Mix of Pliocene and Pleistocene species

deposits as 'post-glacial'. But when his opponents were faced with stone tools found *beneath* the boulder clays, some would be forced to reconsider their views or, at least, to find different ways of expressing them.

2.7 CURIOUS RELICS OF THE TOOL-MAKERS

In the 1870s, the relics of tool-makers were found, several times, in positions that could have been taken as confirmation of Geikie's interglacial theory. Three reports are of particular interest: one from London, one from Yorkshire, and one from East Anglia. The first of these was announced in the same issue of the *Geological Magazine* that contained the last of Geikie's series of papers on changes in climate. Osmond Fisher reported in the June number, 1872, that he had found a worked flint in the lower brickearths at Crayford in London. This flake had been certified as genuine by Evans, the authority on stone tools, and came from deposits that Dawkins thought earlier than other valley gravels in the Thames Valley. In 1876 another worked flake was recovered from the lower brickearths at Erith (Fig. 2.6).[89] Although the lower brickearths had once been dated to pre-glacial times, these two London flakes caused little excitement. Dawkins's views of 1867 had not been popular,

[89] Fisher 1872; Dawkins 1874, 416; 1880, 136.

Figure 2.6. Flint flake from the lower brickearths at Erith (Dawkins 1880, 136, Fig. 27).

and the Crayford and Erith flint flakes were classed with other post-glacial tools from the Thames Valley.

Dawkins had changed his opinion about the lower brickearths since his pre-glacial stance of 1867.[90] In 1869 he described them as 'postglacial passage-beds' between the pre-glacial Pliocene fauna and the northern fauna of post-glacial times. He became wary of glacial terminology in later works, although he would occasionally suggest that the brickearths might also be called pre-glacial. Dawkins had cemented his reputation in learned society with a paper on the classification of Quaternary strata using fossil mammals, published by

[90] On Dawkins's changing views of the brickearths, compare Dawkins 1869, 214 to Dawkins 1877, 610, 612.

Table 2.6. The palaeontological divisions developed by Dawkins (after Dawkins 1872, 411, 422–423; 1874, 413–420). Dawkins believed that the changes in mammal species were connected to a gradual drop in temperature through the Quaternary period.

Palaeontological divisions	Characteristic fauna	Geological deposits
Late Pleistocene	Northern species dominate (reindeer, mammoth, woolly rhino, musk sheep); also southern species (hyena, cave lion, hippopotamus)	Cave deposits and river drifts
Mid Pleistocene	Disappearance of Pliocene deer. Southern species (hippopotamus, *E. antiquus*) accompanied by northern species (woolly rhinoceros, mammoth, musk sheep)	Thames Valley lower brickearths; Clacton-on-Sea; parts of Kent's Cavern
Early Pleistocene	Pliocene species (*E. meridionalis*, Pliocene deer); some Pleistocene immigrants (mammoth); no northern species	Cromer Forest Bed

the Geological Society in 1872, the same year that Fisher had found his flake. Dawkins now preferred to class the brickearths in palaeontological terms as 'Mid Pleistocene' (see Table 2.6). Like Falconer, Dawkins found that the divisions of his three-fold mammal classification had some overlap with Pliocene species, so there was no neat alignment with Lyell's divisions or the pre-glacial, glacial, and post-glacial divisions that had been applied to the deposits of central and north Britain.[91]

A more exciting relic of ancient tool-makers was reported from Victoria Cave, Settle, Yorkshire, in 1873. The committee of the British Association for the Advancement of Science who were conducting the cave excavations included Dawkins, Lubbock, Prestwich, Hughes, and Richard Hill Tiddeman (1842–1917) of the Geological Survey. Tiddeman revealed to readers of *Nature* that a bone, found the previous year beneath glacial deposits, had been identified as a human fibula. He argued that Victoria Cave had preserved a valuable remnant of human occupation in northern England from pre-glacial or interglacial times, all other evidence having been erased by the great ice-sheet that later ploughed down from the north. Tiddeman preferred a destructive ice-sheet to the destructive submergence that Geikie still supported at this time; in other respects, he upheld Geikie's case.

Furthermore, the layer that yielded the fibula also produced the bones of hyena, mammoth, woolly rhino, and other species familiar from the southern

[91] Dawkins 1872, 440; 1874, 413–414; 1878, 161.

river drifts, and this 'hyena-bed' layer seemed to be covered with boulder clay. For Tiddeman, this confirmed Geikie's view that the southern river drifts pre-dated the northern ice-sheet. He later strengthened his case with the observation that warm species (such as straight-tusked elephant and hippopotamus) were peculiar to the hyena bed, but cold species (such as reindeer) were restricted to the later bed above.[92]

James Geikie agreed that Victoria Cave supported the existence of tool-makers before post-glacial times. He incorporated Tiddeman's discovery as a last-minute postscript to the first edition of *The Great Ice Age*. Ramsay also welcomed the evidence. Dawkins, though initially in favour of a pre-glacial date for the fauna and fibula, later changed his mind, perhaps provoked by a disagreement with Tiddeman. He re-identified the human fibula as that of a bear; argued that this was a typical mixed post-glacial fauna, like the bones from other British caves; and dated them to his Late Pleistocene division of mammals.[93]

Dawkins also questioned the care with which Tiddeman had conducted these excavations. He announced that he had found reindeer bones mixed with southern species when digging in Victoria Cave in 1872. This, he argued, cast doubt on Tiddeman's observation of two discrete layers and on his interpretation of a warm period (with southern species) followed by a cold period (with reindeer). Finally, Dawkins described the boulder clay above these layers as a post-glacial deposit that had been derived from the wreck of true, but much older, boulder clay.[94] Geologists would often use a later disturbance to dissolve the pre-glacial positions of bones or tools. In the 1880s, when Henry Hicks (1837–1899) claimed that his cave tools from the Vale of Clwyd in North Wales were pre-glacial, he would receive a comparable response from Hughes: that this supposed Welsh boulder clay was merely post-submergence debris from older glacial drifts.[95]

Back in 1870s Yorkshire, Evans added his suspicions to a growing pile.[96] And even if the fibula and bones from Victoria Cave were contemporary with

[92] Tiddeman 1873, 14–15; 1874, 133–135. In *The Great Ice Age*, Geikie 1874, x described how traces of the Palaeolithic further north had been 'swept out of these regions by confluent glaciers, and by the sea during the period of great submergence.'

[93] Geikie 1874, 510; Ramsay in Dawkins 1877, 612. Geikie 1877a, 381 made sure that Tiddeman's discovery had a place in the second edition of *The Great Ice Age* as evidence of a human presence early in the last *inter*glacial; he disagreed with Tiddeman's pre-glacial date. For Dawkins's change in opinion about the identity and date of the Victoria Cave discoveries, compare Dawkins 1874, 124, 411 to Dawkins 1878, 155, 158–159.

[94] On the origin of the Victoria Cave boulder clay, see Dawkins 1877, 607. See Murphy and Lord 2003 on the arguments between Dawkins and Tiddeman about Victoria Cave and the recent re-examination of these deposits.

[95] Hicks 1885, 1022–1023; 1886a, 18; 1886b, 16; Hughes 1887, 108–113.

[96] Evans 1875, lxxiii.

Figure 2.7. Sydney B. J. Skertchly (1850–1926), glacial geologist. Photographed in 1879. Source: BGS Library Archive: IGS1/639,122.

the Thames Valley river drifts and did pre-date the northern boulder clay, this would still not have been enough to convince Wood of their pre-glacial or interglacial age: he placed this northern boulder clay in his 'Older post-Glacial' period (see Table 2.5). Only stone tools from beneath the Chalky Boulder Clay itself—the great benchmark that had led so many geologists to describe the Thames Valley deposits as 'post-glacial'—might lead geologists to accept an earlier glacial date for the stone tools of the southern river drifts.

Sydney B. J. Skertchly (1850–1926), yet another geologist on the Survey, made just such a discovery in 1876. Skertchly (Fig. 2.7) had been mapping the Fenland area, rich in drifts, since 1869. He announced in a letter to *Nature* that he had come independently to the same conclusion as Geikie: stone tools had been discarded in interglacial times. This date now had the support of his own recent discovery of stone tools *beneath* the Chalky Boulder Clay, lying in pockets of brickearth along the valley sides around Brandon in Suffolk. Skertchly suspected at the time, and would demonstrate the following year, that these deposits corresponded to Wood's Middle Glacial series (see Tables 2.1 and 2.4). This placed them between the Cromer Till and the Chalky Boulder Clay.[97] If Skertchly was right, tool-makers would have occupied

---

[97] Skertchly 1876a, 449; 1877, 142. Skertchly soon added to his original find (of four implements from two localities), and in 1877 had nearly 150 tools from six different sites, which included pits at Brandon, Mildenhall, West Stow and Bury St. Edmund's (Skertchly 1876b, 476; 1879; Miller and Skertchly 1878, 547–548).

East Anglia early in the Glacial epoch before the deposition of the Chalky Boulder Clay, as well as making a later appearance in the post-Chalky Boulder Clay (but, as Geikie might interject, not necessarily post-glacial) river drifts.

Geikie and Ramsay both visited Brandon and incorporated Skertchly's discovery into their books. Geikie took the tools as proof 'that man lived in Britain as early at least as that interglacial mild period which preceded the climax of glacial cold'.[98] He described Skertchly's find as 'by far the most important discovery of Palaeolithic beds which has been made since Boucher de Perthes first detected the flint implements in the ancient river-drifts of Abbeville'.[99] Ramsay was also pleased that the relics of such early tool-makers had finally been found. He wrote that it was 'obvious that his advent into our area was either of pre-Glacial or of inter-Glacial date. I say inter-glacial because Mr. Skertchly has lately discovered palaeolithic flint implements in certain brick-earths'.[100] Tiddeman, equally delighted, took the chance to urge Geikie, who was working on a new edition of *The Great Ice Age*, to erase the submergence of the land and retain the grinding power of an ice-sheet as the single destroyer of Stone-Age traces in the north of England.[101]

Skertchly's discovery was greeted by the Survey geologist H. B. Woodward, in the *Geological Magazine*, as 'startling' to previous assumptions of a post-glacial Stone Age, but he accepted the evidence.[102] Fisher converted to inter-glacial implements on his second visit to the brickearths, in Skertchly's company, and wondered whether the brickearth at Hoxne might belong to the same era. Whitaker also accepted Skertchly's conclusions.[103] Lubbock, though he added the Brandon find to a footnote in the fourth edition of *Prehistoric Times* (1878), remained cautious: 'Other geologists, however, have contested his interpretation of the fact'.[104] Wood echoed this warning in 1880. By then too ill to read his paper to the Geological Society, he wrote that the interglacial age of Skertchly's brickearths 'has not yet been generally admitted by geologists'.[105]

The sceptics had been fuelled by the negative responses of Evans and Hughes at a conference on human antiquity, held at the Anthropological Institute in May 1877.[106] Hughes suggested that the geological beds that held Skertchly's tools were not of glacial age, but post-dated the emergence of the land from the glacial sea.[107] Evans warned the audience 'that our watchword must for the present be "caution, caution, caution"'.[108] Geikie was impressed

[98] Geikie 1877a, xiii.   [99] Geikie 1881a, 263.   [100] Ramsay 1878, 481.
[101] Tiddeman 1876, 505–506; Geikie 1874, 484–485.
[102] H. B. Woodward 1878, 232–233.
[103] Fisher 1879, 288–289; Dawkins 1880, 169, footnote 1.   [104] Lubbock 1878, 424.
[105] Wood 1880, 499.   [106] Geikie 1877b, 141; Skertchly 1877, 142.
[107] Hughes 1878, 164–165.   [108] Evans 1878, 151.

neither by the occasion, nor by the anonymous review of the conference that appeared in *Nature* (the tone and content suggest Dawkins as a likely author). He thought the conference had tolerated too many tired old views and had inspired little new treatment of human antiquity: the argument still circled wearily around the assumption that since the implementiferous deposits of the south of England seemed to post-date the Chalky Boulder Clay, they could be labelled 'post-glacial'.[109] It had not been difficult for geologists to filter the discoveries of supposedly early tools in East Anglia, Yorkshire, and the Thames Valley through their different perspectives of Quaternary time. The day after Geikie delivered this opinion to *Nature*, he expelled more irritation in a letter to Ramsay:

English geologists are so conservative, and some of them are so wooden that there is no getting them to see a logical conclusion. Prestwich ignoring Skertchly's work, reiterates that all Palaeolithic beds must be *post*-glacial because all that he has seen rest upon '*the* boulder-clay'. Now the boulder-clay of East Anglia, (that near Brandon, etc.) has been demonstrated not to be *the only* boulder-clay. There is an *older one*, and there are *two younger ones*: all of which I have seen in actual sections over and over again. The mere occurrence of Palaeolithic beds lying *on* the boulder-clay of East Anglia, therefore, is no proof that these Palaeolithic things are younger than the glacial period.[110]

2.8 CONFLICTING CHRONOLOGIES: GEIKIE *VERSUS* DAWKINS

Geikie's complaints about the inability of wrongheaded English geologists to see his own perspective struck to the heart of many other debates of the time. Such restricted spheres of relevance gave impetus and character to arguments. A clear example of the adamant incomprehension that made the logical conclusion of one geologist unacceptable to another is seen in the tangles between Geikie and Dawkins. In 1877 Geikie referred to Dawkins as 'my able opponent' when he observed how often, in previous years, their different views of the faunal and glacial sequences had been put before 'our fellow-hammerers'.[111] Two years previously, Dawkins had willingly signed Geikie's certificate for admission to the Royal Society.[112] By 1881, however, they

---

[109] Geikie 1877b, 141; Anon 1877.
[110] J. Geikie to Ramsay, 16 June 1877: ICL: KGA/Ramsay 8/412/13.
[111] Geikie 1877b, 141.
[112] James Geikie 1875, Certificate of a Candidate for Election: RSL: EC/1875/22.

reached the sour peak of their conflict and published angry letters in *Nature* that clarified the great barriers between their opinions.

The disagreement between Geikie and Dawkins echoed Dawkins's arguments of 1867 with the stratigraphical geologists, Wood and Whitaker: each believed that he had seen the most correct sequence of Quaternary events and charged the other with overextending his restricted viewpoint. Dawkins, inspired by bones, thought glaciers a foolish basis for a Quaternary time-scale. He argued that the northern ice-sheets had little influence on life in the river valleys of southern England, where animals and tool-makers continued to browse and hunt.[113] He ridiculed the idea of applying a glacial chronology to non-glaciated areas like the Thames Valley, where the tool-makers 'may have been, to use the nomenclature of the glacialists, pre- inter- and post-glacial'. 'For purposes of geological classification over wide areas an appeal to the purely local phenomena of glacier or iceberg is useless'.[114] Geikie asserted: 'Dawkins' papers on the subject are full of the wildest absurdities', and was convinced that his own glacial and interglacial cycles of climatic change had a wider range and influence than changes in fauna, river drifts, or stone tools, which might display different patterns in different regions.[115]

In 1881, at the height of their disagreement, Dawkins was Professor of Geology at Owen's College, Manchester. He had published his *Early Man in Britain* the previous year. Geikie would not become a Professor of Geology at Edinburgh until 1882, and produced his book on *Prehistoric Europe* in 1881. He claimed that he had been so exhausted when he reached the end of his manuscript that he had written the last sentence in his sleep.[116] Although this was, perhaps, an exaggeration—an example of the myths that geologists spun around themselves—the work had evidently cost Geikie much strain and trouble. His book would be reviewed by Dawkins. The review was published in February 1881 in the respectable pages of *Nature*. Geikie would have known, as he read it, that the review would also be seen by a large majority of his peers. Dawkins, unsurprisingly, was uncomplimentary; the tone can be summarised by his remark: 'Dr. James Geikie takes his stand upon the glaciated mountains of Scotland, and attempts to throw the glacial net woven in his previous work, "The Ice Age", over the whole of Europe'.[117] Geikie, angry and upset, responded privately that Dawkins was

---

[113] Dawkins 1880, 171–172, footnote 1.

[114] Dawkins 1878, 160; 1880, 114, footnote.

[115] J. Geikie to Ramsay, 12 December 1871: ICL: KGA/Ramsay 8/412/3. Marcellin Boule 1888, 132, the French geologist and palaeontologist, observed that the large scale climatic shifts of successive glacial episodes provided a more reliable basis for a relative chronology of Quaternary times than fauna, which were regionally variable, adaptable and often migratory.

[116] Newbigin and Flett 1917, 88–91.   [117] Dawkins 1881a, 309.

a vain cocky humbug, who has endured so long simply because no one has been examining the evidence derived from foreign sources. For years past he has been 'asserting' & 'expressing his belief' that I am all wrong, in anonymous letters and notices—the authorship of many of which I can prove, and that of the rest is self-evident. It is monstrous that such a nincompoop in physical geology should be allowed to strut about as an authority.[118]

Geikie wrote to Ramsay in amazement that Dawkins had been allowed to review his book, 'as it was notorious that my views were directly opposed to his, & that he could not therefore give an unbiased notice of my work. But a more unjust and mendacious notice I never saw'. In a disconcerting twist, Geikie learnt that the geological editor of *Nature* who had let the uncomplimentary review through was his brother, Archibald.[119] In his letter of reply to *Nature*, James Geikie responded to Dawkins's accusation—that he had based his interglacial theories and his 'ice-classification', 'on ice, and ice only'—by throwing back the criticism: 'Geologists rightly refuse to accept classifications which are based upon so narrow a foundation as a single series of phenomena, such, for example, as Mr. Dawkins' attempt to classify the Pleistocene by reference to the mammalia alone'.[120]

Their controversy continued through later numbers of *Nature*, a journal that Geikie accused of being a one-sided 'kiss-my-arse of Macmillan & Co' when the editor tried to suppress his next reply to Dawkins.[121] Geikie assured Ramsay: 'if it does not appear in next week's Nature I will send it *with additions* to the Academy & Athenaeum', but admitted to his friend, the Scottish surveyor John Horne, that 'The affair has affected me more than I can tell [...] You will laugh, but it is true all the same, that I can hardly eat or sleep'.[122] Dawkins, undeterred by Geikie's ferocious public response, delivered an identical opinion in his address to the Anthropological Section of the British Association the following year. The tool-makers had, he argued, been around both before and after the great submergence that left the boulder clays, so were both pre- and post-glacial.[123] Geikie's interglacials were pointedly ignored.

Whitaker called for an end to such debates over the glacial or post-glacial age of the earliest tool-makers, pleading that the variation in glaciations

---

[118] J. Geikie to Peach, 17 February 1881: BGS: GSM1/321/68.
[119] J. Geikie to Ramsay, 27 February 1881: ICL: KGA/Ramsay 8/412/15.
[120] Geikie 1881b.
[121] J. Geikie to Peach, 9 March 1881: BGS: GSM1/321/69. On the rest of the controversy in *Nature* see Dawkins 1881a, 1881b and Geikie 1881b, 1881c.
[122] J. Geikie to Ramsay, 27 February 1881: ICL: KGA/Ramsay 8/412/15; J. Geikie to Horne, 1881, quoted in Newbigin and Flett 1917, 94.
[123] Dawkins 1882, 601–602.

between north and south England meant that 'He may be both, without any contradiction!'.[124] But beneath these vocal arguments over terminology lay quieter and greater differences in perception. Faced with conflicting patterns, personal beliefs were buffered by the assumption that these discrepancies were only the distorted perspectives of the past held by their peers. Some were accused of being so narrow in their focus (on a geological speciality, geo-graphical region, or segment of time) that they missed the broader picture entirely; others, so broad, that their picture failed to account for the specific anomaly that filled the vision of their accuser. The lines of argument traversed by Geikie and Dawkins were familiar to all who searched for chronological patterns in Quaternary Britain.

2.9 GLACIAL CHRONOLOGIES AT THE END OF THE NINETEENTH CENTURY

If a series of distinct climatic cycles had left their mark in the sediments and bones of Britain, as Geikie suggested, then his glacial sequence could be used to connect these scattered drifts and bones into comprehensible order. Skertchly suggested in 1878 that stone implements, too, might be teased out from their vague post-glacial cluster and assigned to different glacial and interglacial periods: 'the generally received opinion is that palaeolithic imple-ments belong to one unbroken series—some a little newer or older than others, as the case may be, but still geologically speaking of one date. This, my own researches have proved to be far from the case'.[125] Skertchly described three successive Palaeolithic occupations of Britain and placed them within a glacial sequence that was similar to Geikie's version of 1877 (see Table 2.7).[126]

If Geikie's climatic cycles were not unique to Britain, but had also afflicted the European mainland, then British tools might be compared in their date and character to implements found far across the Channel. In 1878, the year that Skertchly made his suggestions about the British Stone-Age sequence, Geikie was delighted to hear from the German geologist, Albrecht Penck (1858–1945), and learn that his book, *The Great Ice Age*, had opened Penck's eyes to the meaning of superficial drifts in north Germany. Here, Penck had observed evidence of three glaciations with intervening glacial deposits. Al-though sequences similar to those favoured by Geikie were being developed in Europe, a great framework of glaciations and industries would not become

---

[124] Whitaker 1889 (i), 387.   [125] Miller and Skertchly 1878, 534.
[126] Geikie 1877a, 489; J. Geikie to Ramsay, 16 June 1877: ICL: KGA/Ramsay 8/412/13.

Table 2.7. Skertchly drew these connections between the glacial deposits and stone tools of England in 1878. Note the similarity between Skertchly's glacial sequence and the scheme published by James Geikie in 1877 (after Miller and Skertchly in 1878, 551; and Geikie 1877a, 387–393).

Miller and Skertchly 1878		Geikie 1877
Glacial deposits and Stone-Age divisions	Glacial sequence	Geikie's terms for the glacial deposits
Neolithic Period	Post-glacial	
Hessle Boulder Clay	Glacial	Last glacial (Hessle Boulder Clay)
Modern Valley palaeoliths	Interglacial	Third/Last interglacial
Purple Boulder Clay	Glacial	Third glacial (Purple Boulder Clay)
Ancient Valley palaeoliths	Interglacial	Second interglacial
Chalky Boulder Clay	Glacial	Second glacial (Chalky Boulder Clay)
Brandon Beds	Interglacial	First interglacial
Cromer Till	Glacial	First glacial (Cromer Till)
		Pre-glacial (Cromer Forest Bed)

popular in Britain until the twentieth century. A reason is not difficult to find: Geikie was more interested in glacial patterns than stone tools; John Evans, the leading authority on stone tools, did not believe that Palaeolithic stages could be interspersed with glacial periods, and the majority of his peers who worked on the implements would have agreed with him.[127] The glacial-interglacial chronology had few adherents at the end of the nineteenth century.

In the 1890s, the post-glacial date of the southern river drifts and their stone tools gained even more support. T. V. Holmes (1843–1923), formerly of the Geological Survey, recorded in 1892 that he had seen the Chalky Boulder Clay lying *beneath* the oldest spread of valley-gravels in a section exposed by a railway cutting at Hornchurch, north of the River Thames. This was the section that Falconer had wished for in 1863. Holmes was cautious in the wording of his report, but his conclusions were still powerful. Aware of claims for later boulder clays further north, he stated: 'the Thames Valley deposits are (locally) post-Glacial, or newer than the local Boulder Clay'.[128]

In 1896, Clement Reid (1853–1916) reported on a re-investigation of the celebrated site at Hoxne by a committee of the British Association for the Advancement of Science. Reid, who had worked for the Geological Survey since 1874, was best known for his knowledge of the Quaternary and Pliocene deposits on the Norfolk coast and the climatic changes reflected in ancient plant-remains.[129] After looking at the Hoxne sediments, Reid had no doubt

---

[127] Evans 1897b, 568.   [128] Holmes 1892, 370; Murchison 1868 (ii), 582.
[129] Lamplugh 1917, lxii–lxiii.

that the tool-bearing levels lay *above* the Chalky Boulder Clay: tool-makers had only occupied the area *after* the last glaciation of this district.

Reid used the plant remains to reconstruct the post-Chalky Boulder Clay climate: a bed of temperate species and another bed of Arctic species suggested that cold conditions might have recurred; the tools had been made still later. He concluded that stone tools were post-glacial in Suffolk, but conceded that tool-makers might have lived in interglacial or pre-glacial times in other areas. This result was confirmed by the Royal Society's investigations at Hitchin in Hertfordshire: once again, stone tools lay above the Chalky Boulder Clay. The doubts raised by Skertchly's discoveries faded.[130]

Geikie's grand glacial sequence for Britain had no place in these proofs of a 'post-glacial' Stone Age. Holmes and Reid were cautious in their expression, speaking of deposits that were *locally* post-glacial and conceding that the Palaeolithic in *other* areas might be 'interglacial' or pre-glacial. For Geikie, an interglacial for one part of Britain was an interglacial for all; it was nonsense to talk of one region hosting a post-glacial stage and another, an interglacial. The tool-makers did not just have to retreat back into the Ice Age before climatic cycles could be used as a master-sequence for dating and correlating tools, bones, and sediments. The scale of interpretation also had to expand from local sequences to embrace Britain and even, perhaps, Europe.

One final problem faced by Geikie was that few other geologists saw as much detail and division in boulder clays. It was widely agreed that Britain had two different boulder clays, and a belief in their iceberg origin was now unusual. But while Geikie argued that several successive boulder clays, left by several distinct glaciations, could be observed stretching across large areas of Britain, others saw such divisions between boulder clays as the product of purely local phenomena.[131] Whitaker expressed the views of many geologists when he suggested that too much was being read into the glacial drifts: 'as regards the Drift, there seems to me to have been a leaning to over-refinement, and a tendency to over-division of the beds, which leads to the bewilderment of simple-minded persons like myself'.[132]

Geikie was seen as an extreme glacialist in the nineteenth century. He gained little support for his interglacial theories in England but retained a few supporters abroad: Penck, for example, stayed in contact with Geikie and continued to advance the idea of multiple interglacials on the Continent.[133]

---

[130] For the Hoxne conclusions, see Reid 1896, 411; for the Hitchin conclusions, see Reid 1897, 41. Evans 1897b, 577 took care to insert the Hoxne findings into the second edition of his *Ancient Stone Implements* as further proof of a post-glacial Palaeolithic.

[131] Bulman 1891, 341–342; Howorth 1896, 450–454.   [132] Whitaker 1902, 107.

[133] Penck 1897; Newbigin and Flett 1917, 85. Some of the later correspondence between Penck and James Geikie is held by the British Geological Survey (Penck to J. Geikie, 1880–1914: BGS: GSM1/321).

Penck would become more famously associated with Eduard Brückner (1862–1927), another German glacialist who worked on the deposits of the Bavarian and Austrian Alps. And when Penck and Brückner published their fourfold glacial sequence in the early years of the twentieth century they would dedicate their book to James Geikie.[134]

The scheme of Alpine glaciations set out by Penck and Brückner in this book, *Die Alpen im Eiszeitalter*, would provide an influential framework for discussion of British Quaternary sequences. As Geikie's books faded on their shelves, British researchers of the twentieth century would continue to argue over the Glacial epoch and the interglacial slots occupied by different groups of stone tools. But before entering the new century, this chapter on the Ice Age must be matched by an account of the Stone Age. The treatment of stone tools over the four decades that followed the acceptance of human antiquity, the same four decades discussed above, is the subject of the next two chapters.

---

[134] Preller 1894, 29; Penck and Brückner 1901–09; Imbrie and Imbrie 1979, 115.

3

Ancient Dwellers of the Thames Valley

Stone-Age classifications of the nineteenth century are usually dismissed in
a few sentences that refer briefly to Lubbock's 'Palaeolithic' and 'Neolithic'
divisions; the faunal chronology developed by the French palaeontologist,
Édouard Lartet (1801–1871); and the famous industrial classification promoted by the French prehistorian, Gabriel de Mortillet.[1] The reactions of
other researchers to the stone tools of the British river drifts have been hidden
under their shadow. But if the varied and detailed patterns that these
researchers saw in the river-drift tools are painted back into the historical
picture, a clearer perspective is gained of their response to Continental
research and the reasoning behind more comprehensive classifications of
the British Palaeolithic.

Some of those who worked on the stone tools of Britain have already been
introduced: Lubbock, Dawkins, Lyell, and James Geikie played geological
roles in Chapter 2; Evans defended human antiquity in Chapter 1. All five
published synthetic works on the British Palaeolithic. Lyell was one of the first
to draw the new mass of information together in his book on the *Antiquity of
Man* (1863), which sold well but did not impress all his peers. Lyell received
several charges of plagiarism and Greenwell, the Durham archaeologist,
confided to his friend: 'We want Master Evans, a good book on the Antiquity
of Man, there is quite sufficient matter now accumulated to admit of one.
Lyell's book is not satisfactory, there is too much of dubious evidence bought
in, & with all humility I say, he is not master of the subject'.[2]

Other authors followed close on Lyell's tail. In 1865, the first edition of
Lubbock's *Pre-Historic Times*, an account of past and present savages, could
be purchased for fifteen shillings. Evans produced *Ancient Stone Implements* in
1872, a cautious, catalogue-like description of stone tools, full of careful
engravings and priced at twenty-eight shillings. Dawkins presented his ideas
about the Stone Age alongside now-familiar palaeontological arguments
in *Cave Hunting* (1874) and *Early Man in Britain* (1880). Geikie's glacial

---

[1] Daniel 1975, 99–109; Trigger 1989, 94–97.
[2] Greenwell to Evans (undated, probably late April 1874): AMO: JE/B/1/13.

chronology had a central place in *Prehistoric Europe* (1881), the book that widened his rift with Dawkins.

A few new names also appear in this chapter. Colonel Augustus Henry Lane Fox (1827–1900) of Kensington, London, is best known today for his evolutionary typology of weapons and other artefacts, and his ethnological collections in the Pitt-Rivers Museum, Oxford. He took the name 'Pitt-Rivers' when he came into his inheritance. But before Lane Fox used his wealth to embark on a series of famed excavations on the prehistoric sites of Cranborne Chase, he worked briefly on the Palaeolithic gravels of London: at Acton and Ealing.[3] 'Pitt-Rivers' was, as Whitaker joked, 'a fit name for a finder of implements in pits in the River Drift'.[4]

Three other researchers looked more closely at the stone tools that were emerging around London. Worthington George Smith, John Allen Brown, and Flaxman C. J. Spurrell all kept a careful watch on gravel-pits, brick-pits, sewers, and any other exposures in their neighbourhoods that might penetrate the Palaeolithic layers beneath their expanding Victorian city and its sprawling suburbs. Their papers clustered in the 1880s, though Smith gave a later account of Palaeolithic sites in *Man the Primeval Savage* (1894), a book adorned 'with two hundred and forty-two illustrations by the author'.[5]

Smith was a draughtsman as well as a Palaeolithic researcher. He had trained as an architect and harboured dreams of designing cathedrals; but in his twenties, tired of working on drains for his employers, he turned to freelance illustration. His self-engraved notepaper proclaimed him a 'Wood Engraver and Artist' and incorporated an ornate and revealing design: a spray of flowers and toadstools bursting from a pile of quills, pens, and a fine, pointed flint implement.[6] The quills and pens represented Smith's work as a draughtsman; the toadstools, his reputation as a fungus expert. In 1864 the *Journal of Botany* published his graphic account of the effects on the human constitution of the fungus *Agaricus fertilis*, with which Smith had poisoned his family by mistake. He was tormented with headaches, sickness, and dreams, 'in which Fungi, and particularly toadstools, always played a prominent part'.[7] But his relatives all survived the misidentified meal. The flint

---

[3] On Lane Fox, see Bowden 1993.

[4] Whitaker 1889 (i), 341.   [5] Smith 1894, v (title page).

[6] For an example of this notepaper, see the letter from Smith to Evans dated 22 January 1879 (AMO: JE/B/2/80). The life and work of Worthington Smith has been described by James Dyer in '"Middling for Wrecks" Extracts from the story of Worthington and Henrietta Smith' 1959, '"W.G.S." and the Potato Blight Mystery' (1967–68), and 'Worthington George Smith' 1978. See also Smith 1888 on 'Lepores Palaeolithici: Or, The Humorous Side of Flint Implement Hunting'. Smith's reports of four Palaeolithic sites have been used by White 1997 to gather data relevant to the questions of late-twentieth-century archaeologists.

[7] Smith 1864, 217.

implement on this notepaper referred to an interest that had started in the late 1870s, when Smith read Evans's *Ancient Stone Implements*. His attention had been caught by descriptions of two finds from his own district: north-east London, and Smith soon became an authority on the stone-tools in his area.

Brown was a diamond merchant, like his father before him, but the interests of both father and son reached beyond their business. John Brown senior had been one of the founders of the Ethnological Society and had spent his spare time encouraging Arctic exploration.[8] His son undertook explorations closer to home, though more distant in time. Brown studied the geology and Palaeolithic tools around Ealing in west London, inspired by the discoveries made by Lane Fox in his neighbourhood during the late 1860s.[9] Brown published most of his papers on Palaeolithic discoveries in the 1880s. He gathered his scattered articles into a book: *Palaeolithic Man in N. W. Middlesex* (1887).

Spurrell also contributed a small number of noteworthy articles to journals. He was a quiet authority on botany, natural history, early charters and legal documents as well as stone tools. There was family interest amongst the Spurrells in the river drifts of London: Dr F. Spurrell of Belvedere, Spurrell's father, had been a notable collector of fossil bones at Crayford long before his son worked on the Palaeolithic flakes from the site.[10] The names of many more researchers can be found scattered through prominent synthetic works, hidden in thankful footnote references to their discoveries.

3.1 SHARING STONE-AGE RESEARCH: SOCIETIES AND JOURNALS

Most of the books that have been mentioned were constructed and adapted from shorter papers that had already appeared in journals. Some had been published in popular periodicals like the *Quarterly Journal of Science*, the *Natural History Review*, and *Nature*. Discoveries and opinions were also shared in the rooms and journals of provincial learned societies, which had been multiplying rapidly before the 'Palaeolithic' had even received a name.[11] Favourite choices for Palaeolithic researchers were those local societies that welcomed a combination of geology, archaeology, and natural history: Brown, for example, enjoyed membership of the Ealing Microscopical and Natural History Club.

[8] Woodward 1904, xiv.   [9] Brown 1889, 56.   [10] Kennard 1944, 121; 1947, 287.
[11] Morrell and Thackray 1981, 12–14. On the rise of provincial geological societies in the early nineteenth century, see Knell 2000, 33–72.

The rise in local societies and field clubs around London in the later nineteenth century gave Archibald Geikie much satisfaction. He had succeeded Ramsay as Director–General of the Geological Survey in 1882, and hoped 'that henceforth no section of importance will be laid open and covered again without being carefully noted by local observers.'[12] Similar expectations led Whitaker to berate members of the Essex Field Club, frequented by Smith and Spurrell: 'One of the most irritating sites to a field-geologist is a new railway-cutting carefully soiled over, of which no geological record has been kept. The Essex Field Club ought to have saved me from such irritation!'.[13]

In the heart of London sat the grander rooms of the national societies: expensive and much more exclusive than the small, provincial societies. The *Quarterly Journal of the Geological Society of London* and the *Journal of the Royal Anthropological Institute* both took papers on the Palaeolithic. Recent finds and theories could also be read in the less prestigious but more accessible pages of the *Geological Magazine* and the *Proceedings of the Geologists' Association*. Papers appeared in both anthropological and geological sections at the well-patronized, peripatetic Meetings of the British Association for the Advancement of Science.

Archaeological societies published surprisingly few papers on Palaeolithic topics. Spurrell sent several of his contributions to the Royal Archaeological Institute, but when these appeared in their *Archaeological Journal*, they looked out of place next to the usual antiquarian articles which ranged from Roman to Medieval times.[14] Spurrell, though, was already familiar with the Archaeological Institute—he had an interest in old charters as well as stone tools—so this was not an unusual choice for him. Evans, whose antiquarian interests also predated his studies of stone tools, read two of his earliest articles on the Palaeolithic to the Society of Antiquaries of London in 1859 and 1861. Their journal, *Archaeologia*, contained little further discussion of Palaeolithic subjects through the rest of the century, although a few short papers did appear in their *Proceedings*.

Journals devoted exclusively to prehistoric archaeology also existed in the nineteenth century, but enjoyed more success in France than Britain. One of the earliest journals on prehistory was founded in 1864 by de Mortillet: *Matériaux pour l'histoire positive et philosophique de l'homme*. Perhaps inspired by this example, S. J. Mackie began a short-lived, illustrated popular magazine: *The Geological and Natural History Repertory; and Journal of*

---

[12] Whitaker 1889 (i), iv.   [13] Whitaker 1887, 180.
[14] On the scanty discussion of Palaeolithic and geological topics at the Archaeological Institute, see Ebbatson 1994, 63–65.

Pre-Historic Archaeology and Ethnology (1865–1867). Mackie had edited the *Geologist* (1858–1864), a popular monthly magazine that was superseded by the *Geological Magazine*; but he wanted to develop his new prehistory publication into a popular weekly number. At only twopence per issue (the *Geologist* had cost a shilling), this was to provide 'a quick, full, and *cheap* means of intercommunication amongst Geologists, Naturalists, Ethnologists, and those engaged in kindred pursuits'.[15] His subscribers included Lubbock, Ramsay, Godwin-Austen, and Pengelly. Mackie observed in the first issue: 'For the many labourers on that important subject which is now attracting such universal attention—the Antiquity of Man—there needs [to be] a special and ready organ to convey the knowledge of new facts, and to afford the means of discussion'.[16] Only twenty-nine issues of the *Repertory* were published; to discuss the Palaeolithic, researchers were turning instead to the varied organs that were already available in the City and the provinces.

Researchers wrote about the British Palaeolithic in various different ways, but the most common kind of publication was a simple statement of discovery. These short notes or articles might give a description of the tools found, the layer of sediments in which they lay, and any other associated material: bones, shells, or plant fragments. Then there were longer articles and books that speculated about broader patterns: suggesting how particular tool forms or stone-working techniques might be arranged in time and space, and—whether intentionally or not—drawing Palaeolithic sequences into arguments over geological sequences.

A few researchers applied a finer time-scale to the Palaeolithic. Although, on the one hand, tools of similar character could be grouped together, classed as a single 'industry' and regarded as similar in age even if they were separated by millennia; on the other hand, the work of a few ancient minutes could be reconstructed by scrutinizing the scars on a single stone implement, or the scatters of tool-working debris lying frozen in space at a single site. Smith, Spurrell, and Brown each discovered abandoned land-surfaces, or 'Palaeolithic Floors', where the flakes once struck by ancient flint-knappers still lay. They reconstructed ancient methods of manufacture by re-fitting these flakes back together again, and even tried to make tools themselves: Smith and Spurrell were adept Victorian flint-knappers, and Evans demonstrated stone-working techniques at the International Congress of Prehistoric Archaeology in 1868.[17] But it was difficult to reconcile these two perspectives of Palaeolithic tools: as the products of individual members of tool-making groups, and as patterns of industries lying in broader swathes through time and space.

---

[15] Mackie 1865a.   [16] Mackie 1865b, 7.   [17] Evans 1869, 192.

3.2 THE RIVER-DRIFT PALAEOLITHIC OF JOHN EVANS

When Evans joined Prestwich in Abbeville on 1 May 1859 to examine the implements found by de Perthes in the Somme Valley river drifts, he had little idea that he would become one of the most respected authorities on stone tools in Britain, and even in Europe. But by 1861 he had already won the name of 'Flint Evans' at the British Association for his part in the Somme Valley trip and his papers delivered soon afterwards.[18] After his book was published in 1872, *Ancient Stone Implements* became the bible of flint-hunters: a weighty and undisputed classic. It was difficult to argue with what was, essentially, an illustrated catalogue of finds. The second edition, which appeared in 1897, incorporated more descriptions of discoveries made over the intervening years; but the cautious suggestions of their patterning through time had changed very little. Evans had founded these suggestions on two early papers that he read to the Society of Antiquaries.

In his first paper, given in 1859 and published the following year, Evans's main aim was to reinforce the credentials and age of the river-drift tools: he asserted that these had been formed by human action rather than natural processes, and had been found alongside the bones of extinct animals. But he also divided the river-drift tools into three classes, based on their appearance. This classification structured his own thoughts about Palaeolithic patterning for many decades more and moulded the views of his peers.[19] Although his scheme was founded on discoveries from pits around St Acheul and other sites along the Somme Valley, it would also be applied to English finds.

The three classes of John Evans comprised: first, flint flakes; second, pointed weapons with a thick base-end, or butt (found in two varieties, one more acutely-pointed than the other); and third, oval implements with a cutting-edge that extended all round the implement, including the butt (these, he later described as 'sharp-rimmed implements'). Implements of the second and third classes are illustrated in Figure 3.1. Evans wondered if their shape held a clue to the use of these tools. He observed that the butts of some of his pointed weapons had been left smooth, nodule-like and un-chipped: perhaps so they could have been grasped in the hand with no need of an extra handle or haft—which would, in any case, have rotted away

---

[18] Evans 1943, 107. When Evans was awarded the Lyell Medal in 1880 for his 'distinguished services to geological science', the President observed, 'we now can scarcely say where archaeology ends and geology begins, nor whether to rank and value you most as an antiquary or a geologist' (Sorby 1880, 30).
[19] Evans 1860, 289–291; 1872, 560; 1897b, 640–641.

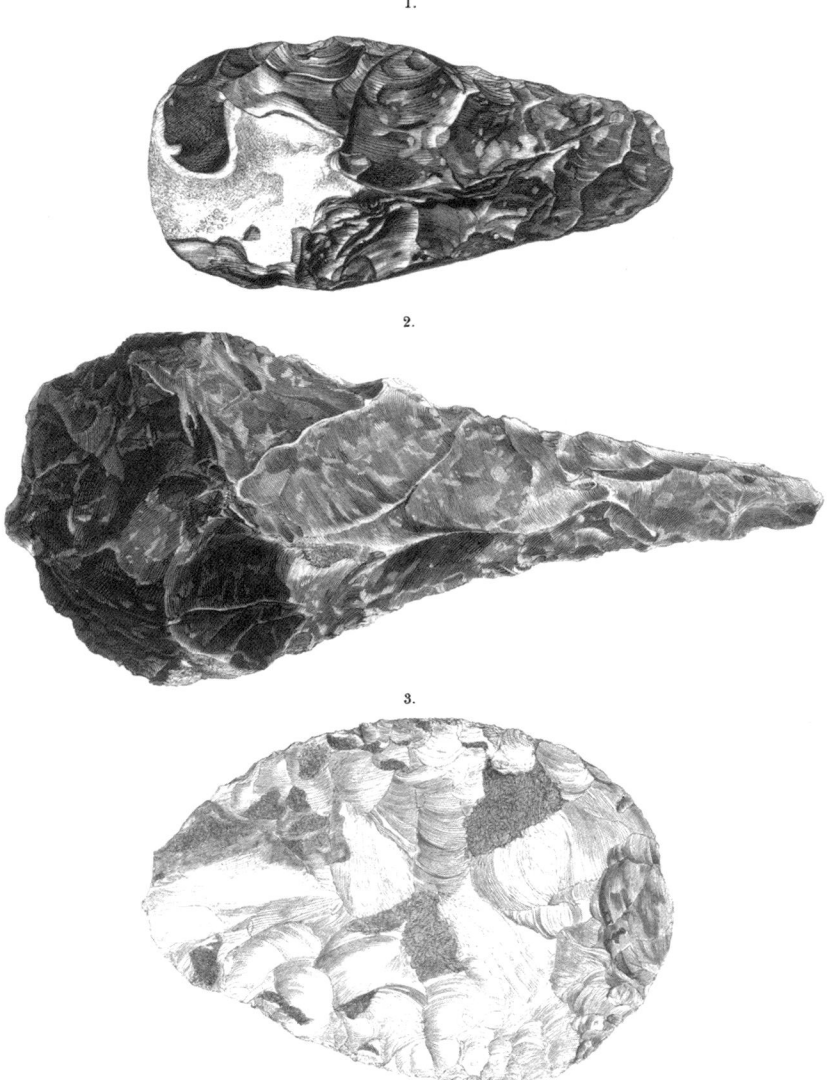

Figure 3.1. These three implements illustrate two of the classes defined by Evans (1860, Plate XV, facing p. 291). The upper two are examples of Evans's pointed weapons (the middle implement is one of the acutely pointed varieties); the lower one is an example of his oval implement with a cutting edge all round or 'sharp-rimmed' implement.

long ago.[20] There were other practical explanations for differences in shape: Evans suggested that variation in his sharp-rimmed oval implements might arise 'from defects in the flints from which they were shaped'.[21] Indeed, he warned that there was so much variety in his second and third classes that they might be said to blend one into the other.[22]

Evans dug back into the work of eighteenth-century British researchers to dilute the French flavour of the new discoveries surrounding the question of human antiquity. The stratigraphy of Hoxne was joined by the implement from Gray's Inn Lane, found long before by Conyers and described by Bagford. This became a famous example of Evans's pointed weapons (see Figs. 1.1 and 3.2).[23] In these early years, the pointed weapons and sharp-rimmed oval implements of his second and third classes were known by a variety of names. In France, they were called *hâches* before de Mortillet introduced the term *coups-de-poing*. Then there were local names: workmen at the pit of St Acheul called them *langues de chat* (cats' tongues); those at Hoxne named them 'fighting stones'. But there was no widely accepted British term until the British Museum *Guide* to their Stone-Age collections advised in 1911 that 'hand-axe' might offer a convenient equivalent to the French *coup-de-poing*. The 'hand-axe' became far more popular than an alternative suggestion made the same year by William Sollas (1849–1936), Professor of Geology and Palaeontology at the University of Oxford: that they be called 'bouchers' after their early discoverer, Boucher de Perthes. The names of Prestwich and Evans were also requisitioned in 1911 for other tool-types, but 'evanses', 'prestwiches', and 'bouchers' never became fashionable.[24]

In his second paper to the Society of Antiquaries, read in 1861 and published in 1863, Evans reported on discoveries of the previous two years, explained that each of his three classes included a variety of typical forms, and made tentative remarks about their patterning through the river drifts. Some types from his first class of flakes, for example, were particularly abundant in the lower deposits of the Somme Valley. As Evans explained to Lyell: 'The implements which, as far as I have hitherto seen, are almost without exception peculiar to the lower gravels, are the broad skilfully made flakes like No. 4 in

---

[20] Evans 1860, 291; 1872, 560–566.   [21] Evans 1860, 291.
[22] Evans 1860, 288; 1863, 75.   [23] Evans 1860, 301–302.
[24] De Mortillet 1883, 148 preferred the term 'coup de poing' to 'hâches' because the cruder pointed weapons, which he referred to his 'Chellean' epoch of the Palaeolithic, were not hafted but used in the hand. The terms used by the French and British workmen for stone tools are recorded by Evans 1860, 291, 300. On the 'boucher', see Sollas 1911, 75; on the 'hand-axe', see Read 1911, 12. The terms 'prestwich' and 'evans' were advocated by Abbott 1911, 469–471 for the cores and the flakes (respectively) that were soon more commonly associated with the 'Levallois' industry.

Figure 3.2. Evans's illustration of the Gray's Inn Lane implement (Evans 1860, Plate XVI, facing p. 301). Published full size in the original, this implement is a little over 16cm long from base to tip. Bagford's description of the same implement had also been accompanied by a crude sketch (see Fig. 1.1).

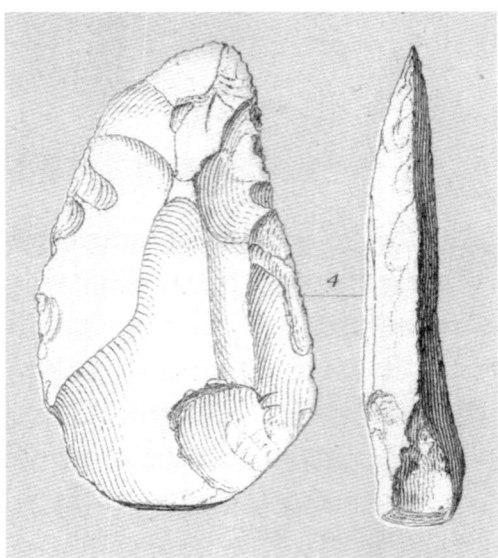

Figure 3.3. This is the flake (No. 4) mentioned by Evans to Lyell in 1862: a carefully prepared type that seemed to be restricted to the low-level gravels of the river valleys (Evans 1863, Plate IV).

my plate' (Fig. 3.3).[25] Flakes shared a common characteristic: one side was flat, where the flake had once joined its parent block of flint; the other side was convex and might display a number of facets. But the type mentioned to Lyell had been shaped very carefully while still on its parent block of flint: the convex side had been prepared with several facets before the finished flake was detached. Many flakes prepared in this distinctive manner had been found in the low-level gravels of the Somme Valley at Menchecourt near Abbeville, and at Montières near Amiens.[26]

Evans also suggested that the implements of his second and third classes (later known as 'hand-axes') might come from different levels in the Somme Valley. The sharp-rimmed oval hand-axes and the flakes (particularly the more finished kind shown in Fig. 3.3) were plentiful in the low-level deposits and were found at sites like Menchecourt and Montières, but the pointed hand-axes were more common at higher levels and were found at sites like St Acheul near Amiens. The connection between hand-axes of distinctive shape and particular pits was noted by other researchers: any pointed hand-axe was soon described as being of 'the St Acheul type', and the sharp-rimmed, oval hand-axes might be referred to as the 'Abbeville' or 'Menchecourt' type.[27] Evans linked these different types, though in very tentative terms, to different

[25] Evans to Lyell, 17 May 1862: ULE: Gen 110/1185–1186.
[26] Evans 1860, 290; 1863, 75.
[27] Evans 1860, 292; Lyell 1863a, 114–115; Prestwich 1864, 254, 258, 261; Flower 1867, 45; Brown 1888, 366; 1889, 61.

positions in the river valley: a distinction that also hinted at a difference in age, because the higher band of drifts was older than the lower band (see below).[28] In 1862, Evans mentioned this pattern to Lyell, who reported the distribution in his *Antiquity of Man*. Lyell made the point about their difference in age with more conviction than Evans, and thought that the overall similarity of shape between these implements suggested a very slow rate of progress.[29]

Evans was drawing on the geological conclusions of his friend, Prestwich, when he made this connection between the height and the age of river sediments. Prestwich believed that Quaternary rivers had cut down into their beds through time, leaving river drifts in broad bands by their sides. As the rivers deepened their valleys, they flowed below their previous levels; the older river drifts clustered higher up the valley sides than the more recent river drifts. Prestwich therefore assigned his high-level group of valley gravels to a more ancient time than his low-level group, which lay nearer the rivers of Victorian times. This river-drift sequence offered Palaeolithic researchers a useful time-scale for the implements that lay within these sediments. Tools from the higher group could, admittedly, have been swept down to the lower group; but their intrusive, 'derived' nature could be identified by the wear and abrasion they received on their downward journey. Height, condition, form, and technique all helped to decide the age of a river-drift tool—but in the early 1860s most of these tools still came from France.

Evans and Prestwich urged members of antiquarian and geological societies to recover more tools from British deposits and to record associated sediments and bones. They suggested profitable areas to search; Prestwich encouraged hunters with the motto *Nil desperandum*, and the hunt began.[30] Keen society members soon added to the numbers of known specimens. Labourers also learned to spot the now-valuable flint implements. Down in their pits, they were in a better position to scan large quantities of gravel than were their customers of higher standing, and they took their new skills with them as they moved between jobs.[31]

After a decade of additional finds, from the gravels of Bournemouth to the basin of the Ouse valley, Evans's original conclusions were in need of expansion. In 1870, Edward Stevens, author of the popular book *Flint Chips*, observed that new types of implement had been found since Evans first set out his classification: he described heart-shaped implements with a cutting-edge all round,

---

[28] Evans 1863, 81–82.
[29] Evans to Lyell, 17 May 1862: ULE: Gen 110/1185–1186; Lyell 1863a, 114–115, 376–377.
[30] Evans 1860, 307; 1863, 58; Prestwich 1861a, 363, 368.
[31] Evans 1872, 479.

shoe-shaped implements with a flat underside, scrapers, and other forms.[32] Two years later, when the first edition of *Ancient Stone Implements* was published, Evans retained his original three classes but he did take up some of Stevens's suggestions and gave a longer account of types within each class.[33] By then, it was clear that English river drifts were producing similar forms to those from France, which supported the suspicion that the two countries had once been joined by an ancient land-bridge.[34]

Evans still thought that the general preponderance of the 'ruder pointed implements in the high-level gravels and of the flat, oval, sharp-rimmed implements in the low-level' seemed to hold good, despite the variety of implements from the two levels, but admitted that the distinction was not as marked as he had formerly expected.[35] Brown stated with more certainty that the pointed implements with heavy, often un-worked, butts (like the hand-axe from Gray's Inn Lane) were among the oldest of the river-drift types.[36] Thomas Rupert Jones (1819–1911), Professor of Geology at the Royal Military College, Sandhurst, agreed that the crudest implements came from the oldest valley-gravels: from high-level sites like St Acheul, ninety feet above the nineteenth-century River Somme.[37] More recent deposits yielded sharp-rimmed, oval hand-axes, the prepared type of broad flakes, and a variety of trimmed flakes, or scrapers. The patterns outlined in the 1860s gained support: it seemed that tool form had changed over time, tool-working had become more skilful, and carefully-manufactured trimmed flakes appeared relatively late in the river-drift sequence.

Some of these labels need more explanation. The term 'implement' tended to be used for stone tools that had been trimmed or flaked ('retouched') to alter their form: these 'implements' were distinguished from simple un-worked flakes. But the same term could also be restricted to retouched tools made on nodules or 'cores' of flint—'core-tools' like hand-axes—and not extended to the retouched flakes ('flake-tools'). In a list of the specimens in his collection, Smith distinguished between 'implements' and 'trimmed flakes', but he noted: 'the latter are true implements, made from large flakes, one side being left unworked, or only slightly worked'.[38] Unless in direct quotation, 'implement' is used here to refer to both core-tools and flake-tools; 'tool' is used as a general term for any stone artefact.

---

[32] Stevens 1870, 40–41.   [33] Evans 1872, 565.
[34] Horner 1861, lxv–lxvi; Flower 1867, 51; Evans 1872, 614–616.
[35] Evans 1872, 616. The wording survived unchanged to the second edition (Evans 1897b, 701).
[36] Brown 1887a, 88–89.
[37] Jones 1877, 24.
[38] Smith, printed leaflet, 1 December 1882: AMO: JE/B/2/80.

3.3 DISCOVERIES AROUND LONDON

London became an important centre for Palaeolithic research in the decades after the acceptance of human antiquity. The pits that supplied gravel, sand, clay, and brickearth for the roads, railways, and buildings of the growing city often pierced implement-bearing deposits. These materials were extracted by hand; the process would not become fully mechanised for many decades, so the workmen missed little of value—and implements could be sold for high prices.

In the late 1860s, Lane Fox was hunting for implements around Acton and Ealing. He had been encouraged by a similarity between the geological position of these districts and the positions occupied by the implementiferous gravels of the Somme, and the Ouse where Wyatt had made his discoveries in 1861. He tracked down some implements in 1869. They belonged to Evans's second class with thick butts: both the pointed and the more oval forms.[39] Lane Fox related his discoveries to Prestwich's division of the valley-gravels, but he also mentioned a sequence observed by Whitaker.

As he mapped the geology of London for the Survey, Whitaker recognised three horizontal spreads in the river gravels of the Thames Valley, three steps (or 'terraces') left by the old river on the valley sides: a high terrace at 50 to 100 ft above sea level; a middle terrace at 20 to 30 ft; and a low terrace at 10 to 20 ft. These three terraces would often, thereafter, be used to indicate the geological position of Palaeolithic tools. Lane Fox assigned his implements to Whitaker's high-terrace gravel.[40] Although Lane Fox saw much variety in the river-drift implements, he believed that Evans's two kinds of hand-axe could indeed be classed as two distinct river-drift types.[41] One class with a roughened end followed by another class with a cutting-edge all round fitted his penchant for progress: they seemed to represent two stages in an evolution from hand-held to hafted implements.[42]

The richest decade for nineteenth-century discovery around London was the 1880s, when three researchers—rarely mentioned today—became familiar to casual labourers digging in their neighbourhoods. Smith, Brown, and Spurrell discovered Palaeolithic assemblages lying amongst horizontal beds of stones that they thought to be the remnants of ancient land-surfaces. Smith found these 'Palaeolithic Floors' under Stoke Newington Common and in pits

[39] Lane Fox 1869; 1872, 458.
[40] Lane Fox 1872, 450–453; Whitaker 1864, 87. On the divisions of the Thames Valley river drifts, see Whitaker 1864, 82–83; 1875, 61; 1889 (i), 389–391.
[41] Flower 1872, 293.
[42] Lane Fox 1867, 618; 1868, 409–410; 1875, 294–295.

around the small village of Caddington on the Hertfordshire-Bedfordshire border.[43] Spurrell described a Palaeolithic 'workshop' from Crayford and Brown recorded Floors around Ealing.[44] Their response to these frozen fragments of Palaeolithic time, captured in the stone scatters left by ancient tool-makers, and their detailed accounts of personal discovery offer an interesting contrast to the blend of British find-spots presented by Evans in *Ancient Stone Implements*.

Evans and Prestwich once had to plead with members of societies to go out into the field and gather tools and information. The atmosphere had changed by the 1880s. Flint implements were now desirable prizes for curio-collectors who wanted pieces for their cabinets; Palaeolithic researchers, though, prided themselves on collecting knowledge as well as implements. Confronted by competition, they became reticent about the location of their discoveries (whether gravel-pits, house-foundations, railway-cuttings, sewer-trenches, or any other entrances to the richer seams of river drifts). When curio-collectors swarmed to a site, the workmen could raise their prices for implements and the number of forgeries could rise. Instead of lists of likely find-spots, like those provided by Evans and Prestwich in the early 1860s, Spurrell tried to keep his discoveries to himself.[45] Smith also recalled defiantly that he had kept secret the location of one site for twenty-seven years: 'Otherwise the members of local scientific, literary, and philosophic societies would have visited the place like a swarm of locusts and totally destroyed all my work, as was the case in North London'.[46]

3.4 THE TRIALS AND TRIUMPHS OF WORTHINGTON SMITH

Worthington Smith had started to hunt for stone tools in north-east London after reading Evans's *Ancient Stone Implements* in the late 1870s. Thereafter his tall figure, clad in a long cloak and wide-brimmed hat, had become an oddity of the neighbourhood as he strode between exposures of ancient deposits, looking for implements in drains, house foundations and pits.[47] Smith became sharp-eyed and successful. He kept Evans abreast of his discoveries and suggested trips they might make together to see the deposits where these tools were coming from. There were the Clapton and Shacklewell gravels, for

---

[43] Smith 1882; 1884; 1894, 64–65, 106–107.
[44] Spurrell 1880a; Brown 1887a, 55.
[45] Spurrell 1884, 112.   [46] Smith 1916, 68.
[47] Smith 1884, 358; 1894, 189; Evans 1897b, 586.

example, where 'The old pits are now nearly filled with new villas & dead cats but still there is something to be seen in the way of sand & gravel'.[48]

Pits were not the only source to be exploited by Smith. He also examined the gravel that lay in heaps dotted around the London suburbs, transported from pits to supply stone ballast for roads and railway lines—and stone tools for the keen-sighted observer. His unusual fixation with these heaps astonished passers-by. One day, as Smith was gazing upon a gravel pile, he recalled:

two labourers (both unknown to me) were sitting close by, when one said to the other, 'Do you see that gentleman, Jack?' 'Yes', said the other. 'Well', said the first, 'if you ever sees a heap of gravel anywhere, it don't matter where, if you keep your eye on that heap of gravel long enough you will be bound to see that gent come and walk about on the top of it'.[49]

Smith realised that parish officials and builders tried to save their pennies by buying the cheapest loads of gravel, so these heaps might lie a considerable distance from their original pits.[50] He managed, nonetheless, to trace many to their sources, where he could discuss the gravel with pit workers.

Fellow enthusiasts were given this advice by Smith on the etiquette of approaching gravel diggers and sifters: 'To make a good beginning with strange diggers a man should not be too well dressed; he should jump gaily into the pit, and at once begin conversing in a friendly and familiar manner with the men'.[51] Though engaging in the pits, Smith was quieter in the company of fellow researchers. They saw him as 'a remarkable personality, very reserved [. . .] so retiring in disposition that his knowledge and friendship had to be sought'.[52] Smith had little money and was not prominent in large national societies, but his energetic research on the Palaeolithic was well known to his peers.

Smith made an awful mistake in his first major paper on the British Palaeolithic, which was published in the *Journal of the Anthropological Institute* in 1879: he revealed the locations of nearly every implement-bearing position he knew of in north-east London. Curio-collectors swept down to these sites and openly offered the workmen five or ten times as much money for tools as they had received from Smith. Fearing what might happen if the workmen discovered their value, Smith had always tried to hide his true interest, talking of bones and fossils as he scrabbled furtively for implements and flakes. The atmosphere around Smith's pits changed when curio-collectors appeared.

---

48 Smith to Evans, 19 May 1879: AMO: JE/B/2/80.
49 Smith 1888, 10.   50 Smith 1880, 319; 1894, 90–91.
51 Smith 1888, 7.   52 Kennard 1947, 280.

Public houses near the pits became a source of stone tools for Victorian gentlemen: workmen could get advances of beer and gin on the security of a now-valuable implement. And as prices rose, so did forgery. Forged tools were already common in the less prolific pits of London, but experienced researchers could spot them. More damaging were the false provenances of genuine tools: collectors gathered at the well-known sites, so workmen who found tools in obscure areas might pass them on to be passed off at more famous pits. The site of Gray's Inn Lane, where discoveries were expected, was salted not just with forgeries but also with genuine Palaeolithic artefacts brought from Stoke Newington.[53]

This story of careful research blighted by trophy-hunting collectors and cunning forgers had formed a familiar part of London life for some while. Forgery of fine-quality implements, beloved by collectors, was a booming industry that reached beyond the world of the pit-workers. As Smith observed, these labourers were not skilled in the use of hammers and punches: it was carpenters and plasterers who made good forgeries. They sold them for small sums to the labourers, who then passed them on for larger sums to the collectors. Smith condemned the foolishness of any collector who let the men know exactly what they were looking for. He was appalled that some even lent them genuine implements: blueprints for forgery. But as collectors became aware of implement-fraud, ever more perfect forgeries were produced. Newly-made implements might be boiled with iron to imitate staining, shaken with stones and sand to simulate abrasion, or brushed until they gleamed with the shine peculiar to certain pits: all according to the expectations of gullible patrons.[54]

The most notorious forger was Edward Simpson, or 'Flint Jack', who duped many collectors of curiosities—and who was regarded as a curiosity himself. At a meeting of the Geologists' Association in 1862 at Cavendish Square, Flint Jack's muddy corduroys, greasy hat, and weather-beaten appearance excited speculation until, at a word from the Vice President, he undid his bundle and exhibited his skills. Flint Jack made a more anonymous appearance as 'a person' in the Association's *Proceedings*.[55]

There was some affection for eccentric characters like Flint Jack: when he ended up in Bedford Gaol, one antiquarian even addressed a plea to readers of the *Antiquary* asking for financial assistance to secure his release.[56] But the collecting ethos that provided Flint Jack and his fellow forgers with a

---

[53] Smith 1884, 377–378; 1894, 251–253, 295.

[54] Smith 1894, 294–297. On forgery, see also Evans 1872, 575–577 and East 1888.

[55] Wiltshire 1862, 224, 226. The fuller account of Flint Jack's appearance at the Geologists' Association was given by Jewitt 1867, 73–74.

[56] Jewitt 1867.

livelihood was distained by Palaeolithic researchers. Smith complained that the habit of collecting only the ornate and elaborate implements at the expense of the ruder or damaged forms—which were far more common— gave a false perception of the past, and that more could be learnt about the ways of Palaeolithic peoples from their rough, everyday tools.[57]

Nonetheless, Smith believed that well-managed collections could make a great contribution to Palaeolithic research. His splendid personal collection included all kinds of flints from the finest hand-axes to the rudest flint chips, most of them from around London. The massive holdings of Evans, in contrast, displayed a representative selection of different forms and techniques from major sites across Britain and many from the Continent. Smith felt honoured to deplete his own collection to feed the needs of this authoritative series. Early in his relationship with Evans, Smith wrote:

It was exceedingly kind of you to offer me one of your Boscombe (?) implements yesterday. I hardly liked to say *yes* as I felt fearful of in any way injuring your collection, which ought to be *added to* by everyone rather than diminished. If however you can spare one at any convenient time I shall be only too glad to have it & I shall esteem it highly.

If you care for one or two *genuine flakes* from some of the new positions I shall be very pleased to send them on to you but possibly you do not care for flakes.[58]

The relationship between local researchers like Smith and elite investigators like Evans is an interesting one. Both were happy to purchase finds directly from workmen or from middlemen who collected merely for profit. But Smith felt awkward about selling material to a fellow researcher, even though he was not a rich man. Money could alter the relationship between colleagues in science, who ranked the respectable and honourable search for knowledge above such worldly concerns. Anne Secord has described a similar awkwardness amongst botanists in Lancashire in the early nineteenth century. In that case, though, it was the lower-class artisan botanists who were loathe to collect in return for payment, preferring to exchange specimens or information with botanists of higher social rank. A request for payment would have damaged an artisan's claim to belong to the learned community of naturalists.[59]

Unlike the lower-class artisan, the middle-class Smith paid a lot of money to assemble his collection. Some of his cheapest pieces had cost five shillings. Implements from sites known for their large, fine-quality pieces might cost as much as £1 each: this was the price commanded by implements of flinty chert from the ballast pit at Broom, near Axminster on the boundary between

---

[57] Smith 1879, 279; 1883a, 274.
[58] Smith to Evans, 26 June 1878: AMO: JE/B/2/80.
[59] Secord 1994a, 288; 1994b, 393–396.

Devon and Dorset. Transactions were not always a straightforward exchange of implements for money: workmen might be paid to look out for flints but find nothing, or only find 'wasters' (waste flakes) which would then have to be purchased. There were other currencies than coin, which made it difficult to estimate the cost of an artefact: for three good specimens, Smith recorded: 'I gave the man £1. for & his wife (previously) a leg of mutton & a bottle of port'.[60]

By 1882, Smith had over a thousand specimens in his collection. He printed a small leaflet about his hoard, which stated proudly: 'None of the Stone Implements enumerated in this list have been purchased from dealers. One hundred and sixty of the examples were found by MR. W. G. SMITH himself. With a few exceptions all the others were found by men specially set on to search for stone weapons by MR. W. G. SMITH'.[61] But in December 1882, Smith offered the pickings of his collection to Evans: 'You are welcome to the largest & best specms. for they could not be in better hands than yours'. The question of money placed Smith in a difficult position. He eventually supplied Evans with the prices he had given to the workmen, but did not include carriage time and was still embarrassed about the situation: 'I am quite willing to *give* a few specms. from any of the places'.[62] On Christmas Day, he admitted: 'It is a pleasure to send the stones, but when it comes to sending *a bill* it is different & I cannot see my way to it'. Again, he asserted Evans's right to the specimens:

I have always designed that the first & best of these things should go to you. Mr Backhouse wished to buy a number from all the positions but I said, no,— Canon Greenwell brought a Mr Robinson here one day but I would not dispose of any—you are the only person who has had more than one or two examples (as gifts). I think now; if you have all you wish for, I will let others have some, if they will pay 10ˢ./ ea—they have cost me at least three or four times as much as that, in time & money.[63]

Despite Smith's fears, he was not classed by Evans as a dealer or curio-collector, but welcomed as a worthy fellow researcher. A few months after this episode, Smith was flattered to be invited to see Evans's collection. He replied, 'I should exceedingly like to spend a *very short time* at your house to glance at the things, especially the Palc. [Palaeolithic] stones & your manner of keeping them together'.[64] A decade and a half later, Evans made full acknowledgement to Smith for his generosity in the second edition of *Ancient*

---

[60] Smith to Evans, 25 December 1882: AMO: JE/B/2/80.
[61] Smith, printed leaflet, 1 December 1882: AMO: JE/B/2/80.
[62] Smith to Evans, 9 December 1882: AMO: JE/B/2/80.
[63] Smith to Evans, 25 December 1882: AMO: JE/B/2/80.
[64] Smith to Evans, 15 April 1883: AMO: JE/B/2/80.

Stone Implements. When Smith read the section on the river-drift implements of the London area, he might have been pleased to read: 'By the kindness of Mr. Worthington Smith most of the important specimens that he has found are now in my collection'.[65] These were illustrated with Smith's own woodblocks, again with due acknowledgement, and Evans referred with respect to Smith's discoveries, his views, and his 'excellent book'.[66]

Many fine implements made their way to Evans's collection from deferential peers. Their names can be found in his book. Most have been forgotten, but the Smiths of Palaeolithic research commanded respect and did not always lose on a Palaeolithic bargain. In 1884, a Mr G. F. Lawrence informed Evans in apologetic tones:

I have found several good ones. A splendid pear-shaped imp*t*. [implement] in Clerkenwell road [...] now in the collection of Mr. W. G. Smith. I was debating whether I should give it to you or not, when he offered me 20 impt in exchange for it & I could not resist the temptation of enlarging my collection knowing at the same time that it was of more use to him than to such an obscure individual as myself.[67]

Smith had been proud to state that he, personally, had gathered much of his fine collection together. His ideas about the distribution of these tools in space and time were also founded on detailed personal knowledge of the deposits around his neighbourhood in north-east London. The 'Palaeolithic Floor' was his most famous discovery: a thin layer of tools that he had first seen in 1878 under Stoke Newington Common. Smith was soon convinced that this Palaeolithic Floor spread for miles, a few feet beneath the feet of unsuspecting Victorian Londoners.[68] When he compared the tools from the Stoke Newington Floor to other discoveries, he concluded that they were the most recent of three distinct classes belonging to three distinct geological ages: three classes that were not restricted to London, but could also be found elsewhere in southern England.

Smith's oldest class of Palaeolithic tools lay deepest in the gravels. Their heavy abrasion suggested that they had travelled far from their original position. These large crude implements (soon called hand-axes) had heavy butts and were stained a deep chocolate-ochreous colour (Fig. 3.4). No worked flakes were found with them. But Smith's second class of finely made hand-axes (Fig. 3.5) was often associated with choppers and large scrapers. Their condition, style, and geological position suggested that they dated to a time between the oldest class that lay below and the fresh, sharp group from the Palaeolithic Floor eight feet above.

---

[65] Evans 1897b, 587.   [66] Evans 1897b, 583–586, 597–602.
[67] Lawrence to Evans, 21 January 1884: AMO: JE/B/2/80.   [68] Smith 1884, 357–358.

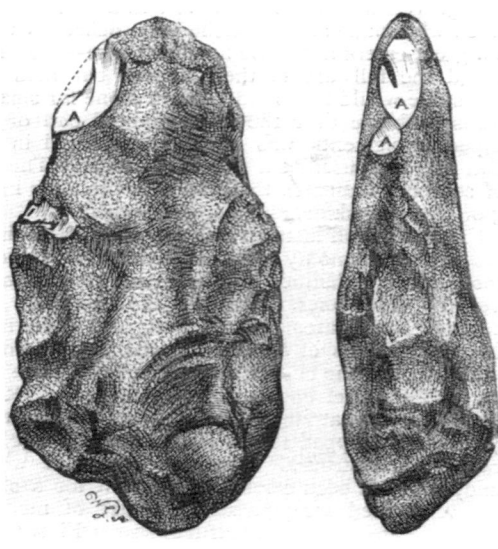

Figure 3.4. Worthington Smith's illustration of an implement belonging to his oldest class (Smith 1883a, 272, Fig. 1). This specimen came from Canterbury (AMO: JE/B/2/80).

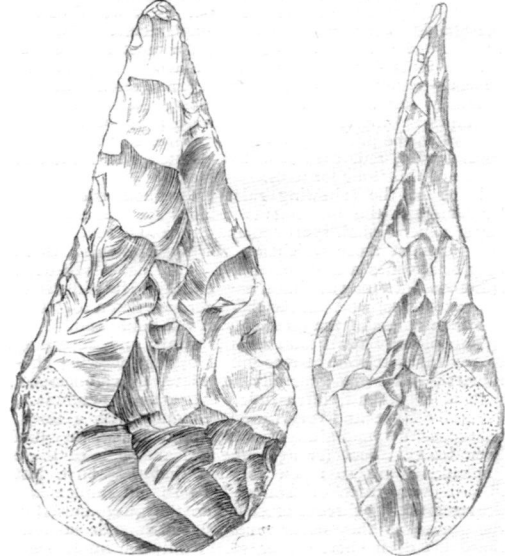

Figure 3.5. The second class of Worthington Smith (1883a, 272, Fig. 3). This specimen came from Lower Clapton (AMO: JE/B/2/80).

　　　Pointed and oval hand-axes could be found in both the oldest class and the second class, but sharp-rimmed oval hand-axes were restricted to Smith's third class from the Palaeolithic Floor. This class was characterized by tools that were smaller, sharper, lighter, and neater than the previous two

Figure 3.6. The third class of
Worthington Smith (1883a,
274, Fig. 8).

classes, and scrapers were now common (Fig. 3.6).[69] Smith saw a clear progression in tool-working skill through his three stages, 'from the large and rude, to the extremely small and neat scraper.'[70] The tentative conclusions of John Evans seemed to be confirmed by the observations of Worthington Smith.

In 1885 Smith moved out of London to live in Dunstable, near Luton. He soon recognized his oldest class and his third class of Palaeolithic tools in the new neighbourhood, but this time was careful about sharing their location with others. Once again, his interest was sparked by some interesting flakes in heaps of gravel brought into the area for ballast. Some of these flakes were slightly worn, as if they had been churned up in distant times with the gravel, and they had a deep ochreous colour; others ranged from white to indigo-black and were as sharp and fresh as the specimens from Stoke Newington. Smith traced them both to brick-pits scattered around Caddington (see Fig. 3.7), where he was delighted to find traces of another Palaeolithic Floor. The workmen had long been fearfully aware of this seam of sharp flakes that cut their fingers to the bone.[71]

[69] Smith 1883a, 271–272; 1883b, 119–129; 1884, 362–365.
[70] Smith 1883a, 273.   [71] Smith 1894, 90–96, 103–106, 113.

Figure 3.7. Worthington Smith at the Cottages site, Caddington, April 1892. (My thanks to James Dyer for supplying the photograph.)

Table 3.1 summarises Smith's suggestions about the Palaeolithic sequence of Caddington, Stoke Newington, and other areas. He assigned most tools from the Thames Valley, including the Gray's Inn Lane hand-axe, to his second Palaeolithic class.[72] He observed that certain distinctive types from his third class could be found across southern England and France: the trimmed flakes with delicate retouch to one face, from the Palaeolithic Floor at Caddington (Fig. 3.8), resembled those from the Floor at Stoke Newington, the brick-earth of High Lodge near Mildenhall and the cave deposits of Le Moustier in southern France.[73] There were also some interesting differences within the third class: Smith noticed that the technique applied to the trimmed flakes from Caddington was different to the technique displayed by the tools found by Spurrell at Crayford in Kent.[74]

[72] Smith 1894, 235.   [73] Smith 1894, 110–111, 169, 224.   [74] Smith 1894, 113.

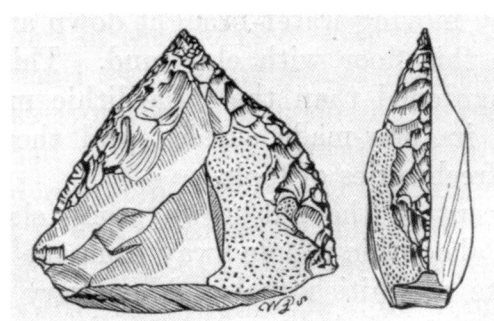

Figure 3.8. This 'trimmed flake' from Caddington was considered by Smith (1894, 110, Fig. 70) to be analogous to types from High Lodge and Le Moustier.

Smith had taken a great interest in tool-working techniques when working on the Palaeolithic Floor at Caddington. Unobstructed by the army of curio-collectors who had trampled over his sites in north-east London, Smith had the time and the patience to match scattered flakes to scars on the blocks of flint from which they had been struck.[75] He worked on 2,259 flakes and blocks from Caddington, gave each a date (painted in oil) and arranged them on tables according to their colour and markings. For more than three years, these flint-covered tables took over two rooms of the house that Smith shared with his understanding wife, Henrietta. He would examine the giant puzzle daily 'as a relief from other work, and at times when I was tired'. He kept hundreds more flakes and blocks in undisturbed grassy places down in the brickfields.

Table 3.1. Worthington Smith's three Palaeolithic classes, their characteristics, and notable sites where each class could be found (after Smith 1883a, 1883b, 1884, 1894).

Classes	Sites	Characteristics
Third class	Palaeolithic Floors at Stoke Newington, Erith, Northfleet, Grays Thurrock, Caddington. Similar in type to those from High Lodge and Le Moustier	Small, neat, thin implements, fresh and sharp; scrapers are common. Trimmed flakes (large, worked on one face). Oval implements with a cutting edge all round
Second (intermediate) class	The majority of the Thames Valley finds, including the Gray's Inn Lane implement	Oval and pointed implements; scrapers are rare. Less fresh in condition than the Floor implements
Oldest class	Found derived at Caddington and Stoke Newington	Oval and pointed implements, heavily abraded and stained; no scrapers

[75] Evans 1897b, 599–600; Boule 1895, 320.

Figure 3.9. 'Man and woman making implement with hammer-stone and punch on anvil-stone' (Smith 1894, 264, Fig. 198).

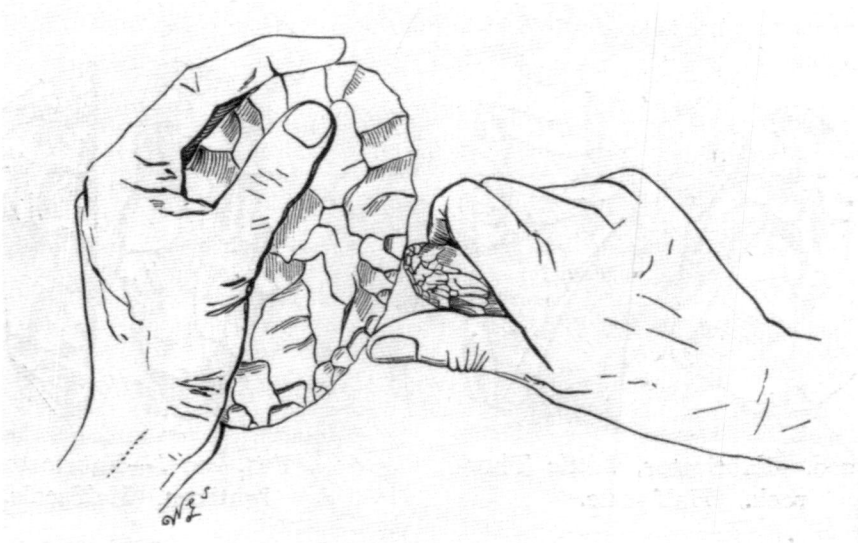

Figure 3.10. Sketch showing how the finest chipping was produced, founded on the methods of the Stoke Newington forgers (Smith 1894, 266, Fig. 204).

Smith slowly matched over 500 flints back together into numerous blocks of conjoined flakes: reconstructing, in reverse order, the techniques once used to detach them.[76] He spoke with knowledge of and respect for their makers: 'men with thinking heads, with correct eyes for beautiful and true forms, and with uncommonly skilful fingers'.[77] Like his peers, Smith wrote in general terms about 'Man' the primaeval savage, but he also discussed the lives of women and children and was unusual in depicting the women making implements. In Figure 3.9, a woman chips the hand-axe while a man holds the tool steady. Smith saw the toolmakers as short, stocky, hairy savages and gave the women whiskers, beards, and moustaches. He believed that they would have been naked or, at most, only slightly protected with ill-dried skins. Other illustrators gave their subjects more furry attire, inspired by different views of the climate and by the skin-working tools that had been found in the caves of France.[78] Smith based some of his ideas about their tool-working techniques on the methods of Victorian forgers from Stoke Newington (see Fig. 3.10). He had made some of them confess and show him their skills. Ironically, these skills had probably been inspired by Smith's earlier work in the area.[79]

3.5 BROWN, SPURRELL, AND PALAEOLITHIC PATTERNS IN THE BRITISH DRIFTS

John Allen Brown of Ealing also worked on Palaeolithic Floors. He was a respected Palaeolithic researcher, like Smith, and was also a reputable geologist. Whitaker, who had found Brown's descriptions of geological sections helpful to his Survey work, described him as 'one of those useful local observers'.[80] Ramsay read the draft manuscript of Brown's book, *Palaeolithic Man in N.W. Middlesex*, corrected the proofs and made further suggestions. In this book, Brown discussed the age and habits of the Palaeolithic tool-makers, which he illustrated by analogy to the activities of 'primitive' living tribes; he also outlined the sequence of events in the Glacial epoch: a subject of particular interest to Ramsay.

Like Smith, Brown saw ancient land-surfaces beneath Victorian London and believed that the very abraded, ochreous implements came from much older gravels than the tools from these land-surfaces. His suspicions were

---

[76] Smith 1894, 126–127, 157.   [77] Smith 1883b, 142.
[78] Smith 1894, 49–59. On images of prehistoric tool-makers in the eighteenth and nineteenth centuries, see Moser 1998, 107–145.
[79] Smith 1887, 87–89; 1894, 267.   [80] Whitaker 1889 (i), 382.

confirmed in 1885 when he examined a series of pits, dug by a builder for sand and gravel, in the high-terrace gravel at Creffield Road, Acton. Here, Brown discovered a Palaeolithic Floor. But he described it as a Palaeolithic workshop because the relics of human work, sharp and unabraded, 'were discovered generally in small heaps or nests, as well as scattered over this area'.[81] Brown noticed a similarity between these tools and those from High Lodge and the cave of Le Moustier: the same analogies that Smith had drawn for his third Palaeolithic class.

Although Brown agreed that Smith's sequence of tools had been deposited over a very long period of time, he preferred not to lock these artefacts into three distinct ages. He admitted that there had been a progression in skill and a specialization in the form of tools through time, but emphasized the continuity of Palaeolithic occupation over a lengthy period in the Thames Valley.[82] The earliest tools, Brown observed, came from Whitaker's high river terrace. He suggested that tool-makers had arrived from Europe after the time of the great submergence when much of Britain was emerging from the sea.[83]

Flaxman C. J. Spurrell, the third in this great trio of London researchers, also worked with energy on the Palaeolithic but left few records: he was so retiring that he rarely reported his findings and had to be pressed into publication.[84] Most of Spurrell's sparse papers describe the Palaeolithic Floors that he discovered in the Crayford brickearths and at Northfleet in Kent. Here, scatters of flakes and discarded implements still rested where they had been knapped and abandoned by the ancient tool-makers, and Spurrell realised that he could fit the flakes back together—if they had not flown too far.[85] Smith declared that Spurrell was the first to trace flakes from Palaeolithic Floors to their parent blocks: 'I shall never forget reading for the first time of this remarkable achievement'.[86] It was Spurrell's work that stimulated Smith to try the same process on his own Palaeolithic Floors.

Like Smith and Evans, Spurrell was a skilled flint-knapper. He thought of the ancient inhabitants of London as he chipped away, wondering if they too had experienced the same practical difficulties and alarms: 'occasionally a flake flew to a great distance: one flew with a fearful whirr a distance of over 60 feet; doubtless this incident occurred to the old men'.[87] Spurrell's knowledge of knapping and patience in re-fitting ancient flakes helped him to reconstruct the techniques and styles of 'the old men.' So did his

---

[81] Brown 1889, 56–57.   [82] Brown 1887a, 62; 1889, 50–51, 63; 1893, 89.
[83] Brown 1887a, 44, 48, 52, 67. Brown 1886, 192 extended the highest point of Whitaker's High Terrace from 100 ft to 130 ft.
[84] Kennard 1944, 162; 1947, 286.   [85] Spurrell 1880b; 1884, 109–110.
[86] Smith 1887, 83; Spurrell 1880b, Plate XXII.   [87] Spurrell 1884, 112.

Figure 3.11. One of Evans's sharp-rimmed implements, exhibiting the twist remarked on by Spurrell in 1883 (Evans 1872, 520, Fig. 450).

detailed observations of individual tools and scatters on Palaeolithic Floors. At Crayford, he was even able 'to see from the disposition of a heap of flakes [. . .], that the operator *sat* on the sand with his legs but slightly apart'.[88] By following every stage of ancient tool manufacture in his mind, Spurrell explained puzzling phenomena such as the curious twist down the sharp side-edge of some hand-axes (see Fig. 3.11):

It is the result of manufacture, and a defect got rid of when the size allows freedom in shaping; it shews also that the implement was chipped on one side, then turned over, with the same end towards the workman, and worked on that. Presuming that the majority were the work of the right hand in flaking, the rarer kind may have been that of the left, or of right and of left handed men.[89]

Spurrell also drew attention to a distinctive method of manufacture on his Northfleet Floor. A flint block was carefully trimmed, and the prepared surface was then detached by a blow to one end of the block, producing a large, sharp-edged flake—a 'turtle-backed' flake—that might then be trimmed into a scraper.[90] Spurrell's description recalls Evans's earlier account of broad flakes from the lower levels of the Somme Valley (see Fig. 3.3). In

[88] Spurrell 1884, 112.
[89] Spurrell 1883, 96, footnote. This twist was often associated with ovate hand-axes. Evans 1872, 502, 520 gave a similar, though less detailed, explanation for the twist.
[90] Spurrell 1884, 113.

later years these prepared flakes would become known in Britain as 'Levallois' flakes; J. Reboux found them in large numbers at Levallois near Paris in the early 1860s.[91]

Working amid such variety of shape and technique, Spurrell was cautious about the possibility of arranging tools into groups along a Palaeolithic sequence. He believed, like Smith, that the height at which tools were found and the degree of abrasion they exhibited were valuable indicators of their age: ancient tools might work their way downhill, but they would find it difficult to ascend.[92] But he was unsure whether tool types and techniques could be reliable time-markers: 'At Northfleet, for instance, five different forms of *hâches* [hand-axes] were found; all made on the same spot, with great diversity of finish, at the same time that it continued to be a flaking floor'.[93]

Despite the differences between the approaches and the questions asked by researchers like Spurrell, Brown, Smith, and Evans, a few broad patterns emerged over the last decades of the nineteenth century for the tools in the river drifts of southern England. The most ancient group seemed to be the crude hand-axes; heavily abraded and stained, derived from the high-level gravels; the oldest class of Worthington Smith. Next came Smith's second class: hand-axes and trimmed flakes that corresponded loosely with Evans's class of pointed implements (some approaching an oval shape). Typical examples had been found at Hoxne, Gray's Inn Lane, and St Acheul. The most recent river-drift tools came from the low-level valley-gravels: Smith's third class from the Palaeolithic Floors, Evans's sharp-rimmed oval implements, the distinctive 'turtle back' or 'Levallois' flakes from Northfleet and Montières, and the finely-made scrapers from High Lodge and Le Moustier.

This was not a definitive or rigid sequence: Evans was tentative; Smith suspected that later tool-makers might have used different flaking techniques in different regions; Spurrell warned that hand-axe form might vary too greatly within a site to offer a reliable indication of age. But Palaeolithic researchers were describing similar patterns across southern England. The distinction drawn by Evans between his two classes of hand-axes and the third class of flakes reappeared with little change in the second edition of *Ancient Stone Implements* (1897), along with his expanded list of types.[94] Evans did, however, make a few concessions inspired by classifications and sequences that had been developed by researchers working on the other side of the Channel, in France.

---

[91] Reboux 1869, 222; de Mortillet 1883, 255–256; Read 1911, 35.
[92] Spurrell 1883, 89–90.   [93] Spurrell 1883, 93.   [94] Evans 1897b, 640–641.

4

River-Drift Men and Cave Men

British researchers often compared their river-drift tools to French finds. The gravel pits of St Acheul, rich in hand-axes, were a popular choice. Another was the cave of Le Moustier, known for its scrapers. By the end of the nineteenth century, British tools might be called 'Mousterian', 'Acheulian', or 'Chellean'. These labels are associated with the French prehistorian Gabriel de Mortillet, and it is often assumed that de Mortillet's classification clarified a hazy image of the British Palaeolithic sequence. There are two problems with this assumption.[1]

First, the picture of river-drift tools gained from Chapter 3 suggests that British researchers cannot be compared to a sponge, waiting to soak up Continental classifications. They did use these French labels, but they did not necessarily adopt all de Mortillet's beliefs as well. Second, researchers other than de Mortillet were building sequences on the Continent in the nineteenth century, and de Mortillet was aware of their research when he developed his Chellean, Acheulian, and Mousterian epochs. So were British researchers, but they were also encouraged by these Continental findings to make another, more fundamental, division of the British implements.

One question that had puzzled British researchers was the connection between tools from their caves and those from their river-drifts; it was difficult to link the isolated pockets of cave sediments to the drifts lying in river valleys. Another question, closely related, concerned the term 'Palaeo-lithic': what tools and what time-period did it encompass? The geologists from Chapter 2 used their sequences of bones, river drifts, and glaciers to answer the first question; but answers to both were also found in the bones and tools of France and Belgium (Map 1 gives the location of the major sites mentioned in the text).

---

[1] Daniel 1975, 109 describes how the de Mortillet system 'became the orthodox system of prehistory until well into the twentieth century'. His account focuses on the later Palaeolithic of France and barely mentions British research on earlier tools from river drifts and caves. Daniel 1975, 100–102 does, however, recognise the influence of Lartet and Christy's definition of the Reindeer period on British research.

The famous division of the Stone Age into two periods—Palaeolithic and Neolithic—was the work of John Lubbock. In the first and second editions of *Pre-Historic Times* (1865 and 1869), Lubbock distinguished between the chipped stone implements of the 'Palaeolithic' period (from the Greek *palaios*: ancient and *lithos*: stone)—which he called 'Archaeolithic' in the first edition—and the polished stone implements of the 'Neolithic' period (*neos* being Greek for new).[2] Amongst Palaeolithic researchers, these new terms replaced old geological indications of age: they mentioned Drift and Surface (or Diluvial and Alluvial) periods less frequently. Geologists, for their part, often favoured the simple term 'Drift' over 'Pleistocene' or 'Post-Pliocene', partly because of uncertainty about which of the two latter terms to use.[3]

Although Lubbock was not the first to make this division between two Stone Ages, he did supply the most popular labels. (Lubbock credited John Evans with drawing the first distinction between the flint implements of the Palaeolithic and Neolithic periods, and there are other claimants.[4]) The definition of the Palaeolithic was not, however, as simple as it might seem. Peter Rowley-Conwy has discerned a doubt in these early years about whether certain chipped stone tools ought to be assigned to the Palaeolithic at all. The trouble had emerged from the caves of southern France. Lubbock was among those who suspected that some of the tools from these caves belonged to a period between his Palaeolithic and Neolithic periods.[5]

4.1 THE REINDEER-CAVES OF SOUTHERN FRANCE

Two years before Lubbock's *Pre-Historic Times* appeared, and a few years more before de Mortillet began to publish his Palaeolithic classifications, Édouard Lartet and Henry Christy (1810–1865) were exploring the caves and rock-shelters that pock-marked the steep limestone cliffs of the Vézère Valley. They started their work in August 1863 and their observations in this valley of

---

[2] Lubbock 1865, 60; 1869, 74; Stevens 1870, 33.

[3] In 1862, Prestwich told Lyell that he preferred 'Pleistocene' to 'Post-Pliocene' (Prestwich to Lyell, 18 May 1862: ULE: Gen 115/4982–4985). In 1872 Lyell was wondering whether to replace 'Post-Pliocene' with 'Pleistocene' (Lyell to Dawkins, 3 June 1872: BMD: WBD: 70045).

[4] Evans 1860, 293; Lubbock 1866, 196. Suggestions about different ages of stone were also made by Worsaae 1859, 889, Lyell 1863a, 373–374, and de Mortillet 1883, 17–18, amongst others. See Laming-Emperaire 1964, 114–122 on the distinction drawn by Jouannet, in 1834, between a time of flaked stone and of polished stone. Laming-Emperaire 1964, 161 also remarks that the title of Boucher de Perthes's book, *Antiquités celtiques et antédiluviennes*, foreshadowed the distinctions drawn by Lubbock nearly two decades later.

[5] Lubbock 1865, 245; Rowley-Conwy 1996.

caves, which lay in the department of the Dordogne, Aquitaine, would convince British researchers that there was an intermediate period of the Stone Age.

Christy was a British collector of ethnological and ancient stone tools; he was also the wealthy Director of the London Joint-Stock Bank. Lartet was a distinguished French palaeontologist of independent means who had briefly studied law before abandoning it for Tertiary and Quaternary mammals. Lartet had been an early convert to human antiquity and enjoyed some eminence among prehistorians: he would preside over the Congress of Prehistoric Archaeology when it met in Paris in 1867.[6]

The best-known publication by Lartet and Christy on the French caves was their massive volume *Reliquiae Aquitanicae* (a title suggested by Falconer, Lartet's friend).[7] This was published at Christy's expense in quarto, on fine paper with plates, at the modest price of three shillings sixpence a part. The first part appeared in the last year of Christy's life, 1865, and Lartet had also departed this world before the seventeenth and final part arrived a decade later. Notwithstanding their early deaths and the comparatively late emergence of the full report, British researchers had noticed their findings in France within months of discovery.

Lartet and Christy shared their discoveries and opinions in person and in print. They welcomed several prominent visitors to their caves in 1864. Lubbock recorded a pleasant and informative visit in the spring, and was stimulated to write an article on Cave Men for the *Natural History Review*.[8] Falconer remembered the terrible heat of May, when he departed from the Dordogne 'leaving M. Lartet to survive, if he can, another day's work at the Caves'.[9] Evans, too, travelled to the Caves and met Lartet and Christy. His report to the Geological Society was published only in abstract, but the original account would be reproduced later in *Reliquiae Aquitanicae*.[10] Various articles on the caves by Lartet and Christy could also be read in both British and French journals in 1864. There was a long account in the *Revue archéologique* and a translated summary in the *Quarterly Journal of the Geological Society of London*. A letter from Lartet summarizing their finds appeared in the *Comptes rendues de l'académie des sciences*; this was reproduced in *Annales des sciences naturelles* and (in translation) the *Annals and Magazine of Natural History*.[11]

---

    [6] On the life and work of Henry Christy, see Topley and Jones 1865 and Harrison 2004; on Édouard Lartet, see de Mortillet 1871, Fischer 1872 and Prestwich 1872.
    [7] Lartet and Christy 1865–75; Prestwich 1872, xlviii.
    [8] Lubbock 1864; 1865, viii.
    [9] Falconer to McCall, 20 May 1864: FMF: HF, 119.
    [10] Evans 1864b, 444; Lartet and Christy 1865–75.
    [11] Lartet and Christy 1864a; 1864b; 1864c; 1864d; Milne-Edwards 1864.

In June 1864, Christy presented his thoughts on the sequence of stone tools in these caves to the Ethnological Society of London. He distinguished three groups of tools: those from the Drift period, when mammoth and rhinoceros were common; those from the succeeding Cave period, when large herds of reindeer lived alongside the cave-dwellers in the Dordogne; and the ground stone axes made far later, during the Surface period.[12] It was Christy's middle division, with an abundance of reindeer rather than mammoth and no clear association with the river-drifts, that seemed to lie outside Lubbock's 'Palaeolithic'. This intermediate age of stone would become known in Britain as the 'Reindeer' period.

Lartet had already distinguished between a time of Mammoth and of Reindeer. In 1861, after working at the cave of Aurignac, he published a sequence of four different periods of fauna in *Annales des sciences naturelles*. This sequence was based on the idea that the successive appearances of different species into Europe during the last geological period had been followed by their successive extinctions. Lartet's paper received more publicity in Britain the following year when it was printed, albeit abridged, in the *Natural History Review*.

The first of Lartet's species to disappear was the Cave Bear: their bones at Aurignac suggested to Lartet that this cave might pre-date the river drifts. He believed that his second group—Mammoth and Woolly Rhinoceros—had arrived in post-glacial times from Siberia, which had become too cold for them.[13] Dawkins, though critical of Lartet's work, took up this idea when he admitted that the arrival of Lartet's northern group from Siberia into Britain 'defines the Postglacial from the Preglacial fauna'.[14] Lartet suggested that these Mammoth and Woolly Rhinoceros disappeared before his third group—the Reindeer—departed from Europe. His fourth group—the Aurochs—was the last to die out.

Lartet's sequence could withstand considerable overlap between species because it was based on a series of extinctions, not appearances. When he dated the Dordogne caves and their tools to his Reindeer period, Lartet explained that reindeer were not restricted to these late times: Prestwich had found reindeer in the river drifts (at Bedford, for example) and Falconer had discovered reindeer in the Gower caves of South Wales alongside a very early fauna.[15] But his critics tended to ignore this overlap and Lartet's fourfold faunal sequence never won much support. The sequence of Palaeolithic (River-Drift), Reindeer (Cave), and Neolithic (Surface/Alluvial) periods, however, became widely used in Britain.

[12] Christy 1865, 364, 371–372.   [13] Lartet 1861, 201, 222–223; 1862, 70–71.
[14] Dawkins 1869, 211, 213–215. See also Dawkins 1872, 445.
[15] Lartet and Christy 1865–75, 8–9.

The first edition of Lubbock's *Pre-Historic Times* (1865) reported that the tools from the Dordogne caves belonged to the Reindeer period of Lartet; tool-makers from the preceding Drift or Palaeolithic period had lived alongside extinct mammals like the mammoth.[16] Lyell observed in 1863 that the caves of southern France, where Lartet had found ancient artwork and reindeer bones, fell between the newer and older divisions of the stone period. In 1867, he followed Lartet's term, the 'Reindeer' period, for this intermediate time to distinguish it from the preceding 'Palaeolithic' period.[17] Evans, in *Ancient Stone Implements* (1872), remarked cautiously that, in general, tools from the caves and the river drifts seemed to belong to much the same time. But he added that certain caves seemed to have been occupied *after* the time of the river drifts, citing good evidence from the south of France and Belgium.[18]

The work of Édouard Dupont (1841–1911) in the caves of Belgium often won a brief mention in British publications alongside lengthier appraisals of Lartet's shortcomings.[19] It was partly through Dupont's researches that the simple distinction between a time of Mammoth and a time of Reindeer survived the demise of Lartet's four mammal periods. Dupont's work on the caves around Dinant, south east of Brussels, began in 1864 and was supported by the Belgian Ministry of the Interior; Belgium was not going to be left behind in the new subject of human antiquity. In 1871, Dupont, now the Director of the Royal Belgian Museum of Natural History, Brussels, published *Les temps préhistoriques en Belgique*. This book was popular enough for a second edition to appear the following year.

Dupont presented his readers with three Stone-Age periods: of Mammoth, Reindeer, and Polished Stone. Even Dawkins, despite his suspicion of Lartet's faunal sequence, had to admit that mammoth had arrived in Europe before reindeer, although he still advised the rejection of a Reindeer/Mammoth distinction on the grounds that these species had enjoyed a lengthy co-existence thereafter. But Dawkins's carping did not harm the popularity of the Reindeer period amongst his peers.[20]

[16] Lubbock 1865, 2, 245. See also Lubbock 1864, 413 and Rowley-Conwy 1996, 941.
[17] Lyell 1863a, 373; 1867, 176–177.
[18] Evans 1872, 426–427.
[19] Lubbock 1865, 239–245; 1869, 304; Evans 1872, 433–434; Jones 1877, 22; Geikie 1881a, 101.
[20] Dupont 1872; Dawkins 1872, 413, 419–421; 1874, 351–352. The existence of an intermediate stage between the Palaeolithic and Neolithic was widely recognised but was not always called the 'Reindeer' period. Reboux 1873, 99–101; 1876, 64–65 described three periods of stone and referred to the first two as 'La pierre éclatée, *époque paléolithique*,' found with mammoth and cave bear; and 'La pierre taillée, *époque mésolithique*,' associated with reindeer. They were followed by the polished stone tools of the Neolithic period.

As the idea of a late race of reindeer-hunters in southern France pervaded British classifications of the Palaeolithic, problems still remained. Some sites, like the cave of Le Moustier, seemed to lie *between* the Palaeolithic and Reindeer periods. The bones from this cave seemed to be earlier than those from the others. Lartet observed that the same species of fauna could be found in all the caves, but reindeer were less abundant in Le Moustier.[21] Evans reported that mammoth were present at Le Moustier, but that Lartet and Christy did not believe mammoth to have been contemporary with the tool-makers.[22] Lubbock was also troubled by the presence of mammoth: if contemporaneous with the tools, this species would place his intermediate Cave period further back in time.[23]

The implements and flakes from Le Moustier were also thought to be early in character. Most of the other reindeer-caves were rich in small, delicate flake-tools, and some also contained distinctive bone-work and artistic engravings: a skill that seemed to harmonize with a late date. The hand-axe-dominated river drifts, in contrast, lacked art or bone-work. In the cave of Le Moustier, however, Lartet and Christy found small, finely-made hand-axes: a type familiar from the river drifts. This cave was also dominated by tools made from large, boldly-struck flakes that resembled those described by Evans from the low-level river drifts.[24]

Evans and Lubbock remarked on this difference in character between the tools from Le Moustier and those from other caves in the Vézère Valley. Evans thought the hand-axes were almost indistinguishable from the Abbeville types in the Somme Valley, and Lubbock identified this cave as the oldest of the Dordogne group.[25] When McKenny Hughes visited the site in 1872, he reflected on these theories about its age: 'The Cave of Le Moustier is supposed to be the oldest in the valley of the Vézère from its containing flint implements of the Amiens type [i.e. hand-axes]. But the great number of flake &

---

[21] Lartet and Christy 1865–75, 5.

[22] Lartet and Christy 1865–75, 166, 174–176, 180.

[23] Lubbock 1864, 420; 1865, 246–247, 257–258. Lubbock 1869, 327 used the same uncertain tone in the second edition of *Prehistoric Times* to refer to the presence in the French caves of mammoth with stone tools, 'remains of which have been found in doubtful association with them'. This follows strangely from his new description and illustration of an engraving depicting a mammoth found at the cave of La Madelaine, made on mammoth tusk (Lubbock 1869, 324–325, Plate 2). It was not long before mammoth became more acceptable companions for the peoples of the Reindeer period, though greatly outnumbered by reindeer.

[24] Lartet and Christy 1864a, 238.

[25] Lartet and Christy 1865–75, 166, 167; Lubbock 1865, 251; Rowley-Conwy 1996, 941. Lubbock had been more cautious about the age of the tools from Le Moustier in 1864, when he argued that, although Lartet and Christy were happy that the hand-axes were identical to River-Drift forms, he suspected that the resemblance was accidental: those from Le Moustier were smaller, and the workmanship was less bold (Lubbock 1864, 418).

fragments about were similar tho generally of a larger average size than those of the other caves'.[26]

The distinctions between the Palaeolithic period and the Reindeer period were clarified by the Le Moustier problem, which stimulated discussion about the elements that were expected, and those that seemed too early. During the nineteenth century the term 'River-Drift man' conjured up a picture of early, hand-axe-making groups living alongside mammoth and woolly rhino, roaming down river valleys and occasionally wandering into caves. The 'Cave Man' of the Reindeer period evoked later groups, akin to the modern Eskimo, hunting reindeer, inhabiting caves, making finer flake- and bone-tools, and blessed with artistic ability. The term 'Palaeolithic' was used in a more restricted sense than today, and Evans was not unusual among his peers in referring to 'the Palaeolithic, or River-gravel Drift, Period' and 'the Reindeer, or Cavern, Period of Central France'.[27]

It was only after the emergence of these broad general patterns—in river drifts and cave deposits, hand-axes and flake-tools, reindeer and mammoth—that cases of overlap like Le Moustier could be identified. Britain had its own problematical analogies to Le Moustier. Greenwell declared that the series from High Lodge—which attracted collectors with its fine examples of trimmed flakes and oval hand-axes, sharp and very dark in colour—was quite distinct from the usual range of river-drift forms.[28] Intrigued by the character of the tools on his Palaeolithic Floor at Stoke Newington, Smith kept a special lookout for bone needles and harpoons, but with no success.[29]

4.2 CAVE MEN AND RIVER-DRIFT MEN IN BRITAIN

What of the caves of Britain? How were their tools linked to these Palaeolithic and Reindeer periods? Although the more recent artefacts from a few British caves were assigned to the Reindeer period, this term tended to be restricted to tools from the small cluster of French caves explored by Lartet and Christy. Many British cave tools were regarded as Palaeolithic in age and character. The sediments in Wookey Hole, Brixham Cave, and the deeper levels of Kent's Cavern, for example, contained bones and tools like those from the river drifts and Le Moustier (see Figs. 4.1 to 4.3).[30] When Evans made his cautious

---

[26] Hughes, April 1872: SMC: TMH (fieldmaps), Notebook B, France.

[27] Evans 1869, 193 (see also Evans 1875, lxvii, lxxii; 1891, 253–254).

[28] Greenwell to Evans, 3 December 1870: AMO: JE/D/1/9 (in an envelope enclosed in Evans's scrapbook of Cave Implements); Evans 1872, 492–493.

[29] Smith 1884, 374, 379.

[30] Evans 1860, 292; 1872, 449, 454, 468–469; Lubbock 1864, 420–428; Dawkins 1874, 351.

Figure 4.1. Evans thought this 'round-pointed lanceolate implement' from Brixham Cave was identical to his pointed class from the Palaeolithic (or River-Drift) period, despite being found in a cave (Evans 1872, 468–469, Fig. 409).

suggestion that some artefacts in caves and river drifts were of much the same age and had probably been made by the same race of people, he was gesturing to the earlier deposits of British caves that belonged to the Palaeolithic (or River-Drift) period, not the later tools of the Reindeer period.[31]

Brown agreed with Evans about the overlap between Cave and River-Drift dwellers, pointing to the similarity between the bones and tools from the *lowest* levels of caves and those from the river drifts. But Brown, Smith, and others saw much variety on each side of this overlap, and drew finer divisions in the tools from caves and river drifts. Brown mentioned distinct groups of implements *within* the River-Drift group, and believed it possible that several waves of tool-makers had entered Britain between the time of the earliest River-Drift peoples and the Neolithic period. Smith believed that his oldest

---

[31] Evans 1872, 426–427, 573–574.

class of tools from the drifts was older than the oldest tools from the caves.[32] Dawkins distinguished between three groups of tools in the caves of Creswell Crags, Derbyshire, whilst Pengelly reported two groups of Palaeolithic tools from the earlier deposits of Kent's Cavern in Devon. Such discoveries suggested that British caves might have been occupied over a vast stretch of time and these tools might have been dropped by several different groups of tool-makers.

Pengelly had returned to Kent's Cavern in 1865 to renew his old excavations. This time he was one of a larger Committee of the British Association: Evans, Lubbock, Dawkins, and Lyell were among fellow members who helped him conduct the dig. Deep in the Cavern, in the oldest levels of gritty, dark-red breccia, Pengelly found implements that he described as 'nodule-tools', because they had been made directly and exclusively on *nodules*. Above, in the light-red clay of the cave-earth, lay more recent flint tools, quite different in character: they had been made on *flakes* that had been deliberately detached from a nodule for the purpose. Pengelly thought that these two types of tool had been made by two different races, widely separated in time. He also believed that there had been a clear progression in tool-working skill through the Palaeolithic.[33]

In *Ancient Stone Implements*, Evans noted a similarity between Pengelly's older group and the pointed implement from Brixham Cave nearby (see Figs. 4.1 and 4.2). He agreed that they differed from the group in the cave-earth above—the sharp-rimmed oval implements (see Fig. 4.3) and the trimmed flakes like those from Le Moustier—and thought that both groups had their analogies in the river drifts. The varied tools from Kent's Cavern embraced a wide spectrum of chipped stone tools. Evans described early River-Drift types; later River-Drift types and boldly trimmed flakes familiar from Le Moustier; and even bone-work and fine scrapers, like those from the reindeer-caves of southern France.[34] It seemed that cave tools could not be compressed into a single period merely because they all came from caves.

When Pengelly suggested that there was a great difference in age between his two groups of Cavern tools, he had drawn on their character and relative position. But he also looked outside Kent's Cavern to find a date for his early group of nodule-tools. It was around this time that Richard Tiddeman and Sydney Skertchly had been attacking the post-glacial Palaeolithic. Pengelly decided to relate his nodule-tools to the glacial sequence and made the controversial suggestion that they had been made during interglacial, if not pre-glacial,

[32] Brown 1887a, 129–130; Smith 1894, 160, 169.
[33] Pengelly 1873, 208; 1874, 149–150; 1877, 59; 1883, 558.
[34] Evans 1872, 446–449, 454–461; 1897b, 495, 514–515.

Fig. 388A.—Kent's Cavern. (6,022)

Figure 4.2. An implement from the breccia of Kent's Cavern, which Evans thought similar to the older River-Drift types (Evans 1897b, 495, Fig. 388a).

Figure 4.3. A sharp-rimmed oval implement from the cave-earth of Kent's Cavern, which Evans thought similar to the more recent River-Drift types (Evans 1872, 447, Fig. 386).

times—long before later occupants left flake-tools in the cave-earth above.[35] The character of tools provided one link between isolated pockets of cave debris and discoveries in other caves or river drifts; geological sequences offered another guide out of the caves. Lartet had found the sequence of bones in the French caves invaluable; Pengelly turned to the glacial sequence. Both sequences were controversial and stimulated conflicting opinions about the relationship between the Cave men and the River-Drift men.

4.3 FURTHER DISAGREEMENT BETWEEN DAWKINS AND GEIKIE

The arguments between William Boyd Dawkins and James Geikie over sequences of bones and glaciations have been described in Chapter 2. It is not surprising that they also disagreed about the patterning of tools. The early opinions of Dawkins were similar to those held by many of his peers; he would, however, describe a different chronological and racial relationship between the tool-makers of the caves and the river drifts after he developed his classification of Quaternary mammals (see Table 2.6). Geikie, in contrast, chose to recount their relationship in terms of his interglacial cycles. Stone tools became a platform for their deeper disagreements.

Dawkins declared in 1866 that the 'reindeer-folk' were 'intermediate in condition' between the 'flint-folk' of the river drifts and the Celtic tribes of Neolithic times. By 'reindeer-folk' he meant the tool-makers of the Dordogne caves; he believed that the tools from British caves, like Wookey Hole, had been made by the 'flint-folk'.[36] But by the time Dawkins wrote *Cave Hunting* (1874), his opinions about the connection between Cave and River-Drift peoples had changed. Their tools were requisitioned for the defence of his 'mixed' northern and southern fauna: his vision of a time when hippopotami had flourished alongside reindeer.

Dawkins now assigned most of the cave and river deposits from middle and northern Europe (including those frequented by his erstwhile reindeer-folk) to his Late Pleistocene division. This division encompassed Lartet's Reindeer period *and* his Mammoth and Woolly Rhinoceros period, an overlap between species that provoked Dawkins to attack Lartet's faunal chronology. He argued that many of these species had lived together, and that the differences described by Lartet between the bones from various caves might reflect

[35] Pengelly 1874, 147; 1877, 64.
[36] Dawkins 1866a, 712. See also Dawkins 1866b, 334–335, 338; 1868, 38.

human selection rather than the range of species living at that time. A reindeer might be killed more easily than a mammoth, or provide a more enjoyable supper than a cave bear.[37]

The Cave and River-Drift peoples were still presented as two different races, but Dawkins now decided that they, too, had overlapped in time. Strangely, he presented his two-race theory as a counter to Evans's supposed belief that these tools were made by a single contemporaneous race. This seems to be a misrepresentation of Evans's case, blurring the distinction that Evans had drawn between the reindeer-caves and 'Palaeolithic' caves, which were grouped together by Dawkins in his Late Pleistocene division.[38] Dawkins made the cautious suggestion that tools of the River-Drift group *might* be earlier than those of the Cave group because of the superposition in Kent's Cavern, but he no longer pressed this chronological distinction. He now proposed that variation between these tools could result from a difference in geographical territories and tribes rather than a difference in the age of their makers.[39]

Like his attack on Lartet's faunal sequence, Dawkins's suggestion of an overlap between tool-makers was closely connected to his arguments for seasonal migrations and mixed fauna. He assigned the Cave-dwellers to his northern group of mammals and suggested that they had followed these animals in their migrations, dropping the same class of tools in the south of France and Devon alike. He placed the River-Drift peoples in hunting grounds further east: in the Thames Valley and the Somme. This confusing account did not explain the conflict between these discrete human territories and the overlapping ranges that Dawkins used to explain the mix of southern and northern fauna, whose bones were found in both caves and river drifts.[40]

When Dawkins published *Early Man in Britain* in 1880, he quietly returned to the dominant view that these two Late Pleistocene races were different in age as well as in the character of their tools. He maintained his old palae-ontological arguments by observing that the Late Pleistocene stage, with its mixed fauna, 'was long enough to allow of a series of migrations of man, or of the development of a new culture in Europe'.[41] He defended his return to a chronological succession of tool-makers with a report of recent discoveries in the caves of Creswell Crags.

---

[37] Dawkins 1874, 351–352, 414–415. See also Dawkins 1872, 424.
[38] Dawkins 1874, 351, 414.
[39] On Dawkins's shifting explanations for the differences between the tools of River-Drift and Cave groups, compare Dawkins 1866b, 344 to Dawkins 1874, 351, 367.
[40] Dawkins 1874, 366–367, 395, 397–398.
[41] Dawkins 1880, 231–232.

At Creswell, Dawkins had seen a stratigraphic sequence of different classes of tools: crude tools in the lower levels, river-drift types in the middle levels, and highly finished flake- and bone-tools in the upper levels. An engraved fragment of rib in these upper layers linked the inhabitants of Creswell to the artistic Cave men of the Continent. Creswell Crags convinced Dawkins that the River-Drift race had inhabited the caves before the Cave race brought their later and higher form of culture to Britain. They seemed to confirm the evidence from Kent's Cavern: that British caves contained two *chronologically* distinct sets of tools.[42]

Dawkins still believed that Cave and River-Drift men had occupied different geographical ranges. He restricted his Cave men to a zone stretching from the Alps and Pyrenees to Britain and Belgium, and associated his River-Drift tool-makers with his southern group of fauna, which roamed across western and southern Europe, through Asia Minor and reached as far as India.[43] The idea that a sequence of different races ended up in the Continental cul-de-sac of Britain became a popular one, each group of tool-makers surviving for a while 'until they became absorbed, and their implements improved, by invading races who had attained a higher level of progress'—as Brown put it at the end of the century.[44]

Both of Dawkins's books had been peppered with pot-shots at Geikie's climatic chronology; the footnotes were particularly explosive. In his own book, *Prehistoric Europe* (1881), Geikie replaced Dawkins's rapid seasonal migrations with the gradual climatic alternations of his glacial and interglacial stages. Although *Early Man* had appeared before Geikie wrote the preface to *Prehistoric Europe*, he chose to focus his attack on the case made by Dawkins in *Cave Hunting* for overlap between Cave and River-Drift races. This choice put him in a good position to fire on Dawkins's theory of seasonal overlap between northern and southern groups of fauna.[45]

Geikie could not stomach the idea that reindeer (northern/cold-loving) and hippopotamus (southern/warmth-loving) had wandered to-and-fro across Europe in a single year. Instead, he described the arrival of Palaeolithic tool-makers to Britain during interglacial periods where they stayed a long while, left their tools in caves and river drifts alike, and were accompanied by a southern fauna that colonized Britain for long spells whenever glacial conditions died away.[46] Geikie was sympathetic to the division between the Reindeer and Mammoth periods developed by French and Belgian geologists

[42] Dawkins 1878, 153–154; 1880, 186, 194–198.
[43] Dawkins 1880, 172–173; 232–233.
[44] Brown 1893, 95.
[45] Geikie 1881a, 16, 115–117.
[46] Geikie 1881a, 65–66, 116, 263–265; 347–350.

(although his own use of the term 'Palaeolithic' embraced both). He observed that mammoth had survived into the Reindeer period, but were far less common than the vast herds of reindeer.[47]

For Geikie, it was the change in climatic conditions during late Palaeolithic times that drove southern species like hippopotamus away; he did not accept Dawkins's suggestion that the hunters of the Reindeer period merely found it difficult to catch such beasts. Geikie believed that the reindeer-hunters, who dropped their delicate tools in the later cave-deposits, had finally deserted Britain as the last glacial period arrived, retreating to the south of France at the height of glaciation and leaving their artistic relics in the Dordogne caves.[48]

The arguments of Dawkins and Geikie illustrate the range of opinions that contributed to the stone-tool sequence of Britain; not all those who worked on the British Palaeolithic had tools at the centre of their thoughts. Not all agreed with the views of Dawkins and Geikie either. Various connections were made between tools and the glacial and faunal sequences. Some French and German researchers described several lengthy arrivals and departures of tool-makers during the Glacial epoch (see Table 2.7), but few of Geikie's English peers accepted an interglacial Palaeolithic.[49] Evans and Smith thought that humans had only arrived in Britain after the last great glacial period.[50] Brown believed that Palaeolithic tool-makers had first lived in the Thames Valley when the area further to the north was covered with ice and the boulder clays were being laid down, influenced by Prestwich's description of a similar scenario in 1887.[51] But even this vision was far removed from Geikie's lengthy glacial periods that affected all of Britain, including the southern river valleys. Most British researchers did not consider a glacial/interglacial sequence to be relevant to a Palaeolithic that they regarded as post-glacial.

The connections between tools and mammals won a little more agreement, despite the criticisms of Lartet's faunal chronology and Dawkins's contradictory views. The warmth-loving hippopotamus and straight-tusked elephant tended to be associated with the distinctive pointed hand-axes of the early, high-level river drifts and a few caves. Mammoth were given a wide range; reindeer were thought to have become more abundant in later, colder times;

---

[47] Geikie 1881a, 16, 101.　　　[48] Geikie 1881a, 111, 115–117, 350–354, 360.

[49] Geikie 1894, 689. Marcellin Boule believed that the Chellean and Acheulian were interglacial. Albrecht Penck held similar views. Boule observed the disagreement amongst British geologists on this subject and noted that his interglacial hand-axe makers were not accepted by some of his French peers: de Mortillet, for example, believed hand-axes to be pre-glacial (Boule 1888, 272, 667, 678).

[50] Smith 1894, 2; Evans 1897b, 697.

[51] Brown 1887a, 122; 1896, 157; Prestwich 1887, 407.

Table 4.1. During the late nineteenth century, British researchers often divided Palaeolithic tools between these three groups. Note that the two earlier groups could be found in caves and river drifts alike.

Groups of tools	Sites
Small, delicately worked flake-tools; bone needles and harpoons; artwork	Reindeer-caves of southern France; Kent's Cavern (late cave-earth); Creswell Crags (upper levels)
Oval implements with a cutting edge all round; trimmed flakes; large, boldly struck, prepared flakes	Kent's Cavern (early cave-earth); Palaeolithic Floors in the London river drifts; Le Moustier
Pointed implements of John Evans; nodule-tools of William Pengelly	Brixham; breccia of Kent's Cavern; high-level river drifts

and both were associated with the fine, sharp-rimmed oval hand-axes and scrapers of the later, low-level river drifts and caves like Le Moustier. Reindeer were generally believed to be most plentiful when the dwellers in the reindeer-caves of southern France were making their finer blades and bone-tools. Table 4.1 gives a brief summary of the divisions of stone-tools adopted in Britain.

4.4 THE PALAEOLITHIC CLASSIFICATION OF GABRIEL DE MORTILLET

It has sometimes been awkward, in previous pages, to compare different ideas about the British Palaeolithic using terms like the 'nodule-tools' of William Pengelly, the 'oldest class' of Worthington Smith, the 'pointed implements' of John Evans, and other individual descriptions. Many of these implements are known today by a standard terminology that retains many of the names popularized by Gabriel de Mortillet. His terms started to pervade British research in the late nineteenth century. Having now looked at the patterns observed by British researchers in earlier decades—the detailed sequences in river drifts and caves and the general division between the Palaeolithic and Reindeer periods—the influence of de Mortillet's classification and terminology can be examined with greater ease, and with less likelihood of imposing anachronistic sentiments on early British schemes.

De Mortillet had been a geologist and mollusc expert before he gained eminence as a prehistorian and anthropologist. He cemented his interest in prehistory in 1864 when he founded the monthly bulletin *Matériaux pour l'histoire positive et philosophique de l'homme*. He was also a founder member of the International Congress of Prehistoric Archaeology, which first met in 1866. *Matériaux* was started on a small budget and cost seven francs a year (eight francs for foreign subscribers). The content illustrates the spread of de Mortillet's interests, embracing prehistory, geology, anthropology, and the question of evolution in the quest to learn more of ancient humans. This bulletin would have made unpleasant reading for many members of the clerical establishment—which would have pleased de Mortillet. He had only recently returned to France, having been exiled in 1849 for radical socialist politics and anti-clerical, anti-monarchist opinions. On his return, de Mortillet's polemical writing on prehistory caught the attention of his peers.[52]

In the late 1860s, de Mortillet was working on the collections in the Museum of Saint-Germain-en-Laye on the outskirts of Paris. The museum held a fine assortment of Palaeolithic artefacts, including the collections of de Perthes and material from Lartet and Christy.[53] De Mortillet observed that the French caves had not all been occupied at the same time, and wondered how best to arrange their tools. Having decided, he presented his stone-tool sequence as an original alternative to Lartet's flawed faunal sequence. He argued that there had been little variation in mammals through Lartet's four periods; and since the species found in these caves had been gathered by humans, they did not necessarily represent a typical sample of the animals living at that period—the same criticism that Evans and Dawkins would later cast on Lartet's scheme.[54] De Mortillet turned instead to stone tools as a guide to time.

Between the late 1860s and 1900, de Mortillet published several slightly different versions of his stone-tool sequence (see Table 4.2). His earlier papers focused on tools from the reindeer-caves but he would widen his scheme to include tools from the Palaeolithic period. He grouped tools of similar character and date into 'industries' and arranged them in progressive sequence. De Mortillet was firmer than many British researchers about the time-slots occupied by his industries, which he described as epochs. Each industrial epoch was defined by typical tool-types and labelled with the name of a site that represented the purest or most famous example of that industrial stage.[55] Though based on observations in France, particularly the Somme Valley and the Dordogne, de Mortillet considered his scheme to be equally applicable to the rest of France, Switzerland, the Rhine area, Belgium, and

---

[52] The life and work of Gabriel de Mortillet has been summarized in obituaries by Cartailhac 1898 and Reinach 1899. The connection between de Mortillet's political views and his linear, progressive, evolutionary vision of prehistoric archaeology is examined by Hammond 1980 and Richard 1999a, 2002. See also Sackett 1981, 1991.

[53] Reinach 1889, 10.

[54] De Mortillet 1869, 583–584; Evans 1872, 433–434; Dawkins 1872, 419–420; 1874, 352.

[55] De Mortillet 1873, 436; de Mortillet and de Mortillet 1900, 240; Reinach 1899, 88.

Table 4.2. The Palaeolithic classifications proposed by Gabriel de Mortillet (after de Mortillet 1869, 584–585; 1872, 464–465; 1883, 21, 131).

De Mortillet 1869	De Mortillet 1872	De Mortillet 1883
Époque de la Madelaine	Magdelénien	Magdalénienne (reindeer)
Époque d'Aurignac		
Époque de Solutré	Solutréen (post-glacial)	Solutréenne (reindeer and mammoth)
Époque du Moustiers	Moustiérien (glacial)	Moustérienne (cave bear, mammoth, woolly rhinoceros)
	Acheuléen (pre-glacial)	Chelléenne (hippopotamus, *E. antiquus*)

Britain. Lartet, in contrast, had considered his own palaeontological scheme to be of regional relevance.[56]

De Mortillet's 1873 sequence started with the *Acheuléen* epoch, moved on to the *Moustérien* epoch, and then ran on through various industrial epochs of the Reindeer period. The *Acheuléen* (hereafter Acheulian) was characterized by the hand-axe (see Fig. 4.4); the flake-dominated *Moustérien* (hereafter Mousterian), by scrapers and points (Fig. 4.5).[57] By 1900, his sequence had three distinct epochs: the *Chelléen* (hereafter Chellean), Acheulian and Mousterian.[58] These Palaeolithic epochs of Quaternary times were preceded by an Eolithic period of Tertiary times, when de Mortillet believed that his hypothetical hominids, the *Anthropopitheci*, had made the crudest of all stone tools in Europe.[59] The reception of similarly ancient Eolithic discoveries in Britain is explored in Chapters 5 and 6.

In the light of de Mortillet's poor opinion of Lartet's work, it is interesting to see that Lartet had already described similar patterns for stone tools from French caves and river drifts in collaboration with Christy. One of the many papers published by Lartet and Christy in 1864 had appeared in the *Revue archéologique*. It was here that they remarked on the different types and ages of tools from the reindeer-caves of the Dordogne. They wrote that the cave of Les Eyzies had flint knives, bone harpoons, artwork, and the same reindeer-rich fauna as the cave of La Madelaine (which would become de Mortillet's 'Magdalenian'). Laugerie Haute had delicately worked lance-heads (de Mortillet's 'Solutrean'). At Le Moustier (de Mortillet's 'Mousterian') many of the tools common at other cave sites were absent, but the lance-head types ('hand-axes' of later authors)

---

[56] Lartet 1861, 222, 231–232; de Mortillet 1873, 447; de Quatrefages 1881, 148.
[57] De Mortillet 1873, 436–437.   [58] De Mortillet and de Mortillet 1900, 21.
[59] De Mortillet 1883, 103–106.

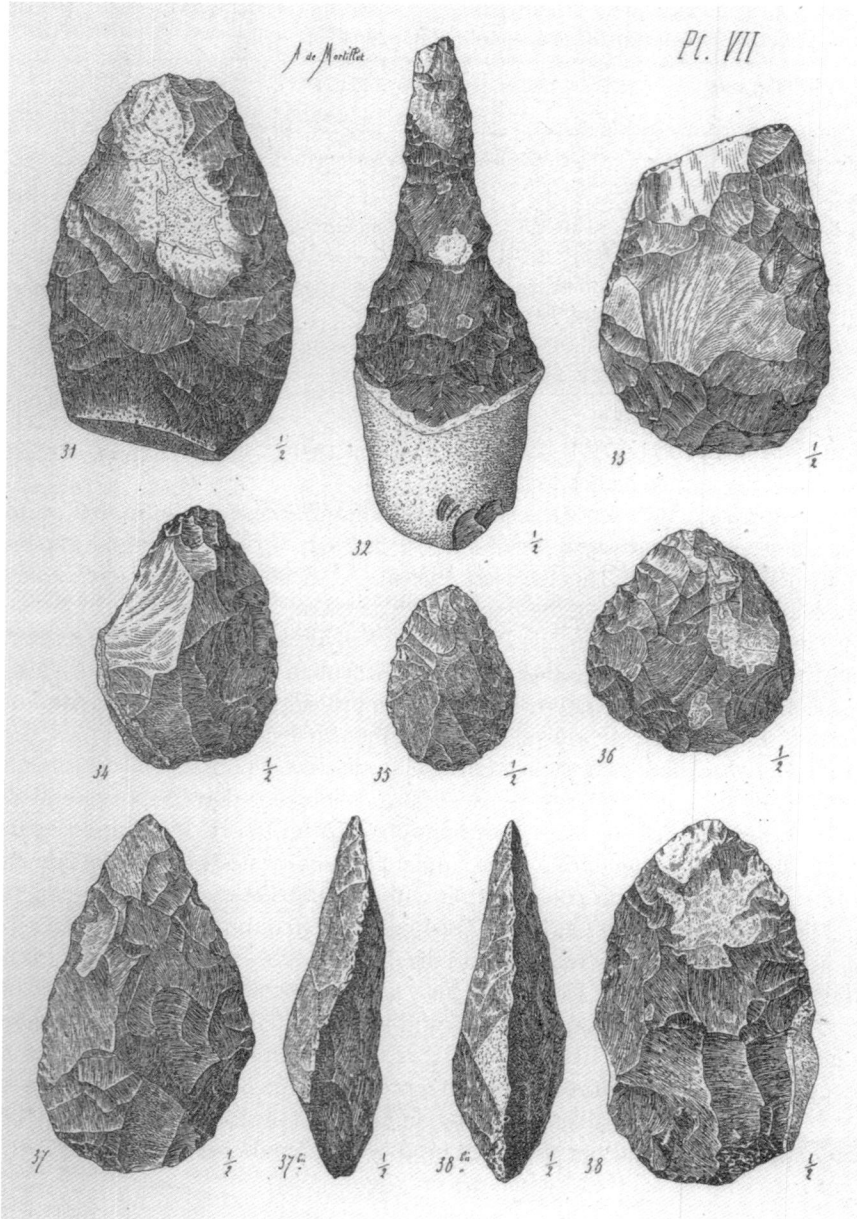

Figure 4.4. Gabriel de Mortillet's '*Instrument chelléen*': hand-axes from the epoch that he had previously called '*acheuléen*' (de Mortillet and de Mortillet 1881, Plate VII. Reproduced by permission of the University of Aberdeen). De Mortillet admitted that some of these hand-axes belonged to the Mousterian epoch (de Mortillet and de Mortillet 1881, description of Plate XI).

Figure 4.5. Gabriel de Mortillet's Mousterian tools: flakes, points, scrapers and blades (de Mortillet and de Mortillet 1881, Plate XI. Reproduced by permission of the University of Aberdeen).

were very similar to those from St Acheul in the Somme Valley (de Mortillet's 'Acheulian').[60]

Lartet and Christy influenced the finer industrial sequences adopted in Britain as well as the general division between the Palaeolithic and Reindeer periods. When Evans was formulating his own sequence in *Ancient Stone Implements*, he referred repeatedly to this 1864 paper. In his section on the chronology of the caverns, Evans followed de Mortillet's arrangement but selected different type-caves. The first two ages Evans described were the Age of Le Moustier and the Age of Laugerie Haute; the primary reference he gave for both was to Lartet and Christy's paper of 1864.[61] It is also worth noting that when British researchers referred to a 'St Acheul' tool-type, this did not necessarily have any connection to de Mortillet's scheme: the site had long supplied a useful analogy to the *pointed type* of hand-axe.[62] Despite the later prominence of de Mortillet, there was no immediate rush amongst British researchers to adopt his scheme after his early papers appeared.

This is not intended to be a hunt for the first to draw such distinctions. Many others besides Lartet, Christy, and de Mortillet had recorded similar patterns. The river drifts held many hand-axes; the reindeer-caves contained bone-work and finely made flake-tools; and between the two, in the earlier caves and later river drifts, lay distinctive flakes, flake-tools, and smaller hand-axes. In very general terms, de Mortillet's scheme did little more than offer a new set of labels for groups of tools that had already been identified in Britain. A loose comparison between the terminology used in late nineteenth-century Britain and the terms of de Mortillet is given in Table 4.3. No exact matches are possible. The Acheulian, for example, has been left out, because it was re-defined by de Mortillet as he adapted his scheme through the decades.

De Mortillet's classification was less flexible than many of the terms used by British researchers: he fixed the character of his industries rigidly to their supposed age. The principles that underpinned his scheme are seen clearly in his decision to change the type-site of his original Acheulian industry from St Acheul on the Somme to Chelles on an eastern tributary of the Seine, near Paris. This decision meant that he also had to change the name of his earliest Palaeolithic industry from the 'Acheulian' to the 'Chellean'. He believed that

[60] Lartet and Christy 1864a, 238–239, 245–246, 253, 255. The similarity between the scheme outlined by Lartet and Christy, and the classification later developed by de Mortillet was also observed by Evans 1872, 435 and Reinach 1889, 94; 1899, 86–87.

[61] Evans 1872, 435–436. Daniel 1975, 103–105 remarked on the relationship between Lartet's palaeontological scheme and de Mortillet's industrial classification, but believed that de Mortillet had a greater influence on Evans, and placed more emphasis on Lartet's palaeontological arguments than his archaeological observations.

[62] Lyell 1863a, 114–115; Wyatt 1864, 187; Brown 1888, 366; 1889, 61.

Table 4.3. A comparison between British industrial terminology of the late nineteenth century and Gabriel de Mortillet's classification.

Industrial terms used in Britain	British sites	British description	Industrial terms of de Mortillet
Reindeer period	Kent's Cavern (late cave-earth); Creswell Crags (upper levels)	Small, delicately worked flake-tools; bone needles and harpoons; artwork	Magdalenian; Solutrean (bone and antler tools appear)
Oval implements with sharp rim (Evans); third class (Smith); turtle-back flakes (Spurrell)	Low-level river drift; Palaeolithic Floors of London; Kent's Cavern (early cave-earth)	Thin, sharp-rimmed oval hand-axes; large, boldly struck, prepared-flakes; many scrapers	Mousterian; later Chellean hand-axes
Pointed implements (Evans); second class (Smith)	High-level river drift; tool from Brixham Cave; (? nodule-tools of Kent's Cavern)	Hand-axes (pointed and more oval forms); scrapers rare; Gray's Inn Lane implement	Chellean
Oldest class (Smith)	Derived in the river drifts	Crude hand-axes; no scrapers	

this alteration was essential, despite the confusion it might cause, because the site of St Acheul was tainted by influences suspected by de Mortillet to indicate a time transitional to the Mousterian industry—it was not pure.

One problem with St Acheul was its bones. De Mortillet thought that animals like hippopotamus and straight-tusked elephant lived in the warm time of the early Palaeolithic when hand-axes were made, and that mammoth and woolly rhino lived in the colder Mousterian epoch. St Acheul, however, contained mammoth as well as straight-tusked elephant.[63] De Mortillet's other problem with St Acheul lay in the character of its implements, which did not fit his idealized Acheulian industry. In 1873, when he placed the Acheulian at the base of his Palaeolithic sequence, this industrial stage was defined by the presence of hand-axes. His next stage was the Mousterian— a time characterized by flake-tools—when the points that had started to appear in Acheulian times became plentiful, many scrapers accompanied these points, and the hand-axes died out.[64] But de Mortillet became convinced that the hand-axe, though variable in form, size, finish, and material, comprized the only implement of the earliest Palaeolithic industry '*elle se compose d'un seul instrument en pierre*' ('it is composed of only one stone instrument').[65] At St Acheul, however, he saw evidence of an advanced technological stage in the scrapers and thin, finely crafted hand-axes.

[63] De Mortillet 1883, 131, 133.
[64] De Mortillet 1873, 436–437.   [65] De Mortillet 1883, 133.

These were the reasons why de Mortillet decided to replace the old impure type-site of St Acheul with the site of Chelles, which he believed had a pure straight-tusked elephant fauna and no tools but hand-axes. The Chellean replaced the Acheulian at the base of the Palaeolithic sequence, and the industry from St Acheul was reinterpreted as a *transitional* stage between the hand-axes of the Chellean and the flake-rich Mousterian.[66] Evolutionary progression, in the character of tools and the skill with which they were worked, was an important part of de Mortillet's vision of the Palaeolithic; a named industry had to display the expected character for its position in his sequence of epochs.

But de Mortillet was faced with more problems in matching the character and age of industries than could be solved by altering their type-sites. Take the Chellean hand-axes, for example, which he figured in one of the plates that accompanied his 1881 publication, *Musée préhistorique* (Fig. 4.4). These incorporated both pointed forms and sharp-rimmed oval forms. The latter were held by many British researchers to be later in date than the former, and de Mortillet noted in the text that some of the hand-axes he had classed as Chellean ought to be brought down to the Mousterian epoch (his Mousterian plate is illustrated in Fig. 4.5).[67] In 1883, de Mortillet reflected that although the hand-axes, characteristic of his Chellean, took many forms, they were nonetheless immediately recognizable as hand-axes. They were like dogs in this respect, he explained, for though there are many varieties of dog they are all clearly dogs.[68] He admitted, however, that there was a close connection between his Chellean and Mousterian: that the one merged into the other. The transitional 'Acheulian' became a useful pigeonhole for tools that did not fit the two idealized epochs that lay on either side.[69]

There were others who found difficulty in accepting de Mortillet's assurance that the hand-axe was the only implement in his earliest Palaeolithic industry, and who provoked him to change the name of his type-site. Ernest D'Acy, a vigorous opponent, argued that a variety of tools, weapons, and flake-tools—which de Mortillet classed as 'Mousterian'—could be found in abundance in de Mortillet's original 'Acheulian' alongside the hand-axes. After de Mortillet made his switch to the Chellean, D'Acy still disagreed with de Mortillet's claim that flakes came from only the more recent levels of sites like St Acheul and Levallois; he attacked de Mortillet's definitions, and countered his belief in progression with the argument that 'Mousterian' types could be found *throughout* the deposits at Chelles and St Acheul alongside

[66] De Mortillet 1883, 132–133, 148, 254; de Mortillet and de Mortillet 1900, 234.
[67] De Mortillet and de Mortillet 1881, description of Plate XI.
[68] De Mortillet 1883, 133.
[69] De Mortillet 1883, 254.

'Chellean' and 'Acheulian' types.[70] D'Acy made the more cynical observation that the rarity of smaller flake-tools at Chelles might be connected to the preferences of the workmen who dug there; they had more interest in conspicuous, finely-finished hand-axes, which had a rich and ready market.[71]

In Britain, researchers were aware of de Mortillet's work on the stone-tool sequence. Rupert Jones, for example, used de Mortillet's classification table of 1872 to illustrate a lecture he gave in 1877 to the Croydon Microscopical Club on the successive groups of Palaeolithic peoples in France and Central Europe. This version of de Mortillet's Stone-Age sequence incorporated the mammoth-dominated Acheulian, Mousterian, Solutrean, and reindeer-rich Magdalenian.[72] But it was not until the late nineteenth century that de Mortillet's terms were in such widespread use for British industries that Evans felt compelled to adopt them, with some reluctance, in the second edition of his *Ancient Stone Implements*. Pengelly's nodule-tools from Kent's Cavern then became described by Evans as implements of the age of St Acheul or Chelles.[73]

De Mortillet was equally aware of the British glacial sequence. He had been forced by the British boulder clays into an interesting contortion to maintain a connection between his glacial and industrial sequences. De Mortillet had dated the Acheulian (soon to be Chellean) industry to pre-glacial times (see Table 4.2). At Hoxne, however, these hand-axes lay above the boulder clay, compelling de Mortillet to stretch his industrial epochs and the behaviour of ice-sheets. He skipped past these problems in 1873 by suggesting that his industrial epochs might have lasted a little longer in some areas than others, and that links to the glacial chronology might have varied between different areas depending on the timing of the maximum extent of the glaciers.[74] In 1881 he decided that these hand-axes, which by then he called Chellean, must date to the end of the Chellean epoch or the beginning of the Mousterian epoch because of their glacial associations.[75] But if tools from a supposedly discrete industrial epoch could continue to be made by tool-makers in the next epoch—if a Chellean hand-axe could be found in the Mousterian—this begins to sound like an anomaly.

As yet, British researchers were little troubled by such concerns. Their expectations of Palaeolithic patterning were not yet firmly tied to other geological sequences (and, in any case, most believed the Palaeolithic to be post-glacial in date). Brown, for example, observed that the pointed flake-tools

---

[70] De Mortillet 1873, 450; D'Acy 1875, 282; 1887, 163, 170–171, 222, 227, 231.
[71] D'Acy in Vauville 1891, 352–353.
[72] De Mortillet 1872, 464–465; Jones 1877, 22–23.
[73] Evans 1897b, 483, 495.
[74] De Mortillet 1873, 447–448.
[75] De Mortillet and de Mortillet 1881, description of Fig. 127, Plate XX.

from Creffield Road were very similar to those from the French cave of Le Moustier, and that de Mortillet had described the same kind of flake-tools from the river drifts of the Somme and Seine as '*pointes mousteriennes*'. Despite these connections between the character of his industry and de Mortillet's Mousterian, Brown had no qualms about concluding, from the fauna, that the tool-makers of Creffield Road had lived *before* de Mortillet's Mousterian epoch (characterized by reindeer and woolly mammoth), and should be classed with his Chellean (associated with straight-tusked elephant, hippopotamus, and a more ancient rhinoceros).[76]

If Brown had followed de Mortillet's idealized scheme wholeheartedly, rather than seeing it as a convenient series of time-specific labels, he might have expressed more surprise that the character of the industry did not match its early date. One of his French contemporaries noticed that British prehistorians attributed less value to classifications that linked the age and character of Palaeolithic tools than did their French colleagues.[77] Not all took Brown's view, and tentative links were starting to be made between the character and the date of industries in Britain. Smith, for example, suggested that the trimmed flakes from his Palaeolithic Floors belonged to the time of the older caverns—to 'the age of Le Moustier'.[78]

But it was on the Continent, rather than Britain, that Brown's Creffield Road situation was most likely to be taken as a discrepancy: an anomaly that required explanation. In 1897, Salomon Reinach (1858–1932), curator of the Museum of Saint-Germain-en-Laye (where de Mortillet had formulated his Palaeolithic classification a few decades before), described an industry from Taubach, near Weimar in Germany. This industry was dominated by small scrapers and points, had no hand-axes, and might have been described in France as Mousterian or Magdalenian. But none of the familiar Arctic mammal species, like mammoth or reindeer, were present. This Mousterian-like German industry was found with a warm straight-tusked elephant fauna, like the bones from the French site of Chelles, and was therefore thought to be contemporaneous with the Chellean epoch of France.[79] Reinach saw this as a problem for de Mortillet's scheme. Criticisms also came from Belgium. Dupont questioned de Mortillet's ideas about a widespread linear progression through different industrial stages. He thought that there had been a parallel and separate development of two contemporaneous industries during his

---

[76] Brown 1887b, 212–213, 215. By 'Chellean,' Brown meant Chellean or Acheulian. He was using a version of de Mortillet's sequence with four divisions: Chellean or Acheulian, Mousterian, Solutrean and Magdalenian.

[77] Boule 1888, 289.   [78] Smith 1894, 110–111, 244.   [79] Reinach 1897, 55, 57–58.

Mammoth period in Belgium: one made by the peoples living around Namur and Liège; the other by those from Hainaut.[80]

In Britain, at the end of the nineteenth century, the patterns of stone tools from the caves and river drifts were more familiar to Palaeolithic researchers. But it would be a while before the tools made by River-Drift and Cave men in the Palaeolithic and Reindeer periods exchanged their varied vernacular labels for the terms of a standard industrial sequence. The character of tools was not yet fixed firmly to specific periods, despite de Mortillet's suggestions, and the British Palaeolithic sequence could fluctuate with the varied indications and interpretations of other geological sequences. Dawkins set the tool-makers off on migrations that mirrored the routes of his Quaternary mammals; Geikie believed that they would have witnessed the slow climatic shifts of his glacial–interglacial sequence.

The eventual adoption of de Mortillet's terms in Britain in the latter years of the nineteenth century could be connected to the controversial reputation of these geological sequences. Geikie's glacial sequence was of little value as a Palaeolithic time-scale, because most English researchers believed in a post-glacial Palaeolithic. The sequence of mammals was often used as a guide to age, but Dawkins had quibbled with and confused the idea that straight-tusked elephant and hippopotamus were earlier than mammoth, and that reindeer were most abundant at the end of the Palaeolithic. There was also the sequence of river drifts, but it could be difficult to identify where Prestwich's earlier high-level river-gravels ended and the later low-level deposits began.

De Mortillet's terms, however, allowed references to be made to particular periods of time with less danger of being swamped by geological argument. It is significant that some British researchers used terms like 'Acheulian', 'Chellean', or 'Mousterian' solely to indicate the age of implements; and others drew analogies to the sites of St Acheul or Le Moustier to describe the character of implements; but many disregarded de Mortillet's connections between age and character. Brown seemed undismayed by the idea of an industry of Mouster-ian-like character dating to the Chellean epoch, although such discoveries were treated as vexing anomalies by some of his contemporaries. There would be louder protests about anomalous industries in Britain when a standard indus-trial sequence was adopted and rigid links were established between the character of industries and their date. Then, British researchers would find themselves in a similar position to de Mortillet at Hoxne: tweaking industrial and geological sequences into alignment with their expectations. The study of the British Palaeolithic was acquiring as many facets as one of Spurrell's turtle-back flakes.

[80] Dupont 1873, 469, 479.

4.5 PERCEPTIONS OF PALAEOLITHIC TOOLMAKERS

The tool-makers must not be left out of the picture. There was curiosity amongst nineteenth-century researchers about their character: How old was the modern human type? Did modern humans emerge from more primitive antecedents? Had there been more than one type of early human? Stone tools were, however, far more common than the skeletal remains of their makers, and de Mortillet had been driven to invent his *Anthropopitheci* of Tertiary times.

New discoveries were welcomed, with hope that they might answer such questions mingled with concern that the find could be an intrusive burial from later times. Ancient skulls were valued as indications of character and type, and skeletons were compared to the bones of distant contemporary races, which were thought to represent lower stages of progress. Since the history of palaeoanthropology has been well studied by historians, only a brief summary of the most notable discoveries, from Germany to Java, is given here.[81]

Early in 1857 an unusual skeleton was found in a small limestone cave in the Neander Valley, Germany. Hermann Shaaffhausen (1816–1893), Professor of Anatomy at the University of Bonn, examined the skull and drew attention to the primitive character of the sloping forehead, the heavy brow-ridges, and the large frontal sinuses. He suggested that the owner of these bones had lived alongside the latest animals of the Drift, long before the time of the Celts, and belonged to a savage, wild race.[82] George Busk (1807–1886), the British anatomist and former Navy surgeon, saw certain similarities to the chimpanzee and gorilla.[83]

Thomas Henry Huxley (1825–1895), however, argued in 1863 that the Neanderthal bones could not be positioned between humans and apes because the brain capacity was so large. He agreed that the skull was the most ape-like of human crania yet discovered, but considered it human, nonetheless, and placed it on the primitive end of the *Homo sapiens* spectrum.[84] Other anatomists suggested that the skull was not even a racial type, but a case of abnormal development. Rudolph Virchow (1821–1902), Director of the Institute of Pathology in Berlin, dismissed the idea that humans had descended from apes. He doubted that the Neanderthal find was fossil at all

---

[81] For more information on the history of palaeoanthropology, see Spencer 1984; 1990; Bowler 1986, Regal 2004, and Sommer 2005; 2006.

[82] Schaaffhausen 1861, 155, 160, 171–172.

[83] Busk in Schaaffhausen 1861, 173.   [84] Huxley 1863, 156–159.

and suggested that the unusual features were the result of disease in a comparatively recent human.[85] De Mortillet, though, who nurtured a linear, progressive view of human evolution, was eager to emphasise the simian character of the Neanderthal skull.[86]

In 1886 the age and distinct character of the Neanderthal specimen were reinforced when two similar skeletons were found in the entrance to a cave at Spy in the province of Namur, Belgium. They were described the following year by Julian Fraipont and Max Lohest, an anatomist and a geologist from the University of Liège. Fraipont and Lohest reported a clear association between the skeletons, extinct animals (which included woolly mammoth, hyena, and woolly rhino), and Palaeolithic implements (such as Mousterian points).[87] For the rest of the nineteenth century, similar bones were seen as a low type of the genus *Homo*, and became generally known as the Canstadt or Neanderthal type. Huxley, pleased at this confirmation of his views, wondered if the same genus (if not the species *sapiens*) might have been represented back in Pliocene or even Miocene times.[88]

Few discoveries had such clear associations as the skeletons from Spy. The five skeletons that had been discovered at the Cro-Magnon rock-shelter in the Périgord in the late 1860s were regarded with suspicion. Louis Lartet, the son of Édouard Lartet, had excavated this site. He had found finely-chipped flint scrapers, mammoth bones, and reindeer in hearth beds that lay just below the skeletons, and concluded that these humans had lived just before the artistic period. Paul Broca (1824–1880), General Secretary of the Anthropological Society of Paris, agreed that the Cro-Magnon group dated to the Mammoth period. Broca argued that their appearance confirmed the existence of several different races of men in Quaternary Europe, and explained that whilst the Cro-Magnon race was tall, robust, and long-headed with a vertical forehead, a different race had been found in the Belgian caves by Dupont.[89]

In Britain, few agreed with the Palaeolithic date ascribed to the Cro-Magnon skeletons, and most concluded that they belonged to relatively recent, though still prehistoric, times.[90] Edwin Tulley Newton (1840–1930), Palaeontologist to the Geological Survey, complained in 1895 to the Geological Society of London that although the appearance of the Palaeolithic

---

[85] Virchow 1872, 10–11. See also Davis 1864, 10.
[86] De Mortillet 1883, 102–103, 248.
[87] Fraipont and Lohest 1887, 662–667.
[88] Dawkins 1874, 241–242; Lubbock 1878, 343–346; De Quatrefages 1884, 60–65; Huxley 1890, 773–775; Smith 1894, 19; Newton 1895, 512.
[89] Louis Lartet in Lartet and Christy 1865–75, 71; Broca in Lartet and Christy 1865–75, 98, 101. On the debate between monogenists and polygenists, and Broca's position, see Blanckaert 1988.
[90] Dawkins 1874, 255; Lubbock 1878, 346; Smith 1894, 32; Hutchinson 1896, 74.

race on the Continent was known from the remains at Spy and the Neander Valley, no one knew what the Palaeolithic race in Britain had looked like. The Victoria Cave fibula had turned out not to be human at all, and other finds were either suspected to post-date the Palaeolithic or were too small to give much idea of their physical appearance.

Newton had prepared an answer to this problem. He described to his audience a fragmentary skeleton that had been recovered from gravels of Palaeolithic age in 1888 at Galley Hill, Northfleet, on the south bank of the Thames. The skull was relatively modern in form, but had large brow ridges and a moderately receding forehead. These features suggested an affinity to the Palaeolithic race of the Continent. The response of his audience was mixed, and reflected the variety of opinions held at this time. Evans and Dawkins both thought it an intrusive burial of later date. Brown (like Broca) saw no reason why more than one Palaeolithic race should not have inhabited Europe. Sollas assigned the skull to the same Palaeolithic type as Neanderthal and Spy.[91] The idea that modern-looking humans had arrived very early into Europe would persuade British researchers to accept other dubious discoveries in later years, the most notorious being the Piltdown skull.

One last important find of the nineteenth century must be mentioned: the fragments of a skeleton recovered from Java in 1891 and 1892 by Eugène Dubois (1858–1940), a Dutch doctor and palaeontologist. Dubois found teeth, part of a skull, and a thigh bone near the village of Trinil. He dated his discovery to the Pliocene, placed its bones mid-way between apes and humans, and called it *Pithecanthropus erectus*—the upright ape-man.[92] *Pithecanthropus* had a thigh bone that suggested some stature and an erect posture, but a skull that was two-thirds the size of a European skull. It was strangely different from the Canstadt or Neanderthal race, which was seen as distinctly human from its brain capacity, even if a low and bow-legged type of human. Virchow disagreed strongly with Dubois's reconstruction of an ape-man from these bones. In Britain there was uncertainty about whether the bones belonged to the same individual, or whether they represented an ape, a low type of human, a pathological human, or a missing link between humans and apes.[93]

There was little consensus about the age, origin, or appearance of the tool-makers as the nineteenth century drew to an end. Some supported an evolution from ape-like stock; some envisaged a very ancient origin for

[91] Newton 1895, 510, 516–517, 525–527.

[92] Dubois 1894, 31. For more on Dubois and his interpretations of *Pithecanthropus erectus* see Shipman 2001.

[93] Hutchinson 1896, 160–162; Sollas 1911, 34–39.

modern humans; and others believed that different races had inhabited Europe. They looked with doubt on the situations in which many of the fossils had been discovered. Spy, the most authentic Palaeolithic type from Europe, was described as a race, not as a separate species.

In the late nineteenth and early twentieth centuries palaeoanthropological theory underwent an important shift, which has been studied in detail by Marianne Sommer. The German anatomist Gustav Schwalbe (1844–1916) worked on the fossils from Neanderthal, Spy, and elsewhere, and classified them as a distinct species—*Homo primigenius*—rather than a race. The idea of a single evolving line of hominids then became widespread, with *Pithecan-thropus* evolving into *Homo primigenius* and on to *Homo sapiens*.[94] This linear view of evolution persisted until Marcellin Boule (1861–1942), the French geologist and palaeontologist, reconstructed a brutish Neanderthal from the bones discovered at La Chapelle-aux-Saints in 1908. Boule had occupied the Chair of Palaeontology at the Museum of Natural History, Paris for the last six years.[95] His interpretation matched his belief that Neanderthals, though relatively late in date, were no ancestors of ours. It then became more common to direct the Neanderthal and *Pithecanthropus* types along separate evolutionary branches. Both were set firmly apart from our own ancestors.[96]

Palaeolithic tools would also come to be distributed along several branches, but that is a subject for later chapters. First, some controversial chipped stones must be added to the earlier end of the Stone-Age sequence. In the late nineteenth and early twentieth centuries, some researchers suspected that tool-makers had arrived in Britain *before* River-Drift times. Chapters 5 and 6 will explore the arguments that developed over these 'Pre-Palaeolithic' and 'Eolithic' periods of the Stone Age.

[94] Schwalbe 1906, 14; Sommer 2006, 210.
[95] Vallois 1941–46, 204.
[96] Boule 1908a, 524–525; Sommer 2005; 2006. On Boule and changing idea about palaeo-anthropological lines, see also Hammond 1982, Spencer 1984, Bowler 1986, 34–35, 68–81, 87–91, Trinkaus and Shipman 1994, Richard 1999b, 268–271, Regal 2004, 49–69.

5

Eoliths: An Earlier Phase of the Stone Age?

In 1912, Edwin Ray Lankester (1847–1929), a well-regarded zoologist, was introducing some interesting and ancient flints to the academic world at a Royal Society *soirée* when he spotted Dawkins in the room. The two men started to quarrel over the stones. Soon afterwards, Lankester described the incident to a friend, explaining 'Dawkins was there and I made him go over them with me'. Dawkins, though now elderly, was still outspoken. He proceeded to attack Lankester's view that the flints in question were very early tools, arguing that they had not been flaked by human hands.

Lankester recalled that Dawkins had 'idiotically said that such conchoidal fractures as they show could be produced by pressure' and had placed the burden of proof on Lankester's shoulders: 'Well, unless you can show that these flints could not possibly be produced by natural agencies, I shall refuse to attribute them to man.' Lankester had responded that this was 'a preposterous & unscientific attitude' and further informed Dawkins: 'neither I nor any one who had studied the subject, attached any importance to his opinion!'[1]

The kind of stones displayed by Lankester in 1912 aroused enthusiasm and irritation in Britain during the late nineteenth and early twentieth centuries. They were claimed as the artefacts of tool-makers who had lived before the Palaeolithic period, in Eolithic times (from the Greek *eos*: dawn and *lithos*: stone). Lankester's flints, which came from East Anglia, belonged to the second major group of eoliths to be discovered in Britain; the first group had been reported from Kent in the 1880s and 1890s. His arguments with Dawkins in the rooms of the Royal Society encapsulate the character of the British eolith debates. Lankester was trying to describe what he thought was an important new Stone-Age industry and was irritated by the suggestion that he should prove they were not produced by natural agencies. Dawkins could see only natural chipping in these stones; neither was convinced by the case of the other and the discussion grew heated. Nowadays, both groups of eoliths are usually regarded as the natural products of geological forces; in these chapters, though, the eoliths will occasionally be described as artefacts to retain the atmosphere of the arguments.

[1] Lankester to Moir, 9 May 1912: BLL: Add.Ms.44968/154.

The eolith debates have again attracted attention in recent years, this time from historians of science. Frank Spencer discusses the eoliths in his work on the Piltdown forgery. For those who believed that Tertiary tool-makers had dropped the eoliths of Kent and East Anglia, the discovery of their possible maker at Piltdown (*Eoanthropus*) was not surprising. Spencer introduces the participants in the eolith debates and explains their differing expectations of the character and antiquity displayed by our earliest ancestors in his article: 'Prologue to a Scientific Forgery' (1988), and in his book: *Piltdown. A Scientific Forgery* (1990). Marianne Sommer examines how eoliths became caught up in theories of human evolution in her article: 'Eoliths as Evidence for Human Origins? The British Context' (2004). The reader is referred to the work of these two authors for the palaeoanthropological aspects of the debates.[2]

The next two chapters explore the treatment of the British eoliths themselves and will touch only lightly on the appearance and affinities of their makers. There are several reasons for devoting two chapters to eoliths. Eoliths influenced perceptions of Palaeolithic industries: many standard Stone-Age sequences included an Eolithic period from at least the 1880s, and the Eolithic was still managing to maintain its position in the 1930s. The eolith debates give another view of the expectations about the age and character of stone tools held in the late nineteenth and early twentieth centuries. The debates also offer a glimpse at the reasons behind a sustained disagreement. These are some of the features examined in this chapter. Looking forward to the subject of the next chapter, the eoliths aroused so much indignation that the strategies to gain their recognition and acceptance became particularly blatant: strategies that were common to both Eolithic and Palaeolithic controversies.

5.1 BACKGROUND TO THE EOLITH DEBATES

The age of the earliest tool-makers in Europe was much discussed in the latter decades of the nineteenth century. Some cautioned that the very earliest artefacts would not necessarily be found in Europe and looked instead to Africa or Asia for the birthplace of humanity, as suggested by Darwin, Wallace, and Falconer.[3] Others reasoned that even if the earliest tool-makers

---

[2] Spencer 1988; 1990; Sommer 2004b. The history of the eolith debates has also been studied in detail by Cremo and Thompson 1998, 85–289.

[3] Murchison 1868 (ii), 579; Wallace 1870, 323–324; Darwin 1871 (i), 199; Dawkins 1874, 425; Smith 1894, 2–3; Hutchinson 1896, 15. Wallace suspected that humans had been around in Europe before Palaeolithic times and accepted Harrison's eoliths (Harrison to Wallace, 20 February 1898: PRM: Misc. Ms. 11).

in the world had not come from Europe, it was still possible that tools older than river-drift types might be found in this continent. After tools had been cast back into geological times by the announcements of 1859, opinions on their age were, for a little while, very fluid.

Expectations of the appearance of Palaeolithic tools also took time to settle. Some suspicious 'tools' masqueraded as true artefacts in the early years of Palaeolithic research. Evans often received parcels of natural stones, sent to him by their discoverers in the belief that these were Palaeolithic implements. Even if the resemblance to natural products was pointed out, this might not dissuade those who believed such relics to have been made during an earlier phase of the Stone Age. Defenders of an Eolithic period might reply that early tool-makers had not yet developed the skills needed to produce Palaeolithic implements. Such primitive peoples would have made much cruder tools; the oldest tool-users would not have made them at all, but used naturally fractured stones. This was one of the conundrums that fuelled the eolith debates: the earliest tools in Europe were *expected* to be primitive and to resemble the products of nature.

The first major claims for eoliths were made in France, soon after human antiquity had been announced, but when the extent of that antiquity was still unclear. In 1863, the French geologist and palaeontologist Jules Desnoyers (1801–1887) reported cut-marked bones from deposits of Pliocene age at St Prest. He did not find any tools, but cut-marks of this age suggested very early human activity and aroused interest.[4] Falconer, that same year, advised a colleague working at Lexden: 'Let me recommend to you to give a glance over the surface of all the bones to see if any of them present knife or cut marks, or grooves made by a scraping implement, some instances of which have lately been turning up in deposits of very considerable antiquity'.[5]

In this expectant atmosphere it was not long before 'tools' were recovered from even older deposits. In 1867 the Abbé Louis Bourgeois (1819–1878) presented stone tools from Miocene deposits at Thenay, near Pontlevoy, to delegates at the International Congress of Prehistoric Anthropology and Archaeology in Paris. De Mortillet was amongst those who accepted the Thenay specimens.[6] More claims followed. De Mortillet acknowledged the Portuguese eoliths of Miocene and Pliocene date found by the geologist Carlos Ribeiro at Otta, near Lisbon; he welcomed the French Upper Miocene 'tools' found at Puy Courny by Jean-Baptiste Ramès (1832–1894), a pharmacist and geologist from Aurillac.[7] In his synthesis, *Le Préhistorique. Antiquité de*

---

4 Desnoyers 1863, 1077.   5 Falconer to Fisher, 2 November 1863: ULC: 7652II/NN/33.
6 Bourgeois 1868; 1873, 81–82; Wilson 1899, 325–326.
7 Ribeiro 1873; Rames 1884, 399–402; Lantier 1945, 132. On Ribeiro, see Carneiro 2005.

l'homme (1883), de Mortillet extended his sequence of Stone-Age industries back into Tertiary times and devoted a large section of the book to Tertiary tool-makers.[8]

De Mortillet coined terms for the makers of these Tertiary tools, matching hypothetical human precursors to each industrial epoch. Thus *Anthropopithecus bourgeoisii* emerged as the maker of the ancient Thenay flints, *Anthropopithecus ramesii* left the more recent Puy Courny eoliths and *Anthropopithecus ribeiroii* manufactured the Portuguese finds from Otta. This list of evolving hominids armed de Mortillet against those who were suspicious of Darwinian transmutation, such as Armand de Quatrefages (1810–1892), Professor of Anthropology at the Muséum d'Histoire Naturelle, Paris.[9] Meanwhile, a growing group of sceptics were gathering around the eoliths. Evans became one of their champions. He was dubbed 'the little St. Thomas' by Professor Giovanni Capellini (1833–1922) of Bologna University for his constant expressions of doubt at the international conferences where eoliths were displayed and debated.[10] Evans would also cast his suspicions on British eoliths when they were discovered in Kent.

5.2 THE KENT EOLITHS AND EXPECTATIONS OF HUMAN ANTIQUITY

The tale of the British eolith debates begins with Benjamin Harrison (1837–1921) who lived in Ightham, a village in Kent, twenty-five miles south east of London. Harrison ran the Ightham village shop. This sold the usual range of groceries and other necessities: candles, tobacco, and so on. Odd piles of flints and fossils lying in the garden might, however, have raised a suspicion that Harrison was an unusual shopkeeper.[11]

If invited upstairs to the room above Harrison's shop in the 1890s, you would enter a curious museum. Large solid wooden shelves covered the walls, filled with boxes. These once held grocery goods, but now bore large labels announcing sites, artefact types, and geological deposits. Photographs, maps, and sketches of flints were pinned to the shelves. A framed picture of Prestwich sat prominently amongst the boxes and a likeness of Evans was also present. More flints were displayed on a large table. Harrison had collected

---

[8] De Mortillet 1883, 25–126; see particularly the table on pp. 28–29. On the reception of early French eoliths, see also Newton 1898, 66–67 and Grayson 1986, 81–90.
[9] De Quatrefages 1881, 89–128, 152; de Mortillet 1883, 105; Hammond 1980, 120; Richard 1999a, 96, 101; Sommer 2004b, 212–213.
[10] Newton 1898, 66.   [11] Harrison 1928, 23.

them all in time spared from the shop, mostly on Sundays and Bank Holidays. Lubbock had won many thanks from his fellow Palaeolithic researchers for his government work on the Bank Holiday Act, passed in 1871; Harrison had been grateful when these came into force in 1872.[12]

Harrison wrote to Lane Fox (now Pitt-Rivers): 'I am but a tradesman but I have applied myself to thoroughly working out the Antiquity of Man & his geological position as illustrated in my own area'.[13] Over the last few decades of the nineteenth century, Harrison collected many box-loads of stone tools. Although his name is now associated most strongly with crude, dark-stained eoliths that he found strewn over the surrounding hills, he was also an avid collector of Palaeolithic tools, as Evans noted in *Ancient Stone Implements*.[14]

In the summer of 1879, a mutual friend had introduced a diffident Harrison to Joseph Prestwich, now Professor Prestwich, who lived nearby on the hill above Shoreham village. Prestwich had occupied the Chair of Geology at Oxford University for the previous five years and had retired from City work in the wine business. He was to become Harrison's most valued supporter in the matter of the eoliths. But on this first visit, Harrison felt encouraged to stretch back the age of his Palaeolithic finds. As they chatted about the local geology, Prestwich mentioned the high-level gravels from the Somme Valley, where the early hand-axes had been found, and gestured at the equivalent heights they might have occupied in the Kent landscape. Harrison realised with excitement that some of his own implements (true palaeoliths, not eoliths) had come from much higher levels. Thinking that greater height meant greater antiquity, since ancient rivers had cut *down* through their valleys over time, Harrison suspected that his finds might be older than the oldest of river-drift tools.[15]

Harrison was not averse to an earlier date for the Palaeolithic. He had been stirred by Skertchly's supposedly pre-glacial tools in the late 1870s and had travelled to the Survey's Museum in Jermyn Street, London, to see them. When Harrison suggested to Evans that some of his own finds around Ightham village seemed older still, he received the disheartening reply: 'As to the implements found by Mr. Skertchly, I for one see no reason for attributing them to any pre-glacial antiquity. I believe them to belong to the same geological period as

---

[12] This description of Harrison's museum is taken from two small blue-and-white photographs in the Ashmolean Museum, Oxford (AMO: JE/B/2/42) and from the description and photograph in Harrison 1928, 20. On Bank Holidays, see Harrison 1928, 38, 71.
[13] Harrison to Pitt-Rivers, 29 February 1896: SSW: P-R, L1518.
[14] Evans 1897b, 608–609.
[15] Evans 1897a, xiv; Prestwich 1899, 236–238, 248; Harrison 1928, 83–84.

the others found in their immediate neighbourhood, and should give the Ightham specimens the same antiquity'.[16]

James Geikie was more encouraging than Evans. Harrison owned a copy of *Prehistoric Europe*, the gift of a friend. In the same year that his book was published, 1881, Geikie had written to Worthington Smith who was 'staggered' by the contents of this letter. Knowing that Harrison was interested in early implements, Smith passed the letter on and Harrison read for himself Geikie's crucial lines: 'They [Palaeolithic implements] will yet be found in such deposits and at such elevations as will cause the hairs of cautious archaeologists to rise on end. I hope other observers will take a hint from you and search for Palaeolithic implements in places which have hitherto been looked upon as barren of such relics'.[17]

These words were prophetic; Harrison was soon finding tools in Kent that appeared to be very ancient indeed. In 1885, he moved his hunting territory from the high-level drifts up to the Plateau above and began to collect rich-dark-brown, heavily worn flints (later known as eoliths) from deposits that seemed to have no connection with the river drifts below (see Fig. 5.1 for an illustration). It was rather late in time that Harrison and Prestwich fully accepted these as true artefacts: Harrison in 1886 and Prestwich in 1888.[18] Prestwich also came to believe that the associated sediments—the Kent Plateau drifts—were pre-glacial or early glacial in age. If his geological reasoning was correct, and if Harrison's artefacts were truly associated with the Plateau drifts, and if these stones really *were* artefacts, then Harrison would be the discoverer of the oldest implements in Britain. All three premises would be scrutinised closely by his peers.

Before following Prestwich into the Geological Society and hearing his case for the ancient date of Harrison's finds, some geological background is required. The spreads of gravel that lay high above the river valleys were not well understood and had perplexed many geologists. Whitaker's Geological Survey memoir, *The Geology of London*, included a whole chapter on 'Deposits of Doubtful Age', and the Plateau gravels appeared among this uncertain crowd. Whitaker's opinions are worth noting. His fellows took them seriously, and this particular memoir was published in 1889: the same year that Prestwich presented his first major paper on the Kent discoveries.

Whitaker observed that the Plateau gravels could be classed neither with glacial drift nor with river drift, but in some cases they appeared to be relics of a once-vast spread of marine deposits that survived only in isolated patches.

---

[16] Evans to Harrison, 7 May 1882, quoted in Harrison 1928, 98.
[17] J. Geikie to Smith, 2 May 1881, quoted in Harrison 1928, 91; Harrison 1928, 90, 92.
[18] Harrison 1928, 133.

Figure 5.1. Three eoliths painted by Harrison in watercolour (Harrison, August 1902: PRM: Misc. Ms. 11). Their respective provenances are 'S Ash', 'Pit 1898' and 'Maplescombe 795'.

He thought it unclear where the very high terraces of river gravel ended and the Plateau gravel began, but suspected that the Plateau deposits post-dated the boulder clay of glacial times.[19] Prestwich would follow neither the marine origin suggested by Whitaker, nor the subaerial theories of wind and rain proffered by other geologists. He believed that the Plateau drifts had been transported over the heights of the now-eroded Weald by glacial activity long before the intervening valleys had even been excavated.[20]

As Harrison picked up his high-level palaeoliths and eoliths in Kent, Prestwich had been reviewing the geological evidence for an early arrival of tool-makers into Britain. Once a vehement defender of a post-glacial age for the river drifts, Prestwich announced in 1887 that the discoveries of Richard Tiddeman, Sydney Skertchly, and Henry Hicks had led him to reconsider his earlier opinion of a post-glacial Palaeolithic (see Chapter 2). He wrote that the high-level river-valley gravels of the Somme, Thames, and other areas, and the caves of Tiddeman and Hicks: 'date back to Glacial or Preglacial times, not in the sense of being anterior to the Glacial epoch, but in the sense of belonging to that part of the Glacial epoch when the great ice-sheet was advancing'.[21]

Prestwich seems to have meant, by this slightly confusing statement, that Palaeolithic groups had occupied the south of England when glaciers were starting to spread down from the north. His, then, was not a wholehearted adoption of the interglacials promoted by James Geikie. This change of view by the venerable Professor Prestwich was welcomed nonetheless by Hicks and appeared in his own article on pre-glacial man the following year.[22] Harrison's discoveries, high on the Kent Plateau, hinted to Prestwich at an even earlier

---

[19] Whitaker 1889 (i), 296–298. Whitaker's term 'plateau gravel' seems to have included both the Plateau and Hill gravels of Prestwich Whitaker 1875, 57.

[20] Prestwich 1891, 156–159; 1892, 250.

[21] Prestwich 1887, 407.   [22] Hicks 1888, 29.

appearance for the first British tool-makers: earlier than the time of Hicks's Welsh caves or the older river drifts.

5.3 PRESTWICH, HARRISON, AND THEIR PRESENTATION OF THE CASE

By 1888 Prestwich had finished the second volume of his *Geology. Chemical, Physical, and Stratigraphical.* He retired from the Geological Chair at Oxford the same year, and offered to describe Harrison's finds from around Ightham.[23] Henceforth, Harrison worked 'largely under his direction'.[24] It was Prestwich who wrote and delivered the major papers on the early palaeoliths and eoliths of Kent, and Harrison who gathered the specimens and listened. This division of roles suited their personalities as well as their positions in learned society.

Prestwich, now in his seventies, was known as a grand old man of Quaternary geology who had been in the vanguard of Palaeolithic research back in 1859. Brown, in his account of Prestwich's first paper on Harrison's high-level finds, described him to the Middlesex Natural History and Science Society as 'the accepted leader in this branch of Geological investigation'.[25] Prestwich would have been familiar to members of the main scientific societies in the metropolis and particularly respected in geological spheres.

Harrison, in contrast, had not applied for membership to any of the learned societies. Many were prohibitively expensive and he had little money to spare. His first copy of Evans's *Ancient Stone Implements*, invaluable to the Palaeolithic researcher, was borrowed from Smith; his second was the gift of Evans himself. Personally, Harrison was on good-humoured terms with fellow researchers like Smith, Evans, Lubbock, and Spurrell; publicly, he was hesitant, diffident, and also rather deaf, which made him even more self-effacing. He was happy for Prestwich to take the lead in presenting his finds of early palaeoliths and eoliths in Kent to their learned peers. Prestwich did so, in three papers delivered between 1889 and 1891.[26]

Prestwich gave his first paper to the Geological Society on 6 February 1889: 'On the Occurrence of Palaeolithic Flint Implements in the Neighbourhood of Ightham, Kent, their Distribution and Probable Age.' Armed with a pointer and maps he sat next to Harrison, who strained to hear and caught only a

[23] Prestwich 1888; Evans 1897a, xv.
[24] Harrison 1928, 128–130.   [25] Brown 1889, 54.
[26] Prestwich 1889; 1891; 1892. On Harrison's character, see Harrison 1928, 59–60, 82, 94.

fraction of the argument.[27] Prestwich focused his case on the unusual position of these implements. He pointed out their localities to his audience and announced that Palaeolithic implements, often assumed to be confined to river drifts and caves, had been found by Harrison in drift beds at heights up to 600 ft above sea level.

In the past, such discoveries had been regarded as specimens of the usual post-glacial river-drift implements, dropped by their makers as they had walked over these high grounds.[28] Now, Prestwich suggested a greater antiquity for these palaeoliths. He compared the geological origins and the age of different pockets of drift: the post-glacial river drifts; the older drifts from the higher or hill-levels; and the still more ancient, possibly pre-glacial, patches of drift on the Chalk Plateau itself.

The different characters of the implements from each of these drifts helped Prestwich reinforce his case for the greater age of high-level finds. In this paper, he described all of them as 'Palaeolithic.' He drew attention to a group of crude, heavily worn, brown-stained specimens that he thought had come from a deposit on the Chalk Plateau, although they could also be found in the river-valley gravels and the higher or hill drifts (they bring to mind the oldest class of Worthington Smith). The village of Ash, 500 ft above sea level (and six miles from Harrison's shop in Ightham), provided Prestwich with a type-site for this most ancient group of flints.[29]

Prestwich argued that the valleys below had cut through the Plateau drifts in post-glacial times, which suggested a great antiquity—'possibly Preglacial'—for the Ash specimens. The condition and the rudeness of these tools also indicated an extreme age. The Ash group included palaeoliths, deep-brown and worn, as well as the cruder stones later described as eoliths; but Prestwich treated them both similarly.[30] A week before, Prestwich had arranged with Harrison for specimens to be sent in advance to the Geological Society by rail to illustrate his paper, requesting about twenty or thirty:

> 1st—from the river gravels;
> 2nd—from the hill gravels;
> 3rd—from Ash and Bower Lane;
> keeping each bundle separate and packed in one box.[31]

The audience, having heard Prestwich's case and seen these specimens, offered their opinions. Evans questioned Prestwich's geological conclusions and his division of the implements into types, particularly his 'Ash' type. Hicks welcomed the Kent finds as confirmation of his own ideas about the

---

[27] Harrison 1928, 143.   [28] Prestwich 1889, 270, 283.
[29] Prestwich 1889, 283–288.   [30] Prestwich 1889, 291–293; Harrison 1928, 145, 153.
[31] Prestwich to Harrison, 31 January 1889, quoted in Harrison 1928, 143.

Table 5.1. Prestwich's divisions of the stone tools from Kent, with the deposits in which they were found, and their heights (after Prestwich 1889, 272, 284–289; 1891, 129; 1892, Fig. 1).

Height/age	Physical geology of Kent	Stone tools
High/old	Chalk Plateau, covered in Red Clay-with-flints, 400–800 ft O.D. Pre-glacial?	Plateau implements (Eolithic)
	Higher or hill-gravels of Ightham, 400–500 ft O.D. Glacial? High-level valley-gravels, up to *c.* 100-ft above current river level. Post-glacial	The Hill group has fewer large, pointed forms; more small, oval forms; and is ruder than the high-level river-drift forms. These are classed together.
Low/young	Low-level valley-gravels, sloping down to current river level. Post-glacial	Palaeolithic implements of a more recent period

glacial or pre-glacial age of the Palaeolithic. Three years earlier Evans had questioned his discoveries in the Welsh caves; Hicks now rose to defend the Kent finds and his own beliefs from similar criticism.[32] Whitaker remarked on the difficulties of dating and correlating different drifts and queried whether Harrison's palaeoliths, which had been found on the *surface*, were truly associated with the supposedly ancient drifts. He also objected to the term 'pre-glacial', which had little precision at that time (see Chapter 2). Finally, Harrison described how he was:

unexpectedly called upon to take part in the discussion, or at least, to say a few words on the subject.

Possibly had I heard the subject matter of the Paper I might have briefly touched upon certain points, but my deafness precluded me from catching one-tenth part of what was said, and whether the verdict was for, or against, or not proven, was a matter of uncertainty to me.

I, therefore, under the circumstances, begged to be tenderly treated, and merely invited the Fellows of the Society to pay a visit to my Palaeolithic Paradise, and examine for themselves, when I would gladly act as pilot.[33]

Harrison would receive many visitors to his shop and his Palaeolithic paradise over the following years.

Prestwich bolstered his arguments in a second paper read to the Geological Society early in 1891: 'On the Age, Formation, and Successive Drift-Stages of the

---

[32] Hicks 1886b, 19; Prestwich 1889, 294–296.

[33] Harrison (undated), printed sheet addressed to 'Professor Seeley, and Gentlemen': AMO: JE/B/2/42.

Valley of the Darent; with Remarks on the Palaeolithic Implements of the District, and on the Origin of its Chalk Escarpment.' The Darent Valley was well known to Prestwich, who saw it every day from his house, and he used his valley in this paper as a case-study to clarify the position and age of different spreads of tool-bearing drift in the area. He concluded from the geology of the Darent Valley that the Chalk escarpment and its drifts had a glacial origin, and that the intervening valleys had been excavated later by glacial and river action. Since Prestwich used the term 'glacial' loosely, this argument was not inconsistent with his suggestion of a '*pre*-glacial' age for the Plateau implements.[34]

Alongside his lengthy discourse on the geological position of river-drift palaeoliths, Prestwich described the Ash types from the Plateau. Several features of the Ash group of flints seemed to indicate that they were older than the river-drift group. They were exceedingly crude in character; Prestwich declared: 'they point to the very infancy of the art.' Natural fragments of flint, rudely worked at the edges, were plentiful and some were difficult to distinguish from natural forms. Then there was the absence of hand-axes, familiar from the river valleys: no 'highly-finished spear-head forms' had been found, although, strangely, a few well-made oval types had been identified. Their physical condition also suggested a great age: the flints from the Ash group seemed worn down by time. Prestwich thought that some scratches on their surfaces resembled the lines (striae) carved by glaciers, which would mean that they had passed through the height of the Glacial epoch—but he did not press this interpretation.[35]

The audience was more taken aback by Prestwich's case for the geological origin of the Plateau drifts than by his suggestions of a great age for their implements. Prestwich had departed from customary opinion in assigning the denudation of the Weald to neither marine nor subaerial action, and his glacial alternative was not well received at the Geological Society. Archibald Geikie, who was then President, observed that although glacial *conditions* must have affected the southern counties of England, no *ice-sheet* had penetrated further south than the Thames: for no true glacial drifts existed this far south. (Whitaker had made a similar point after Prestwich's first paper.) Prestwich defended a localized, southern ice-area in the Wealden hills. Other critics questioned his association between the Plateau specimens and the Plateau drift; Prestwich reminded them that a lack of pits and excavations in the Plateau drift had restricted much of the search to the surface.[36]

Prestwich delivered his most detailed description of the Plateau implements to the Anthropological Institute in 1891 in a third paper: 'On the Primitive

---

[34] Prestwich 1891, 128, footnote, 156–159.   [35] Prestwich 1891, 134, 135.
[36] Prestwich 1889, 296; Prestwich 1891, 161–163; Anon 1891, 37.

Characters of the Flint Implements of the Chalk Plateau of Kent, with reference to the Question of their Glacial or Pre-Glacial Age.' By this time, Harrison had over a thousand Plateau implements in his collection.[37] The object of this paper was 'to enquire whether the character of the implements is in accordance with the early glacial or pre-glacial age, to which I would assign them'; Prestwich was satisfied that the geological question of the age of the plateau drifts had been covered fully by his previous two papers.[38]

This account of the Plateau implements elaborated on the indications of age that Prestwich had already given to the Geological Society. Their deep-warm-brown staining suggested that they had been buried within the similarly ochre-ous plateau drifts for a considerable time. Their worn appearance seemed to reflect the tumble of a journey that Prestwich thought they had taken within the drift, as it travelled from the now vanished heights of the Wealden uplands to the Plateau. His Plateau group was also distinct from the river-drift palaeoliths in type: they were cruder, and different in form. The differences between the two groups are summarized in Table 5.2. Prestwich explained away the few cases of overlap: river-drift types among the Plateau group might have been dropped later by River-Drift men as they roamed over the Plateau drifts; crude implements among the river-drift palaeoliths in the valleys below were, Prestwich declared, quite unlike the Plateau specimens.[39]

The relationship between the rudeness of stone tools and their age was not quite as straightforward as it might seem. The skills required to make a finely finished river-drift hand-axe would, in the opinion of many researchers

Table 5.2. A comparison between the character of flints from the Plateau and those from the river-drifts. The crude, worn Plateau group appeared to be much older than the River-Drift implements (after Prestwich 1892, 254–255).

Plateau group	River-drift implements
Made from naturally fragmented flint, the edges roughly trimmed into shape	Made from larger flints
Very worn	Usually not very worn
Only slight trimming	Skilled workmanship
Some implements of definite patterns, but many show no special design and are primitive in form	Rude specimens belong to the same types as more finished specimens
Only a few well-made specimens	Pointed forms predominate, next come the sharp-rimmed oval forms
Almost all are stained a deep brown	Variety of colours; some are stained

[37] Prestwich 1892, 247, footnote 1; Harrison 1928, 160.
[38] Prestwich 1892, 249.   [39] Prestwich 1892, 252, 253, 257.

besides Prestwich, have taken some time to develop; the earliest artefacts made by humans were expected to be much cruder. But the skilful tool-makers of later times could, and did, produce crude tools *as well.* Evans cautioned Harrison that rudeness of character was no sure test of relative age, because rude implements had been found associated with well-finished forms at Amiens and elsewhere.[40] Prestwich had warned Dawkins, over a decade before, that older implements were not necessarily cruder than later ones. He pointed out that Dawkins's rude group from Creswell Crags was made from quartzite: however skilled the tool-maker, quartzite could not be worked as finely as flint.[41]

For Prestwich, though, the Plateau specimens were not just cruder versions of the river-drift palaeoliths; they were a distinct series with marked differences in character and type that separated them from the river-drift tools. He divided these specimens into three groups: a first group of forms little modified; a second group with identifiable tool-types (mostly scrapers); and a small third group of forms common in the river-drift implements. Harrison classified them according to possible use: his Plateau types included the 'single curve scraper' and the 'drawshaves or hollow scrapers'.[42] Evans was the first to respond to Prestwich's paper at the Anthropological Institute. He regretted that he saw no evidence of human agency in any of these Plateau types, although he welcomed Harrison's Palaeolithic discoveries.[43]

5.4 THE RESPONSE TO THE KENT PLATEAU SPECIMENS

Harrison described how Evans 'closed his observations with the following sentence, "Before we accept these" [the rude implements]—looking at Prestwich—"we must think twice,"—looking at me—"we must think thrice, and"—looking round the whole meeting—"we must think again".[44] Other notable figures had also come to hear the paper. They rose to add their opinions and advice. Prestwich knew well what to expect by now. Dawkins, who thought the Plateau series no earlier than the river-drift implements, reminded the audience that the same people could have used ruder forms alongside highly finished implements. Prestwich agreed that crude

---

[40] Evans to Harrison, 24 March 1891, quoted in Harrison 1928, 161.
[41] Evans 1878, 177–178.
[42] Harrison 1892, 164–165, 265–266; Prestwich 1892, 259–260.
[43] Prestwich 1892, 270–271.
[44] Harrison to Newton, 3 June 1908, quoted in Harrison 1928, 165.

forms could indeed be found amongst more recent implements, but repeated that the Plateau forms were different in character and type to the river-drift forms and were confined to a distinct area.

Pitt-Rivers, more supportive, had long expected that implements older and cruder than those from the river drifts would eventually be discovered. He suggested that when these Plateau types, hitherto found by Harrison on the Plateau surface, had been found *within* the Plateau drifts, this would add greater certainty to their case. Brown echoed the need to find the flints *in situ* beneath the current land-surface. He welcomed most of the Plateau speci-mens as genuine artefacts, accepted Prestwich's geological conclusions about the origin and date of the Plateau drift, and wondered if his own derived series of worn, ochreous implements from the Thames Valley might have come from a Plateau drift too.[45]

The case made by Prestwich in these three papers and the discussions that followed strike a familiar chord. They recall the tone and the lines of argument that he and other defenders of human antiquity had used in 1859.[46] The case for and against a *greater* human antiquity addressed the same three aspects: the geological age of the (supposedly) eolith-bearing deposits; the geological context of the eoliths; and their human or natural origin. The geological age of the Plateau deposits had been confused by the notorious ambiguity and variety of drifts and Prestwich's glacial interpretation was not popular.[47] Even if the drift was pre-glacial in age, this term was imprecise. The context of the eoliths was equally uncertain: most had been found on the surface and might therefore have been dropped by the River-Drift men. Harrison and Prestwich needed to discover such specimens within the Plateau drift to satisfy some of their critics. Finally, their origin was disputed: even if eoliths were found within the drifts, those who thought the eoliths had been formed by natural processes would not be satisfied. In the aftermath of Prestwich's three papers, supporters and critics of the eoliths would address these three aspects—age, context, and origin—in different ways, according to their differing interests and expectations.

James Geikie, who had supported controversial Palaeolithic finds before, wrote to Harrison in 1892: 'Yes, Palaeolithic man is *old*,' and he included a cautious footnote reference to the Kent eoliths in the third edition of *The Great Ice Age* (1894).[48] Harrison was pleased to find another tentative

---

[45] Prestwich 1892, 272–276.

[46] Prestwich 1860b, 58; Lubbock 1862, 247; Newton 1898, 64.

[47] Prestwich 1889, 296.

[48] J. Geikie to Harrison, 14 March 1892, quoted in Harrison 1928, 175; Geikie 1894, 640, footnote 1. When he came to write his *Antiquity of Man*, James Geikie was more sceptical about the eoliths, but his suspicions were mingled with excitement about the recently discovered East Anglian eoliths and he gave them an open verdict (Geikie 1914, 5–6, 307).

mention of his eoliths in Ramsay's sixth edition of *The Physical Geology and Geography of Great Britain* (1894), added as further proof that humans lived in Britain before the Glacial epoch.[49] Those who were more doubtful about a pre-glacial or interglacial date for the Palaeolithic tended to be more critical of the age claimed by Prestwich for Harrison's discoveries in the hills and on the Plateau of Kent.

Prestwich had been advised many times to ascertain the geological context of the Plateau specimens. In October 1894, an excavation was conducted in the Plateau drift at Parsonage Farm, Stanstead, under the direction of Benjamin Harrison. The British Association funded the project; the excavation committee included Evans as well as Prestwich and Harrison. Although Harrison found worked flints in this pit, Evans remained unconvinced. In later years Harrison sank more pits and used his own meagre means to pay for the work.[50] He even had a small leaflet printed describing his 1902 excavations, *Eolithic Flint Implements from the Plateau of North Kent*, which announced: 'In these pits no Palaeolithic implements were found, only Eolithic, thus suggesting that the latter are of greater age than the former, and not contemporary with them'.[51]

Smith, who was more interested in palaeoliths than eoliths, did not find it difficult to list instances of Palaeolithic implements in sediments that had been claimed as purely Eolithic. But Harrison and Prestwich, who promoted the eoliths as a group of tools with their own coherent identity, emphasised their distinction from Palaeolithic tools in age, association, and character. Prestwich had suggested that associated river-drift implements might have been dropped by a later race as they wandered across the Plateau. Harrison seems to have adopted a different solution.[52] According to an archaeologist working half a century later:

undoubted hand-axes have been found on the highest points of the Kentish downs, in the deposits. I once had a fine small one but it was lost in the blitz of 1940. I have been told, I think by his son, that Benjamin Harrison found many such from the plateau gravels but concealed them because he thought they would detract from the interest of the eoliths from the same deposits.[53]

The attempts by sceptics to draw Harrison away from eoliths and back to palaeoliths were unsuccessful. Upon receipt of a box of plateau specimens in 1892, Evans wrote to Harrison:

---

[49] Ramsay 1894, 291–292; Harrison to Pitt-Rivers, 29 February 1896: SSW: P-R, L1518.
[50] Harrison 1895; Harrison 1928, 194–195, 209.
[51] Harrison, undated leaflet on 'Eolithic Flint Implements from the Plateau of North Kent': AMO: JE/B/2/42.
[52] Prestwich 1892, 257; Smith 1908, 53; Harrison 1928, 193.
[53] Crawford to Oakley, 3 October 1957: NHM: DF140/6.

So far as results are concerned, everyone will accept the ordinary forms of palaeolithic implements as having been found at the high levels, and I am doubtful as to the desirability of complicating the question with a second race of men and a set of implements of extremely questionable character. I admire your perseverance, and am only sorry that I cannot go further in accepting your evidence.[54]

Worthington Smith made the same point with more humour:

It puzzles me why you esteem your disputed, kicked-about deformities so much, when you have genuine implements and flakes from the same positions. They hob-nob together here, and, no doubt, there are as many hump-backed and bandy-legged works of the Evil One here as with you.[55]

Supporters of the eoliths often compared their trials to those faced by Boucher de Perthes when his Palaeolithic implements had been attacked by sceptics in the mid nineteenth century. Even in the 1860s, a few British researchers had regarded finely finished hand-axes, described by Evans from the river drifts, as the products of natural forces.[56] In 1896, Harrison referred to his eoliths as 'Palaeolithic *Outlaws* in not having been accepted in full by our chief authority on stone weapons, but looked upon as the result of natural agency rather than the work of man'.[57] By 'our chief authority,' Harrison meant Evans. There seemed little likelihood of conciliation between those who regarded the eoliths as the work of tool-makers and those who thought geological processes had been at work on these stones. The arguments of each side were constructed upon such different premises that there was little engagement between their respective views.

Opponents of the Kent eoliths scrutinized Harrison's specimens for features they thought diagnostic of human workmanship, and found them wanting. The techniques used by ancient tool-workers had been studied since the 1860s: to fight the cunning of Victorian forgers as well as to feed an interest in more ancient tool-makers.[58] In his second paper of the early 1860s on the river-drift tools, Evans had provided the Society of Antiquaries with a simple guide to distinguish artificial flakes from those formed by natural fracture. He described a bulge that could be seen on the underside of a flake struck by a human blow, lying at the point of impact where the flake had been hit. This 'bulb of percussion' was the epicentre from which ripples of fracture had spread through the flint, ripples that had detached this flake.[59]

---

[54] Evans to Harrison, 29 October 1892, quoted in Harrison 1928, 184–185.
[55] Smith to Harrison, 13 October 1895, quoted in Harrison 1928, 201.
[56] Whitley 1865, 33–39.
[57] Harrison, extract from the Ightham Parish Magazine, 1896, enclosed in Evans's Palaeolithic Scrapbook: AMO: JE/D/1/3, p.16.
[58] Murchison 1868 (ii), 605–606; Evans 1872, 575–577.   [59] Evans 1863, 76.

Brown described the bulb of percussion as shell-like in form; they had even been mistaken for small fossil shells. But they were seen by Palaeolithic researchers as a signature of the springy blow of a human hand: reassurance that natural processes had not created their discoveries. Frost-fracture, pressure deep in the ground, casual collisions of pebbles in a river, and other geological forces rarely left such a signature. Evans and Smith argued that the Kent specimens did not look like artefacts—they had neither bulbs of percussion, nor the sharp cutting edges characteristic of tools. Prestwich and Harrison, though, expected their specimens to be crude and bear features that differed from Palaeolithic tools.[60]

As sceptics tried to deconstruct the eoliths and push them back among other natural stones, supporters were trying to reconstruct their roles in the lives of Eolithic tool-makers and position them within a grander scheme of Stone-Age tools. Harrison thought the eoliths lay at the early end of a great evolving sequence of industries. His ideas had been stimulated in part by the theories of Pitt-Rivers on industrial evolution. Edward Burnett Tylor (1832–1917), anthropologist and old collecting comrade of Henry Christy, received from Harrison 'some sketches of Eoliths ranging on to Palaeoliths as illustrated by my finds on the Plateau, led to do so by seeing Gen. Pitt Rivers Evolution of Cultures'.[61] Harrison had also been encouraged by the remarks made by visitors to his museum who had referred to Pitt-Rivers's evolutionary sequence of implements and then declared: 'But you [Harrison] have a much stronger case—You have the Alpha of Man, he starts from Delta.'[62]

In his efforts to divorce Harrison from his attachment to the eoliths, Evans had asked him in exasperation: 'Has the absolute uselessness of such flints as tools never struck you, nor the fact that if the edge of a flint is chipped by hand it may just as well be made to present an acute as a right angle'.[63] To try and find a function for his eoliths, with their strangely blunt edges, Harrison examined ethnographic reports of tribes who displayed a wider and more exotic range of habits than his circumspect Victorian peers. He suggested that some forms might be 'body stones' like those used in the East Indies and Australasia to 'rub the hard skin of the feet, remove scarf skin &c.' (see Figs. 5.2 and 5.3 below).[64] Smith's caricature of a 'Combination Shin-scraper'

[60] Brown 1887a, 81; Harrison 1928, 176.
[61] Harrison to Tylor, 14 November 1907: PRM: EBT, Harrison 11.
[62] Harrison to Evans (undated): AMO: JE/B/2/42.
[63] Evans to Harrison, 13 November 1894, quoted in Harrison 1928, 195. Evans directed this remark at the finds from the Parsonage Farm excavations.
[64] Harrison to Wallace, 10 January 1898: PRM: Misc. Ms. 11; Harrison 1904, 18.

Figure 5.2. A satirical drawing by Worthington Smith of a 'Combination Shin-scraper' eolith from Caddington, March 1906, which he sent to Harrison. The inscription reads: 'AM I NOT AN EOLITH AND A BRICKBAT.' (My thanks to Derek Roe for supplying the negative, which is reproduced by permission of the Institute of Archaeology, University of Oxford.)

(Fig. 5.2) illustrates his friendly relationship with Harrison and his delight in drawing as well as his opposition to Harrison's views.[65]

Ethnographies supplied a fertile source of suggestions about the functions of tools—Palaeolithic as well as Eolithic—but the handiwork of distant tribes was also compared to different stages of progression in the Stone-Age sequence, just as their bones had helped anatomists to pronounce on the physical characters of fossil tool-makers. The River-Drift period, for example, was often connected to the technological stage associated by the Victorians with the Tasmanian and Australian aborigines. The Cave tools were connected to those of the Eskimos who, like the peoples of the Reindeer period, led an Arctic life and made fine bone-tools.[66] Pitt-Rivers was not alone in thinking that 'the existing races, in their respective stages of progression, may be taken as the *bona fide* representatives of the races of antiquity'.[67] Note that when Dawkins linked his Cave race to the Eskimos, he meant this as a blood

[65] Derek Roe 1981, 215–219 discusses humorous sketches of the eoliths and the relationship between Smith and Harrison.

[66] Dawkins 1866b, 341–342; 1880, 233, 244–245; Lane Fox 1867, 616; 1875, 301–302; Tylor 1865, 195; 1869, 25; 1893, 147; 1898; Brown 1887a, 34, 75.

[67] Lane Fox 1867, 618.

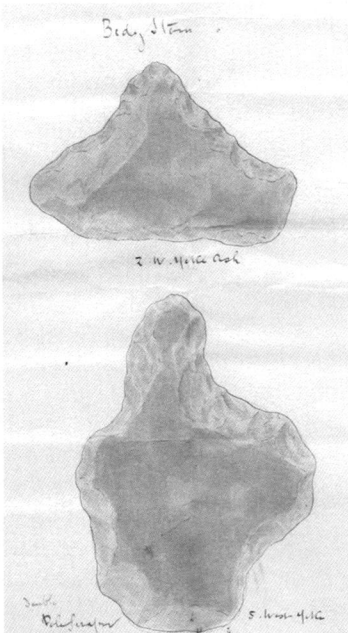

Figure 5.3. Watercolour illustration of two eoliths by Harrison, which he sent to A. R. Wallace on 10 January 1898 (PRM: Misc. Ms. 11). No. 2 (top) is identified as a 'Body Stone' and No. 5 (bottom) as a 'Double Pole Scraper'.

relationship—not as a mere analogy.[68] The Eolithic culture stage tended to be associated with the crude tools made by the recently extinct Tasmanians. Harrison remarked to Tylor: 'I can plainly see my rude specimens will just fit with your Tasmanians & feel sure there is plenty of material to work upon'.[69] For Evans and Smith, who did not even accept the premise that these *were* artefacts, such speculation was irrelevant.

These arguments all circled around the same stones from Kent but the ammunition of each side was not fired directly at the opposition. Supporters of an Eolithic period were expecting crude tools and found them; their opponents took this rudimentary appearance as evidence for their natural origin. Prestwich and Harrison believed in the flints and started to arrange and interpret them; their critics disregarded any suggestion of Eolithic lifestyles. The battle dragged on into the twentieth century, when similar differences in expectations would colour arguments over flints from East Anglia. But here, at the end of this examination of the first of the great British eolith debates, the

---

[68] Dawkins 1866a, 713; 1874, 358–359; Brown 1887a, 140. On the use of 'primitive' peoples as relics of stages in human evolution, see Stocking 1987, 144–185, 282–283, Bowler 1992 and Sommer 2005.

[69] Harrison to Tylor, 26 April 1898: PRM: EBT, Harrison 3. On the Antipodean connections of the eoliths, see also Prestwich 1892, 258 and MacCurdy 1910, 534.

last word is given to Harrison, who summarized the arguments over the Kent eoliths in a poem he sent to an eolith-supporting friend. Harrison's humour comes across well in his verse, but he was not unusual in choosing to present his ideas in this manner. Light-hearted occasional verses allowed their authors to share private opinions with friends, to tell stories, and to explore ideas whilst avoiding the commitment of a more public scrutiny.[70]

Eolith, Palaeolith—Nature or Man. That Little Chocolate Flint

> How often we hear of the wrangles they have
> over one little ochreous stone,
> How they say they can tell all its history long
> from its chocolate staining alone,
> How some gravely proclaim it was made by a man
> or at least by an anthropoid ape,
> While others maintain that in glacial moraine
> it was licked by the ice into shape.
>
> Some assert it was born on the Wealden Heights,
> it was chipped, it was fashioned and used;
> Attached to a handle, in hundreds of fights
> it hammered, it battered, it bruised,
> Till its owner grew tired of the Weald and removed
> on some very remote quarter-day,
> And having no room in the furniture van
> for his weapon, he chucked it away.
>
> It rolled down the slope in a tertiary drift
> from primitive Sussex to Kent,
> Its progress was slow, for a million or so
> were the years on the way that it spent.
> But it got there at last and its troubles were past,
> (like the days on the Wealden Heights)
> Though it bid very fair to come in for a share
> of a second long series of fights.
>
> For others declare that its story is false
> that it never came northward to Kent;
> That the place where t'was found as it lay in the ground
> was the place where its life had been spent;
> That it hadn't rolled down in a tertiary drift
> ere the glacial period or since
> But had stayed all the time on the top of the chalk
> in a layer of red clay and flints.

---

[70] Rudwick 1992, 39–40, 56–57; Sommer 2004a, 60, 72–74; R. O'Connor 2007 forthcoming: *The Earth on Show*, Chapter 2.

If only that Chocolate Stone could explain
what the dickens it did in the past
That Sages might cease from exciting their brains
and the hatchet be buried at last
Whether EOLITH, PALAEOLITH—NATURE OR MAN
could they but of that question dispose
Those eminent men might relinquish the pen
till a new controversy arose.[71]

5.5 THE PRE-PALAEOLITHS OF EAST ANGLIA

A new controversy arose before Harrison's 'Chocolate Stone' spoke any soothing words. On 3 October 1909, James Reid Moir (1879–1944) of Ipswich found some crude flints in a stone heap near his house, at the brickfield of Messrs Bolton and Laughlin. They came from the eroded base bed of the Red Crag: a deposit that was not only older than the Chalky Boulder Clay, but also pre-dated Wood's Middle Glacial series (described in Chapter 2). This late Pliocene stratum was even more ancient than the pre-glacial Cromer Forest Bed (see Table 5.3). Moir was convinced that he had discovered evidence of very early human activity, and suggested that the scratches on his flints had been carved during an early Pliocene glaciation, long before the ice-sheets of the period known as the Glacial epoch arrived.[72] He proclaimed his beliefs forcefully. The second round of the British eolith debates had begun.

Moir had become enchanted by prehistoric flints after the discovery of a Neolithic arrow-head during a round of golf. Lewis Moir, his father, expected his only surviving son to help run the family's successful outfitting and tailoring business; Moir braved the displeasure of his formidable, six-foot father and found scraps of time to devote to his new interest. He had many fellow enthusiasts around Ipswich. Dr W. Allen Sturge (1850–1919) had lived nearby since 1907 after retiring from his former life as physician to Queen Victoria and other members of the Royal Family on the Riviera. Sturge had considered carefully his retirement options before selecting an East Anglia retreat to indulge his passion for flint implements. He became the first President of the Prehistoric Society of East Anglia (hereafter the PSEA) when it was established in 1908, and allowed Moir to roam through his vast

[71] Harrison to Lewis Abbott, 18 May 1898: BM(F): Misc. Doc Files, Eoliths. See also Harrison 1928, 218–219.
[72] Moir 1911, 17–18, 22.

collection of flints.[73] Smith was another member of the PSEA and Spurrell was Vice-President for a short while before ill-health prompted his resignation.

Moir had been examining exposures of deposits that he thought likely to yield implements when he made his first discovery. If his father had read *The Times* on 17 October 1910 and turned to the letters page, he would have been annoyed to read of more stones found by Moir in remarkably old deposits beneath the Red Crag. Moir reported that they seemed to predate the last glaciation, but exhibited 'flint chipping far in advance of any Eolithic work'.[74] Back in July, Moir had shown members of the Geologists' Association over the section where he had made his original discoveries.[75] His flints were becoming widely known.

Moir's finds acquired a powerful local support-group. After he read his first paper on the sub-Crag flints to the PSEA in 1910, a Special Committee was appointed to decide whether they displayed human or natural work-manship. The committee members gathered together a few days before Christmas at Sturge's house. He was one of the members; the others were also prominent in the PSEA: Colonel Underwood, an ex-Army man; W. G. Clarke, a reporter from Norwich; Nina Layard, who had conducted Palaeolithic excavations in Ipswich a few years before; and Dr Frank Corner, who had a large collection of prehistoric artefacts. They examined the flints together and accepted a large number, mainly on the basis of their elaborate chipping and the presence of what they thought were bulbs of percussion.[76]

Colonel Underwood had also taken an interest in Benjamin Harrison's discoveries. His son and biographer, Edward Harrison, recalled the following anecdote. After a long walk together over the hills, Underwood and Harrison arrived at Shoreham village in search of refreshment, but the innkeeper's wife was too busy to attend to them:

'Damn!' ejaculated Colonel Underwood with vigour, as the disappointed travellers turned to leave the inn.
They had proceeded only a few yards when a little girl overtook them.
'Please sir', she asked, 'are you Colonel Underwood?'
'Yes I am', was the answer, 'how did you know me?'
'Oh, please sir, my father was in the bar when you said "damn", and he was once in your regiment, and he said, "That's Colonel Underwood's damn", and so mother said if you will come back she will be very pleased to get you some tea'.
Harrison said afterwards that it was a very good tea, too, that was obtained for them by 'Colonel Underwood's damn'.[77]

---

[73] Clarke 1919, 12; Keith 1944, 733–737; Boswell 1945, 66.   [74] Moir 1910, 8.
[75] Slater 1911.   [76] Underwood *et al.* 1911, 37.   [77] Harrison 1928, 215.

Figure 5.4. E. Ray Lankester
(1847–1929), Moir's supporter (National
Portrait Gallery, London X19879).

Ray Lankester (Fig. 5.4), the zoologist and palaeontologist, had also been informed by Moir of his find.[78] Now in his sixties, Lankester had recently retired as Director of the Natural History Department of the British Museum. Moir suspected that he would be interested in the flints: Lankester had followed the arguments over Harrison's eoliths and had defended them at the 1906 meeting of the British Association in his President's Address. He had also been an authority on the Suffolk Bone-Bed in which Moir had made his pre-Crag discoveries. Lankester agreed from the appearance and geological context of Moir's flints that they were much older than the river-drift palaeoliths, which were still widely regarded as post-glacial in age.[79] Moir had made a sound decision; Lankester would devote considerable energy to the defence of these finds over the next two decades.

The sub-Crag flints, or 'pre-palaeoliths' as they were soon known, drew attention away from the Kent eoliths. They gained sounder geological credentials than their Kent cousins when Moir found more flints at the base of undisturbed Red Crag deposits. The geologists William Whitaker and John Marr (1857–1933) established the sub-Crag age of the site beyond a doubt in

---

[78] Lankester to Moir, 15 May 1910: BLL: Add.Ms.44968/1–2. On Lankester, see Bowler 2004.

[79] Lankester 1906, 22; Goodrich 1930, xi; Moir 1930, 140; 1935, 22, 26. On Lankester's contributions to British biological research, see Lester 1995.

Table 5.3. A summary of the geological deposits mentioned in the text, with associated Palaeolithic, Pre-Palaeolithic, and Eolithic finds (after Lankester 1913; Moir 1913a, 317; 1935, 26).

Geology	Palaeolithic, Pre-Palaeolithic and Eolithic finds
River drifts (post-glacial)	Palaeolithic: Chellean; Acheulian; Mousterian
Chalky Boulder Clay	Skertchly's Palaeolithic finds of the 1870s
Middle Glacial Sands and Gravels	
Cromer Forest Bed (pre-glacial)	Moir's 'implements,' discovered in 1918: considered to be transitional between the sub-Crag rostro-carinates and the Chellean hand-axes
Red and Norwich Crags	The Norwich test specimen, discovered in 1911: Lankester's favourite example of the rostro-carinate type of sub-Crag 'implement'
Suffolk Bone-Bed	Moir's sub-Crag Pre-Palaeolithic, discovered in 1909
Kent Plateau drifts	Harrison's Eolithic

April 1911.[80] Marr was the long-serving Lecturer in Geology at the University of Cambridge. He had been encouraged in his geological interests while at school in Lancashire by Richard Tiddeman, the Survey Geologist who had claimed an interglacial or pre-glacial age for the Palaeolithic back in the 1870s (see Chapter 2), and who was then working in the north of England.[81] Marr was also an experienced field geologist (he named his trusty geological hammer 'Faith', observing that it could remove mountains).[82] With a proven geological age and context, much of the discussion over the East Anglian pre-palaeoliths centred on the question of whether these were naturally fractured flints or true artefacts that could be assigned to their own section of the Stone-Age sequence.

5.6 FLINT-FRACTURE RESEARCH AND FURTHER ARGUMENTS

The history of flint-fracture research has been studied by Donald Grayson in his paper: 'Eoliths, Archaeological Ambiguity, and the Generation of "Middle-Range" Research' (1986). Grayson describes the race for ever-more-convincing proof of the origin of eoliths and examines how the task of distinguishing human from natural work changed through the course of the eolith debates.[83]

---

[80] Whitaker and Marr 1911.   [81] Woodward 1916, 289–290; Oldroyd 2004b.
[82] Boswell, autobiography, 1942–48, p. 41: ULL: D4/1.   [83] Grayson 1986.

Figure 5.5. S. Hazzledine Warren (1872–1958),Moir's opponent. Photographed in 1907 (GSL: Portraits of Fellows of the Society, Vol. 5, p. 35).

The reader is referred to Grayson's paper for a comprehensive account of these arguments. Flint-fracture research is studied here to introduce the early-twentieth-century critics and their reasoning: the researchers whom Moir and Lankester would regard as the opposition in the second round of the eolith debates.

Evans, who had died the year before Moir discovered his first pre-palaeolith, once remarked that questions about the origin of the eoliths had changed over time. Initially, it was the disputed specimens brought forward by their defenders that were scrutinized. The attributes of *human* flaking had attracted most attention in the early decades of the eolith debates. But long before the end of the nineteenth century the critics became uneasy. They began to fear that natural forces might be capable of producing the features they had once taken as proof of human workmanship, such as bulbs of percussion. The study of *natural* processes of flint fracture was then added to the task of deciding whether chipped stones had a human or a natural origin.[84]

A few years before Moir made his first Pre-Palaeolithic finds in East Anglia, Samuel Hazzledine Warren (1872–1958), a geologist from Loughton, Essex (Fig. 5.5), was exhorting researchers to study the principles of natural flint-fracture. The only son of Stephen Warren and Hannah Mary Hazzledine, Samuel Hazzledine Warren rarely used his first name. His friends knew him as

[84] Evans 1878, 150; de Mortillet 1883, 80–83; Abbott 1897, 91–95; Grayson 1986, 87, 92.

'Hazzie' or 'Haz'. In 1903, Warren had left the family business of wholesale provision merchants and moved to Essex where he remained for the rest of his life, devoting his considerable leisure time to Quaternary geology and Palaeolithic research.[85] He believed that arguments over the eoliths hinged on their origin, and explained: 'It is not the geological position of the eoliths that is primarily in dispute, but whether they are, or are not, of human fabrication. Local field evidence cannot help us here; it is a knowledge of the fracture of flint under different conditions that we require'.[86]

Three papers on flint-fracture research were often cited in the eolith debates of the early twentieth century. Warren described his experiments in 1905; Boule presented an account of natural fracture in torrents, also in 1905; and the Abbé Henri Prosper Édouard Breuil (1877–1961) reported field observations of natural fracture under pressure in 1910.[87] Breuil had been introduced to the Palaeolithic of the Somme Valley in the late nineteenth century by a family friend, the geologist Geoffroy d'Ault du Mesnil (1842–1920), and he also learnt much from Victor Commont (1866–1918), the respected Palaeolithic researcher.[88] He spent most of the first two decades of the twentieth century working on later prehistory: the Bronze Age of the Somme Valley and Paris basin, and the reindeer-caves of France and Spain. Although he had been ordained as a priest in 1900, Breuil was better known as an authority on cave art.[89] Boule's early interest in geology had, ironically, been inspired by Ramès, the discoverer and defender of the Puy-Courny eoliths.[90] All three researchers would contribute their knowledge of flint fracture to Moir's East Anglian finds. When they gave these papers, though, they were more concerned with a series of eoliths from Belgium, and Warren was still worried about the Kent eoliths.

The story of the Belgian eoliths and the rise and fall of their discoverer, Aimé Rutot (1847–1933), has been told by Raf De Bont (2003). Rutot was an engineer and a geologist who had become a curator at the Royal Belgian Museum of Natural History in 1880.[91] He successfully publicized the Belgian eoliths in the late nineteenth century, and was described by Dawkins as the 'chief exponent of the Eolithic cult, as it may be called, on the continent'.[92] Rutot's offerings to the

---

[85] Oakley 1959, 144–146.
[86] Warren 1905, 337.   [87] Boule 1905, Warren 1905, Breuil 1910.
[88] Breuil 1921a, 162; 1937a, 61; 1948, 65; 1949, 12; Brodrick 1963, 26–30, 131.
[89] Breuil 1937a, 54; Lantier 1961, 651; Vaufrey 1962, 158–161. On Breuil, see Brodrick 1963, Cohen 1999, and Coye (ed) 2006.
[90] Lantier 1945, 132.   [91] Stockmans 1966, 3–5.
[92] Dawkins 1910, 237. De Bont 2003 examines the strategies employed by Rutot to gain credibility for his finds. He links their rise in prominence to Rutot's central position within an academic trade network of Eolithic flints.

Eolithic cult had been unearthed from a series of deposits that seemed to date to early Quaternary times. Each supposed industry had its own name. The most recent Eolithic industry of Belgium was the 'Mesvinian,' which was regarded by Rutot as a predecessor of the Palaeolithic Chellean. The Mesvinian was preceded by the 'Reutelo-Mesvinian' (later the 'Mafflian'); the simpler specimens of the 'Reutelian' fell further back in time. Later, Rutot would place Pliocene precursors at the beginning of this sequence.[93] With the exception of the Mesvinian (which will appear again in Chapters 7 and 8) these industries were regarded with suspicion in England and France.[94]

Boule's 1905 paper on eoliths was aimed at Rutot. He reported that objects similar to Rutot's eoliths had been produced by a cement works, at Guerville near Mantes, where flint nodules from the chalk had been churned around in the water, much as torrents might have churned flints in the past. Disconcertingly good likenesses of eoliths were produced in the process.[95] Boule published this article in *L'Anthropologie*: he had taken a prominent role in the production of this journal from its earliest years.

L'Anthropologie, founded in 1890, was an amalgamation of three older journals: the *Revue d'anthropologie*, the *Revue d'ethnographie*, and de Mortillet's old *Matériaux pour l'histoire primitive et naturelle de l'homme*. The new journal appeared every two months at an annual subscription of between twenty-five and twenty-eight francs, depending on the location of the subscriber. *L'Anthropologie* gained a sound reputation for its mixture of prehistoric, geological, palaeontological, and anthropological articles. Sturge described it as a 'great serial', 'unique in its kind in the world and quite unapproached by anything published elsewhere'.[96] Boule's opinions were powerful. They encouraged suspicious looks to be directed at Rutot's supposed industries from Belgium.[97] In England, Harrison was provoked to visit his own local brickyards where natural flints were ground up with the loam. A few had been flaked and he observed: 'one or two even show bulbs', but concluded that they were different to humanly struck flakes.[98] Warren, however, felt encouraged by Boule's example to publish his own flint-fracture researches.[99]

Warren's cautious, informative articles began to appear in the last years of the nineteenth century. His interests were not confined to the eoliths. Some of his early discoveries of palaeoliths from the London area appeared in the second edition of Evans's *Ancient Stone Implements* and Warren would

---

[93] Rutot 1898; 1900; 1903. On Rutot's series of Eolithic industries, see also Grayson (1986, 94) and De Bont (2003, 608).
[94] Sollas 1911, 111.   [95] Boule 1905, 262.
[96] Sturge 1908, 11.   [97] Lantier 1945, 133; De Bont 2003, 610, 614.
[98] Harrison to Wallace (undated): PRM: Misc. Ms. 11.   [99] Grayson 1986, 104.

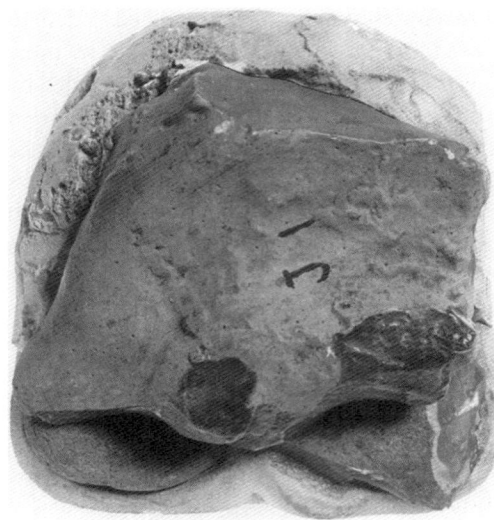

Figure 5.6. This cast of a Kent-type eolith was made by Warren to demonstrate how it was formed by sub-soil movement: 'Double Notch mounted to show how it was found' (PRM: Accessions Book No. X, p. 454; PRM: 1940.12.161.1).

play a quiet role in many of the major Palaeolithic debates of the twentieth century.[100] In 1905, however, Warren addressed the Anthropological Institute 'On the Origin of "Eolithic" Flints by Natural Causes, Especially by the Foundering of Drifts'. The same society had heard Prestwich's paper on the Kent eoliths in 1891.

Warren offered a persuasive account of how eoliths could be created by natural processes and presented the results of his experiments on the fracture-reactions of flint under pressure. He did not deny that eoliths fell into distinct types, but believed that similar geological conditions—not intelligent design—had given rise to similar forms. The chipping on the edges of the Kent eoliths had, he argued, been created as stones ground together in the drift under pressure: a process that he termed 'soil abrasion.' Warren even made casts to show how the Kent eoliths had been chipped under pressure (Figs. 5.6 and 5.7). Many of Rutot's supposed tools had, in contrast, been formed by battering and flaking in river torrents, as Boule had demonstrated, which gave them a different character.[101]

Dawkins regarded Warren's publication as 'absolutely conclusive' evidence for the natural origin of Rutot's eoliths. The audience at the Anthropological Institute, however, gave his paper a negative response. Belief in the human origin of the Kent eoliths still prevailed.[102] Harrison referred light-heartedly

---

[100] Evans 1897b, 603.   [101] Warren 1905, 338, 340–342.
[102] Dawkins 1910, 237; Warren 1905, 360–361.

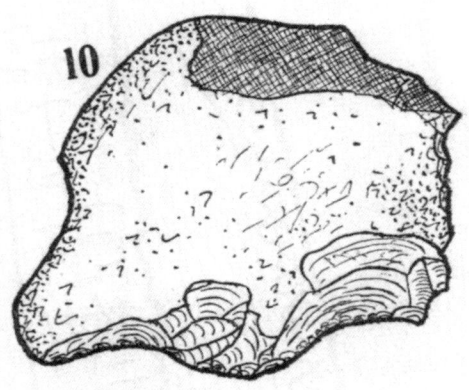

Figure 5.7. Warren's illustration of the original Kent-type 'Double Notch' eolith, cast in Figure 5.6 (Warren 1920, 240, Fig. 10). By permission of the Geological Society of London.

to the two attacks of 1905 in a letter to the sceptical Evans about his latest eolith specimen from Ash: 'It has decided Eolithic chipping on it and according to Mr Hazzledine Warren would *not* be fit to be put into the witness box—and according to Mons Boule might be the product of a cement mill'.[103]

Five years later, Breuil delivered another blow to Rutot's Belgian eoliths. In 1910, he described an Eocene site at Belle Assise in France, where he had found evidence of natural flaking *in situ*. The stone fragments were still preserved in the place where they had been fractured by geological forces. Breuil published his paper on the Belle Assise flints in *L'Anthropologie*, the same journal that printed Boule's paper in 1905. Many of the naturally produced flakes from Belle Assise had bulbs of percussion and fine trimming (retouch) at their edges. Some are illustrated in Figure 5.8.[104] Before, such features had been taken as evidence of human workmanship; now, researchers were more wary.

The papers of Boule, Warren, and Breuil proved that natural fracture could create extremely persuasive 'artefacts' and recommended that greater caution was needed when referring to attributes of human workmanship. Their message provoked a mixed response in Britain. Sollas was roused by Breuil's Belle Assise paper to insert a last-minute note in his book *Ancient Hunters* (1911), announcing that the eoliths 'must now be regarded as the story of an exploded hypothesis'.[105] Sollas would annoy Moir and Lankester over the next decade with his fluctuating opinions of the East Anglian pre-palaeoliths. Sturge, as President of the PSEA, expressed doubts about Breuil's conclusions in the first volume of his new Society's *Proceedings* (hereafter *PPSEA*) and

[103] Harrison to Evans (undated): AMO: JE/B/2/42.
[104] Breuil 1910, 399–402.   [105] Sollas 1911, 68.

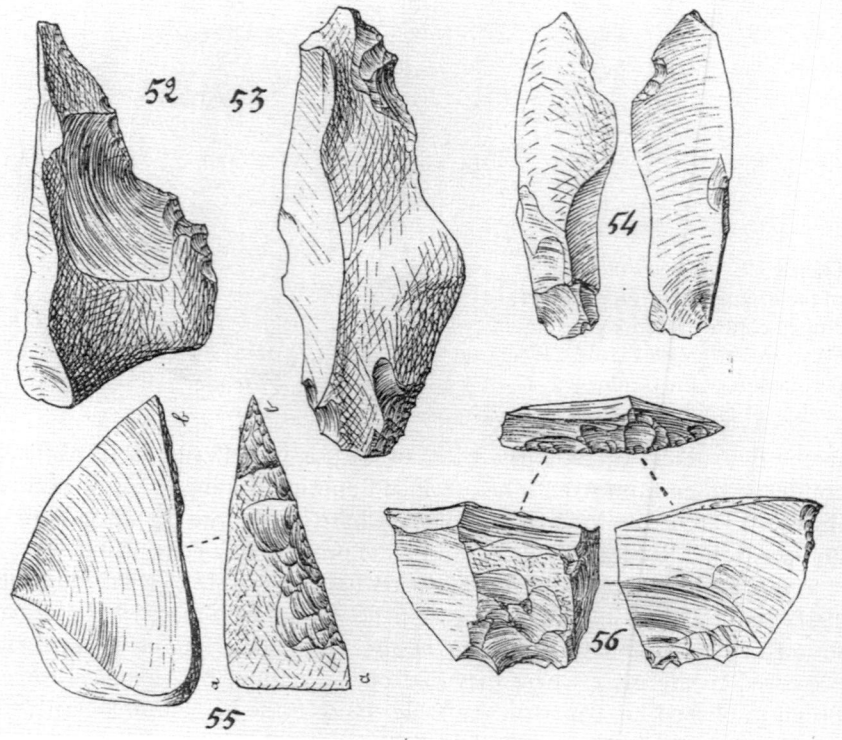

Figure 5.8. These flakes were amongst the illustrations that accompanied Breuil's paper on the natural flakings from Belle Assise (Breuil 1910, 400, Figs. 52–56). Breuil drew particular attention to the bulb of percussion that can be seen in Figures 54 and 55, and to the retouch in Figure 55.

referred the reader to the first article in the volume: Moir's first substantial paper on 'The Flint Implements of Sub-Crag Man'.[106]

Moir had witnessed the attacks made by Boule, Breuil, and Warren on the eoliths of Belgium and Kent. He was, as Grayson has observed, well prepared for an adverse response to his pre-palaeoliths.[107] An account was given at the beginning of this chapter of a quarrel between Lankester and Dawkins at the Royal Society *soirée* in 1912, when Dawkins dismissed the pre-palaeoliths, suggesting that they had been produced by natural processes. Moir had suffered a similar experience when he was confronted by both Dawkins and Warren, who, 'after seeing my specimens when on exhibition at Burlington House, London, told me that they were undoubtedly the result

---

[106] Sturge 1908, 13; Moir 1911.   [107] Grayson 1986, 106.

of natural forces and nothing else'.[108] Boule added his sceptical voice from the pages of *L'Anthropologie*, agreeing with the views of Dawkins and Warren in his review of Moir's first paper in the *PPSEA*.[109]

Lankester was also aware of the papers by Boule and Breuil. He had even described the latter as 'a perfectly fair and good attack on reputed "Eoliths".'[110] But when Warren and Dawkins directed similar arguments at the pre-palaeoliths, Lankester and Moir felt driven to undertake their own studies of flint fracture to represent their position in what they saw as a one-sided field of research. In an echo of Warren's earlier plea, Lankester wrote to the editor of *Nature* in 1912 about the 'great need for a thorough study of flint': its origin, varieties, and fracture. He expressed the same opinion privately to Moir, who had already started experimenting.[111]

When Moir came to publish his results he was criticized by Warren and by Frederick Haward (1871–1953), who also worked on flint fracture.[112] That they objected to Moir's conclusions is not surprising, but their attacks on his approach are more revealing. Moir disagreed with their opinions about the origin of the pre-palaeoliths, but he also made different assumptions about the point and procedures of flint-fracture research. Grayson assesses their respective approaches as early examples of 'middle-range' research.[113] Here, they are examined for the differences in perception and intention that helped to fuel and maintain disagreement between supporters and critics of the pre-palaeoliths.

Haward was an engineer who had worked on railway construction in America. He was also one of the founder members of the PSEA: the first society devoted to prehistory in Britain.[114] (Rutot was their first Honorary Corresponding Member, which indicates the early stance taken by this society towards their local pre-palaeoliths.) Haward's doubts about eoliths, like those of Warren, had gathered after seeing evidence of natural fracture in the field. Haward first saw flaking by natural contact ('contact-flaking') in East Anglia; Warren made a similar observation on the Isle of Wight. These experiences

---

[108] Moir 1911, 19.   [109] Boule 1912a, 428.

[110] Lankester to Moir, 2 June 1911: BLL: Add.Ms.44968/41–42. See also Lankester to Moir, 30 May 1910: BLL: Add.Ms.44968/5.

[111] Lankester 1912a, 331; Lankester to Moir, 30 May 1910, 1 July 1911, and 12 October 1912: BLL: Add.Ms.44968/7, 47–48, 44969/2–3.

[112] On Moir's flint fracture researches, see Moir 1911, 1912a.

[113] Grayson 1986, 106–115. Middle range research is an archaeological approach that was popular in the early 1980s. It involves the direct, present-day observation of human activities as they created material reflections or 'signatures'—rubbish—to help interpret the processes that had produced similar patterns in the past. Archaeologists might, for example, observe a group of hunters butchering a kill and then record the fate of the fragments after the hunters abandoned the site to learn the kinds of patterns indicative of ancient kill-sites.

[114] Thomas 1954, 88.

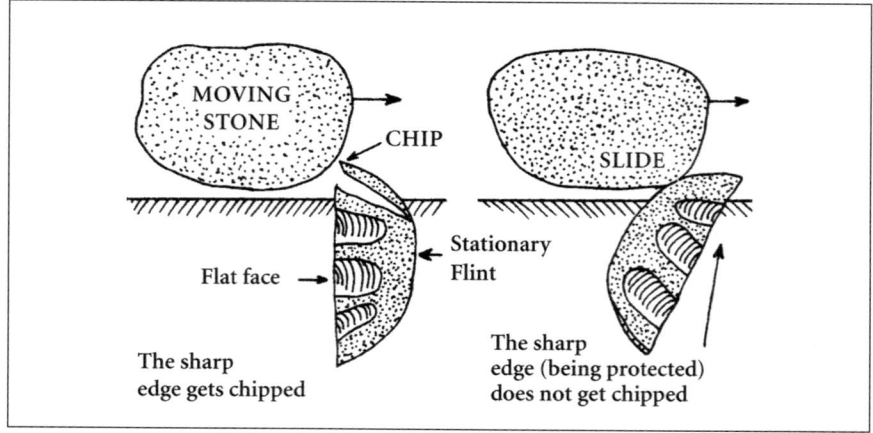

Figure 5.9. A demonstration of the conditions in which the kind of natural flaking described by Haward as 'chip and slide' might occur (redrawn from Haward 1914, 352). Moving stones are forced against stationary flints by natural agencies, flaking the sharp edges of the flint.

stimulated them both to conduct experiments on the fracture of flint.[115] Warren's early attacks had been directed against the Kent Plateau eoliths. When they became eclipsed by the sub-Crag debate, his fracture experiments were aimed directly, though courteously, at Moir.[116]

It is difficult to disentangle the contributions of Warren and Haward to the debates over the pre-palaeoliths. They both took a rigorous approach to flint fracture, which they studied through a combination of experiment, observations in the field of flakes produced by geological forces, and comparison of natural flakes with those knapped by humans. Friends and colleagues, they shared their ideas as well as their outlook and came to similar conclusions independently. Haward, for example, introduced the concept of 'chip and slide' (chipping by lateral movement under vertical pressure, a stationary flint being flaked by a block moving over it: see Fig. 5.9). Warren incorporated this concept into his account of the principles responsible for flint chipping: his second paper on flint fracture to appear in the *Journal of the Royal Anthropological Institute*.[117]

[115] Warren 1900, 411–412; 1913, 38; 1923a, 162; Haward 1912, 185; 1919, 120.

[116] Warren 1913; 1914a; 1923b; 1924a.

[117] Haward 1912, 188; Warren 1914a, 412. Warren 1914a, 426 observed that although Haward had given a name to the principle of chip and slide, he had first suggested this idea in 1905 (Warren 1905, 340, 349–353).

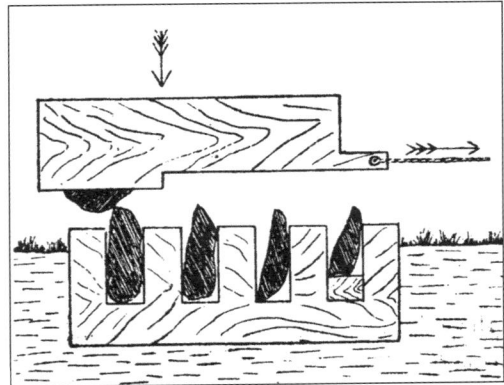

Figure 5.10. Warren's sled experiment (Warren 1914a, 426, Fig. 11). Warren dragged a weighted sled, with pebbles attached to its base, over the projecting edges of flints held in a wooden frame.

Warren's second critique of eoliths at the Anthropological Institute was published in 1914. This time, he included pre-palaeoliths in his comprehensive attack. Warren discussed natural edge-chipping (with a detailed explanation of how the form and condition of a stone could influence the result of natural flaking); the natural agencies that were involved (heat, concussion, crushing, chip and slide, each producing its own characteristic pattern); and the general principles of fracture along planes of least resistance.[118] His investigations into flint fracture had ranged widely since he had last addressed the Institute on the subject. Haward also published a paper in 1914 on 'The Problem of the Eoliths' in the *PPSEA*. He began with praise for experimenters: 'The fullest credit must be given to our members, Mr. Reid Moir, F.G.S., and Mr. S. Hazzledine Warren, F.G.S. (and others also) for their efforts to solve this problem by experiment'.[119] Moir's efforts, however, did not pass without criticism.

For Haward and Warren, the most problematic feature of Moir's experimental work was his attempts to simulate natural geological conditions. Moir countered Breuil's attack of 1910, for example, by simulating the conditions at Belle Assise—using a letter-press filled with sand and flints. He concluded that pressure did not act vertically through sand: a covering of sand protected flints from fracture. (Sand also covered Moir's sub-Crag flint bed at Bolton and Laughlin's brickfield in Ipswich, where he had found many pre-palaeoliths.) In another experiment, Moir simulated the violent action of the sea on a beach. Then he managed to simulate the power of a river dragging pebbles down its reaches by shaking ten broken flints in a sack. He concluded

---

[118] Warren 1914a.   [119] Haward 1914, 347.

from this experiment that natural forces removed flakes at various angles, unlike human blows, which were delivered at a constant angle.[120]

It was Lankester who first suggested to Moir the flaw: 'With regard to your experiment with flints in a sack—it seems to me that the conditions of a torrent or rush of sea ice-bearing water are different'.[121] Haward pointed out: 'Nature does not confine stones in a sack' and suggested, furthermore, in reference to another of Moir's experiments, that the tremendous weight of a slowly coursing glacier could not be reconstructed by scraping stones over a concrete floor by hand.[122] Warren was similarly dismissive of Moir's belief that he could recreate natural conditions.[123]

Haward and Warren were not trying to reproduce nature. They were trying to gain a better understanding of flint and how it fractured by tracing the connections between different varieties of chipping and the natural conditions of the flint as it lay in the ground. Moir's efforts drove Haward to exclaim: 'It is ridiculous to suggest that a few laboratory experiments, made under artificial conditions, are equivalent to the action of Nature, whose forces are so varied.' He suggested that such misguided attempts might arise from a lack of 'extensive geological experience'.[124]

Warren complained that this kind of confusion about the aims of experiments had led to mistrust of flint-fracture research on both sides of the debate. He argued that Nature could only be observed in the field, where a convincing variety of flakings could be recovered—as Breuil had demonstrated at Belle Assise.[125] Moir disagreed. He defended his sack and letter-press experiments, and continued to describe them in his publications.[126] It seems that Moir never fully grasped that Warren and Haward were searching for general principles rather than reconstructions of nature, which was probably due as much to his dogmatic character as his lack of geological knowledge.

Moir did venture into the field to look at natural fracture. Inspired by the example of Belle Assise, he examined an Eocene stratum in Suffolk at the Bramford pit owned by a Mr. Coe. Moir compared the features of these naturally fractured Eocene flints to those of his own flints, produced by experiment, and saw great similarities between two. He described those features he thought most significant. The natural flints exhibited 'pseudo'-bulbs of percussion with no striking platforms, and there was no retouch. There were

---

[120] Moir 1911, 18–20; 1912a 173–174, 181; 1919, 15–18; 1939, 6–7.
[121] Lankester to Moir, 5 July 1911: BLL: Add.Ms.44968/50.
[122] Haward 1914, 347–348; Moir 1912a, 183.
[123] Warren 1913, 38; 1914b, 547.
[124] Haward 1914, 349.
[125] Warren 1914a, 413–414; 1914b, 547; 1923a, 155; 1923c, 39; 1925, 302.
[126] Moir 1919, 15–18; 1939, 6–7; Grayson 1986, 111.

no definite forms, only 'fantastic shapes' fractured by natural processes that could not be divided into tool-types. Characteristics of true tools were at the front of Moir's mind; natural flaking, to him, meant random, unguided forces. Tools were at the back of Warren's mind when he was in the field; he searched for distinct patterns of natural flaking that could be interpreted with the help of general principles learnt from experiment: the two central pillars of his natural-flaking arguments.[127]

The attributes of humanly struck flints also aroused disagreement between Moir and Haward. Moir's pre-palaeoliths had been criticized for lacking bulbs of percussion: one of the traditional signatures of a human blow. In an attempt to convince his critics that these flints were nonetheless the product of human activity, Moir turned to the work of the Brandon Flint Knappers. Brandon, a small town in Suffolk, was one of the last places in England where gun-flints were still made. Most of their flints were exported to South America to service antiquated flintlock weaponry. Back in 1876, whilst mapping the area, Skertchly had recorded their flint-knapping techniques in his Survey notebook and mentioned several knappers by name. After he found tools beneath boulder clay nearby, a reporter from *The Eastern Daily Press* set the scene for Skertchly's discovery with a description of the low, flint-built shops of the knappers in the side-streets of Brandon with heaps of flints piled before their doors.[128]

Moir explained how the peculiar technique used by the Brandon Flint Knappers—a particular 'quartering' fracture with which they dressed their flints—produced a flat, non-conchoidal surface.[129] Thinking back to his Pre-Palaeolithic tool-makers, he surmised: 'The men who made these implements had acquired the art of being able to fracture flints with a blow and leave a flat surface showing practically no bulb of percussion. It was needful to do this when making the rubbers for dressing skins'.[130]

Unfortunately for Moir, back in 1897 Haward had been 'taught the secrets of the art of flint chipping by several of the Brandon "Flint Knappers" with whom I kept in close touch for years'.[131] Haward combined these secrets with experiments and observations of natural-flaking sites to build his case against pre-palaeoliths. He argued that in industries of all ages, the characteristics of bulbous fracture produced by percussion were all the same, since they were all produced by the same force: most flakes stuck by humans *did* have a bulb of percussion although, admittedly, an iron hammer produced finer bulbs than

---

[127] Moir 1914; Warren 1923a.
[128] Greenwell 1870, 419; Skertchly 1876, Notebook 5, 133–138: BGS: NGRC 1009; Anon 1876.
[129] Moir 1913a; 1915a, 25; Lankester 1914, 7.
[130] Moir 1913a, 313.   [131] Haward 1919, 120.

a stone one. The flints from the Norfolk Basement Beds and the Ipswich Sub-Crag Beds, however, had been fractured by great natural pressure that could remove large flakes with an almost flat bulbar swelling, which was different from human-struck bulbs.[132]

These disagreements over flint-fracture and the pre-palaeoliths were part of a broader conflict between the different disciplines that converged on this subject. Warren spoke of principles of flint fracture; Moir, of artefacts. They rarely agreed. But it was unusual that either had entered the field of flint-fracture research at all. Warren regretted the peculiar position held by investigations into natural agencies of flint flaking. Although the study itself was a purely geological subject, its application was purely archaeological, so geologists avoided what Warren called: 'this no-man's-land of unexplored phenomena. Thus the problem of distinguishing human flaking from natural that is at the foundation of prehistory "fell between two stools" and became the Cinderella of Science, casually cloaked in any rags and tatters of misleading half-knowledge'.[133]

Judging from his opinions above, Warren may well have been thinking of Moir when he spoke above of 'misleading half-knowledge.' Despite his contributions to Palaeolithic research, Warren was regarded by most of his contemporaries as a geologist; Moir saw himself as an archaeologist. Boule also remarked on the dilemma of the different disciplines that approached the pre-palaeoliths, and predicted in 1912: '*chacun l'appréciera de son point de vue personnel, au moyen de son expérience en matière géologique ou archéologique*' ('everyone will appreciate it from their personal point of view, through their experience of geological or archaeological matters'). He explained that geologists, although they were happy to assert their opinions on the age and context of Moir's Pre-Palaeolithic finds, avoided the question of whether these flints were humanly worked—which was proclaimed mostly by archaeologists.[134] Warren might have thought archaeologists dangerously ill-equipped to pronounce on such a subject.

Lankester had sent Moir into the realm of flint-fracture research to try and counter the arguments of the opposition on their own ground. But Moir failed to hit the right tone to convince the sceptics, despite all his experiments, observations in the field, and analogies to living flint-makers. This was partly due to a misunderstanding on the part of Moir and his lack of geological experience. But disagreement was also sustained by differences in expectation. The East Anglian debates nurtured attitudes and conflicts similar to those that had surrounded the Kent debates. Warren and Haward, like Evans and Smith, did not believe in the artefactual nature of these flints. Moir, like Prestwich

---

[132] Haward 1919, 122–127.   [133] Warren 1940, 13–14.   [134] Boule 1912a, 428.

and Harrison, saw in the same flints the work of ancient tool-makers, and introduced archaeological lines of argument—describing tool-types and suggesting their functions—that his critics thought irrelevant. The next chapter balances the sceptical deconstructions of Warren and Haward with a study of some reconstructions of Pre-Palaeolithic classifications and sequences offered by Moir and Lankester. But this chapter ends with a picture of two opponents firing missiles past each other, never managing to achieve a direct hit.

6

The Pre-Palaeolithic of East Anglia

The attacks of Warren and Haward had dominated the realms of flint-fracture research, and the defensive forays of Moir and Lankester had not strengthened the cause of the pre-palaeoliths as much as they might have hoped. But Lankester was grooming Moir in another approach: how to promote the pre-palaeoliths as respectable archaeological specimens. He advised, 'You can always state what are the accepted views & pros & cons put forward by geologists as to relative ages of deposits. But *you* yourself are & must be more & more, an expert & critic of the worked flints themselves.'[1]

Moir and Lankester would identify distinct groups and types in their motley assortment of pre-palaeoliths. They would reconstruct lengthy industrial sequences, degenerating back to the Kent eoliths on the one hand and progressing to the river-drift palaeoliths on the other. Similar approaches had characterized earlier attempts to order the tools of the river drifts and caves (see Chapters 3 and 4), but Moir and Lankester were not simply aiming to understand how their industry fitted into a broader Stone-Age sequence. The existence of different tool types and evolutionary stages within the Pre-Palaeolithic provided them with another opportunity to convince their critics that these flints were as reputable as any of their Palaeolithic descendants, and could be classified according to similar principles.[2]

This chapter addresses these archaeological reconstructions and then turns to the style and tone of this work. Presentation could merit as much attention as content; but though Lankester took care to select arenas that would enhance the respectability of his stones, he was often irritated by the reaction of the audience, and Moir could become enraged. For these two believers in the pre-palaeoliths, the scientific ideals of disinterested objectivity sat uneasily alongside the belligerent outbursts that punctuated their practice. For the historian, their unusual forthrightness offers a vivid glimpse of the social tactics behind Stone-Age research.

---

[1] Lankester to Moir, 3 March 1912: BLL: Add.Ms.44968/123.
[2] Sommer 2004b, 214–216 also discusses the classifications of the eoliths.

6.1 MOIR AND HIS PRE-PALAEOLITHIC SEQUENCE

Perspectives on the Stone Age were often coloured by the assumption that the earliest tools would be crude, and that later tools would manifest more skill. These assumptions have been seen in the accounts of the Kent eoliths given by Prestwich and Harrison, explored in Chapter 5, and they were to emerge once again in the arguments over the East Anglian pre-palaeoliths.[3] Just as Evans had warned Harrison that rudeness of character was no sure test of relative age, Haward now cautioned supporters of the pre-palaeoliths, '*Roughness of workmanship is* [...] *no proof of antiquity*', and observed that production quality would also be influenced by the nature of the raw material or by the difference between the skill of a master-knapper and an apprentice.[4] But though crudeness was no proof of antiquity, the earliest industries were nonetheless expected to be crude and unaccompanied by more advanced forms. Warren argued:

> If the eoliths were human implements it appears to me that we should expect that their characteristics would be independent of associated geological forces, but would be dependent upon the relative ages of the deposits containing them. As a matter of fact we do not find that earlier deposits consistently contain more primitive eoliths and later deposits more advanced eoliths.[5]

Moir, however, disagreed. Back in 1910, he had reported to *The Times* that his sub-Crag finds could be divided into at least two distinct types, their different age indicated by the character of their flaking and the appearance of their surfaces (patination).[6] The following year, he provided a longer list of sub-Crag types. Each was named after its supposed function, and Moir described 'Rubbers, Pounders, Beak-shaped Implements, Choppers, Implements with a cutting edge at one end, Scrapers, Struck and Trimmed Flakes'.[7] He also presented the sceptical Warren with different groups of Pre-Palaeolithic types from different geological layers. Some came from the base of the Red Crag, and had been formed by heavy blows. Those lying intermixed within the Middle Glacial Gravels above, Moir separated into at least four different groups according to differences in flaking, form, patination, and staining; he suspected that the rare 'Harrisonian' specimens from these Gravels had been derived from a still more ancient deposit. His youngest Pre-Palaeolithic group came from the Chalky Boulder Clay, the highest of the

---

[3] Prestwich 1891, 134; 1892, 254–255; Harrison to Tylor, 14 November 1907: PRM: EBT, Harrison 11.

[4] Haward 1914, 350.   [5] Warren 1914b, 551 (see also Warren 1914a, 433).

[6] Moir 1910, 8; Slater 1911, 13–14.   [7] Moir 1911, 22–23.

three layers: they resembled the earliest Chellean implements of Palaeolithic times (see Table 5.3 for a summary of this geological sequence).[8]

Moir believed that this Pre-Palaeolithic sequence would arm him against Haward and Warren. Surely, he argued, the uniform action of natural causes could not have produced such variety.[9] But, once again, Moir was failing to engage with the flint-fracture arguments of his critics: Warren had not maintained that Nature flaked in a uniform *manner*; he had observed that natural fracture followed uniform *principles*, with the character of flaking varying according to different kinds of raw material or geological forces.[10] While Moir devoted his energy to the creation of a Pre-Palaeolithic sequence, a sceptical opposition pointed to the disturbed geological context of these finds and to the dubious basis of Moir's criteria of relative age.[11]

Even Lankester was, initially, rather sceptical of this sequence. Commenting on a draft of an early paper by Moir, he criticized the link to the much-debated Kent eoliths with the remark: 'A rough implement is often later than a more elaborate one'.[12] At this early stage in their arguments, Lankester preferred to concentrate on establishing the human workmanship of the pre-palaeoliths, '& not to trouble too much about further questions, such as the various "groups" in Mid-glacial, Boulder Clay &c although they are really of very great importance & interest. Once we have *fixed* the bare fact that such early worked flints *really exist*! That is what people still deny!!'[13]

Moir had, nonetheless, succeeded in arranging his pre-palaeoliths along a progressive sequence that convinced some of his wavering colleagues. Although the drawbacks to a simplistic correlation between the crude and the early were well known, the notion that there had been a progressive improvement in skill through time was persuasive. Moir could not convert Haward or Warren, but he was justified in his belief that a sequence of precursors to the Chellean (the earliest of the accepted Palaeolithic industries) would boost the credibility of his pre-palaeoliths. In age and form they seemed to plug the giant gap between the ancient eoliths of Kent and the more recent palaeoliths of the river drifts.[14] Moir thought that the Kent specimens would have won more support had a convincing intermediate group been

---

[8] Moir 1913a, 308–311; 1913b, 369–370, 374.

[9] Moir 1913b, 370.

[10] Warren 1914b, 551.

[11] See Haward (1914) on why patination was no guide to the age of Kent plateau eoliths.

[12] Lankester to Moir, 29 December 1910: BLL: Add.Ms.44968/16. The draft was probably for Moir 1911. On Lankester's scepticism, see Lankester 1912b, 330–331; 1914, 16; on Lankester's change of opinion, see Moir 1919, 41–42.

[13] Lankester to Moir, 21 January 1914: BLL: Add.Ms.44969/106–107.

[14] Harrison 1928, 293.

identified earlier.[15] He took care to ensure that his own finds did not suffer from a similar disadvantage.

6.2 LANKESTER AND HIS 'ROSTRO-CARINATE'

In 1913, at a meeting dedicated to Palaeolithic tools and their precursors, the President of the Geological Society, Aubrey Strahan (1852–1928), Assistant Director of the Geological Survey, asked four questions of those defending the eoliths. The first three followed the now-familiar lines of enquiry: 1) were they geologically *in situ*? 2) what was the geological age of the deposits? 3) were they humanly worked? The fourth question, however, touched on a more sensitive topic, asking, 'Could such a sequence of types of implements be established in this country as to enable geologists to use implements as zone-fossils in the deposits of the Human period?'[16]

This fourth question was also aimed at those working on the Palaeolithic tools of the river drifts, and at the exciting discoveries made recently in the Thames Valley (see Chapter 8). Moir, though, had been working on the same problem, and Lankester was also contributing to the answer through his work on the so-called 'rostro-carinate', or 'eagle's beak' implement: a 'type-fossil' from the Pre-Palaeolithic zones (see Fig. 6.1). Lankester exalted this distinctive, repeated form above other Pre-Palaeolithic types, such as 'scrapers' and 'hammer-stones'.[17] He brandished the rostro-carinate at the opposition to defy their suggestions that the pre-palaeoliths had been produced by natural agencies of flint fracture. His attempts at casting Moir into the flint-fracture debates may have been unsuccessful, but surely the opposition would see convincing proof of human workmanship in the rostro-carinate? Lankester declared to Moir, 'The adherence to one type—the rostro-carinate—is a great argument for the unbeliever. I want to ram and stuff him with that one type. Once he has admitted that as human, the rest can be discussed with greater advantage'.[18]

Robert Marett, the Oxford anthropologist, captured Sir Ray Lankester's enthusiasm for the rostro-carinate in a short poem. He took up the story of this new type-fossil in the following two verses, which give a mock-heroic aura to the search for Palaeolithic precursors:

[15] Moir 1912a, 172.
[16] Strahan in Anon 1914, ii.
[17] Lankester 1912b, 287. The importance of the rostro-carinate in the arguments of Lankester and Moir is also recognised by Sommer 2004b, 214–215.
[18] Lankester to Moir, 18 March 1913: BLL: Add.Ms.44969/59–60.

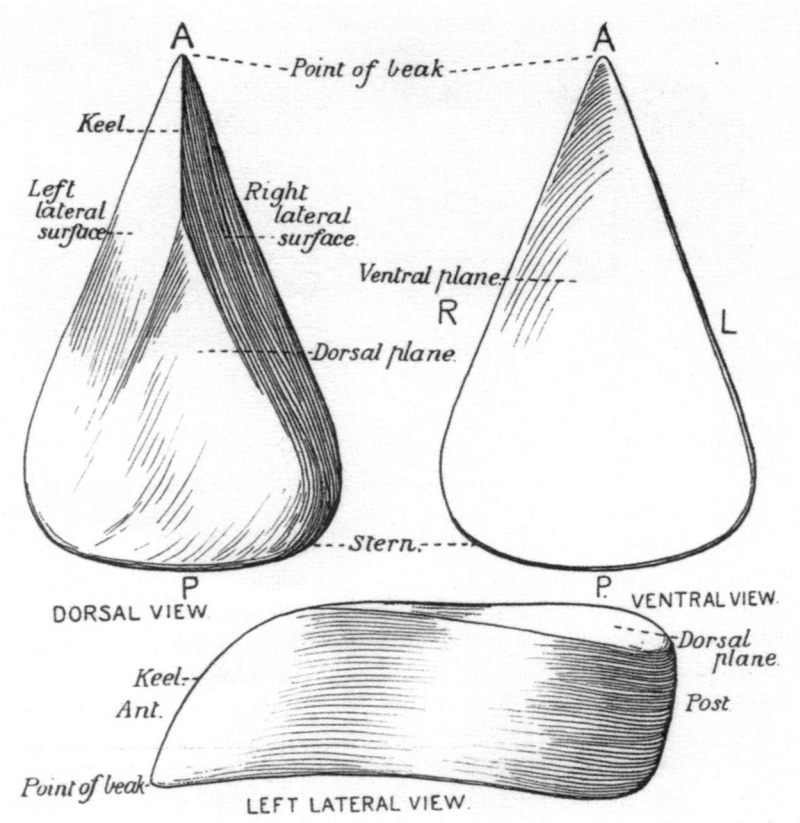

Figure 6.1. 'Diagrams showing the ideal form aimed at by the makers of the Rostro-carinate flint implements'; 'A, anterior; P, posterior; R, right; L, left' (Lankester 1912b, 294, Fig. 1).

Then Mr. M...r took on the job for all that he was worth,
And soon he lighted on a flint in twenty feet of earth;
A very King of Flints it was, of Brobdingnagian bulk,
Looking something like the fore-part of a badly battered hulk.

They lifted it in triumph, and they bore it to the Knight,
He gazed, and suddenly his eyes glowered with second-sight.
Then hoarse he cried, as one through whom the Voice of Time did speak:
'I greet thee, Rostro-carinate! For I recognise thy Beak!'[19]

[19] Marett's *Oratio Capitalis pro Rostris*, quoted in Moir 1935, 34.

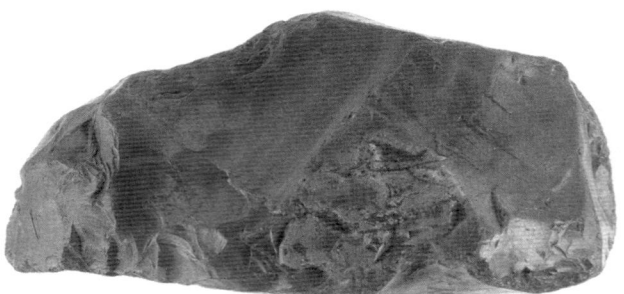

Figure 6.2. The Norwich test specimen rostro-carinate from beneath the Norwich Crag.

Lankester's attempt to slip his rostro-carinate, like some dubious new animal species, into the Stone-Age sequence recalls his background in zoology. But his efforts were not particularly out of place: Palaeolithic classifications owed a historical debt to palaeontological methods. His attention to the rostro-carinate also recalls the devotion of an earlier generation to Palaeolithic hand-axes.

Lankester's masterpiece was a *Description of the Test Specimen of the Rostro-Carinate Industry found beneath the Norwich Crag*, published as a Royal Society Occasional Paper in 1914. This eighteen-page paper on the rostro-carinate included fourteen high-quality figures and three plates, all devoted to a single discovery: 'the best preserved specimen of this class of implements', found by Clarke of the PSEA at Whitlingham pit and known henceforth as 'the Norwich test specimen'.[20] This remarkable type-fossil, illustrated in Figure 6.2 above, is still stored at the British Museum.[21]

The first task that Lankester set himself was to defend the human workmanship of rostro-carinates; the geological age and context of the pre-palaeoliths was of secondary interest. This reversed the order of Prestwich's strategy in his papers on the Kent eoliths, discussed in Chapter 5. As Lankester explained to Moir:

In that paper [Lankester 1914] I insisted, as you & every one else can read, that the main point for me is the human workmanship of this & the other rostro-carinate implements. The occurrence of this & of others in a given geological stratum, and further the exact *age & relations* of such stratum are *separate* questions, and must be discussed by geologists on definite geological evidence. That is what I wrote & still hold.[22]

[20] Lankester 1914, 1.
[21] BM(F): Norfolk, Whitlingham. I am grateful to Roger Jacobi for kindly showing me this marvellous rostro-carinate.
[22] Lankester to Moir, 3 September 1917: BLL: Add.Ms.44970/15–16.

There were several good reasons to justify this approach. The geological context of the East Anglian discoveries was less complex than the problems that had faced Prestwich on the Kent Plateau, and the East Anglian discoveries were not dominated by surface-finds. Lankester had also acquired his reputation in zoology, so did not inspire the respect afforded to Prestwich in geological circles.[23] His strategy was relatively successful. Few dismissed the rostro-carinate lightly, and this type-fossil attracted a number of serious, considered attacks by the opposition. Haward, for example, presented a detailed explanation of how the Norwich test specimen could have been flaked by natural processes along lines of least resistance.[24] But others were converted by Lankester's work on the rostro-carinate: Victor Commont died just when Lankester 'was looking forward to his "coming out" on our side about rostro-carinates'.[25] (Commont was best known in Britain for his diligent work on the Somme Valley sequence, described in Chapter 7.)

Even Lankester's previous adversary, Sollas (Fig. 6.3), changed his mind. In earlier years, Sollas had often clashed with Lankester over the pre-palaeoliths. When Lankester brought his finds to Oxford, where Sollas was a Professor, he had dismissed them with the words, 'We have here choppers which do not chop—borers which do not bore,' to which Lankester had retorted, 'You, Sir, are not a borer which does not bore.'[26] Sollas had attacked the rostro-carinates at a meeting of the British Association at Birmingham in 1913 and again at a meeting of the Geological Society in 1914 (the one mentioned above), arguing that these 'implements' had been formed by the blows of pebbles driven inland by the sea.[27] By 1915, however, Sollas had amended his views: the second edition of his book, *Ancient Hunters and their Modern Representatives*, conceded that rostro-carinates had been made by human hands, not by natural forces.[28] This revision led Lankester to muse, 'Sollas is a hopeless creature. I suppose he has heard from Commont, as I have, that he accepts the Norwich type specimen'.[29]

For Warren, who was one of the reviewers of Sollas's second edition, this revision left him puzzled at 'an apparent inconsistency with our author's attitude to the "rostro-carinates" in general'.[30] Lankester, exultant, marked out to Moir how best to press forward their advantage: 'The successful

---

[23] Lankester's early work on the East Anglian Crags was more palaeontological than stratigraphical in content Lankester 1865; 1870.
[24] Haward 1919, 133–135.
[25] Lankester to Moir, 21 May 1918: BLL: Add.Ms.44970/44–45.
[26] Moir 1935, 32.
[27] Sollas 1913, 788–790.
[28] Sollas 1915, 119; Anon 1914, v–vi.
[29] Lankester to Moir, 30 March 1916: BLL: Add.Ms.44969/174–175. See also Sommer 2004b, 218.
[30] Warren 1916, 79.

Figure 6.3. William Sollas (1849–1936), Professor of Geology at Oxford (GSL: P57/Sollas).

method to follow now, is to *insist* on this. "*There are rostro-carinate imple-ments.*" The *further point* now to be decided & agreed upon, is "What is the *age* of this specimen, and of that?" '[31] Flushed with such success on the question of human workmanship (the rostro-carinate having proved more persuasive than Moir's flint-fracture experiments), Lankester became more amenable to Moir's work on the connections between his flints and the implements of Palaeolithic times.

6.3 FROM ROSTRO-CARINATES TO HAND-AXES

While Lankester promoted the putative human workmanship of his rostro-carinate, Moir worked on its evolution. He made a case for the rostro-carinate ancestry of Chellean implements, and may have hoped that these reputable descendants might reflect some glory back onto their less respectable ancestors. Moir believed that the rostro-carinates from the Sub-Crag Bone-Bed evolved into the finer specimens from the Middle Glacial Gravel, before becoming scarce and degenerate in the Chalky Boulder Clay where the typical implements looked more like early Chellean palaeoliths.[32] He devoted two papers to the evolution of rostro-carinates into early Palaeolithic hand-axes: the first was

---

[31] Lankester to Moir, 18 April 1915: BLL: Add.Ms.44969/161.
[32] Moir 1913b, 373; 1914, 402–403.

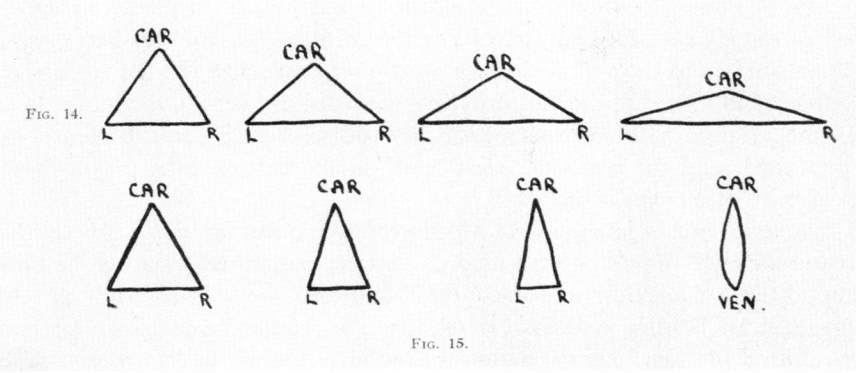

Figure 6.4. Moir used this illustration to explain the difference between two Pre-Palaeolithic sequences (Moir 1919, 41–42, Figs 14 and 15). His Figure 14 illustrates the depression of the keel in the so-called 'Batiform' manner; his Figure 15 shows the 'Platessiform' progression. 'Car' indicates the position of the carina (keel), and 'ven' the ventral side of the flint.

delivered to the Anthropological Institute and published in 1916; the second was presented to the Royal Society in 1917 and published in 1920.[33]

In the past, Lankester had countered Moir's view that rostro-carinates had evolved into Chellean hand-axes. He had asserted in 1912 that 'there is no evidence of any transitional connection' of the East Anglian 'Icenian' industry with the river-drift implements.[34] Elaborating, Lankester had explained that 'The Chellian and Acheuilian and Moustierian types are essentially *depressed* or flattened like a leaf. The Sub-Crag type (*rostro-carinate*) is essentially *compressed* from side to side'.[35] A few years later, when Moir was about to submit his paper to the Royal Society, Lankester reconsidered his old opinion. He solved the question of how to link the compressed rostro-carinate to the flattened hand-axe by turning to the example of a fish, describing the transition as 'Platessiform' because it echoed the evolution of the plaice. Another fish-inspired line was added before Moir's paper went to press: the 'Batiform' sequence for palaeoliths that were triangular, rather than rhomboidal, in section and echoed the evolution of the skate (*Raia batis*), rather than the plaice (see Fig. 6.4).[36]

[33] Moir 1916; 1920a.
[34] Lankester 1912b, 330. 'Icenian', Lankester's term for the sub-Crag industry, was derived from the Roman name for an East Anglian tribe Moir 1935, 99.
[35] Lankester 1912b, 331.
[36] Moir 1920a, 346–348. Moir 1919, 42 cautioned, 'But in the giving of these descriptive names there is, of course, no wish to convey the impression that the makers of the ancient flint implements described, had a knowledge of the manner in which the evolution of the plaice and

On the Platessiform line, a rostro-carinate was transformed into a palaeolith of *rhomboidal* section by extending the cutting edge from the keel on the dorsal surface to the *ventral* surface on the opposite side, the flat underside being flaked away. On the Batiform line, the hand-axe kept to the same plane as the original rostro-carinate form, the dorsal keel became flattened or 'depressed' and the two side edges took on the cutting role, producing a palaeolith of *triangular* section.[37]

Lankester was adamant that Moir's research into the evolution of the rostro-carinate should be presented in a strategic manner. Although he now supported a transition to Palaeolithic hand-axes, he warned Moir not to mention an Eolithic end to this sequence. In earlier years Lankester had supported the Kent eoliths; he had hailed Harrison as the 'courageous and indomitable discoverer of pre-palaeolithic man'.[38] Now, after the discovery of pre-palaeoliths in East Anglia, he wanted to distance these new finds from disreputable associations, and wrote to Moir:

As to your paper [Moir 1920a] on the transition flints from Rostro-carinate to Chellean-type Palaeoliths, I am *convinced* that it is most important *not* to go into the separate (though related) question of the development of rostro-carinates from simplest beginnings (which I beg you *not* to call 'Eoliths'—endless misery & confusion & lies are set rolling by that word 'Eolith'). It is really necessary in a controversy like this to hammer in one point at a time. What you can (& I hope will) do *now*, is to show that there are most convincing intermediate forms between typical rostro-carinate & typical Chellean (or Acheullian), and that there is no explanation of their occurrence (in such number as you can produce) excepting that they were actually flaked by the prehistoric men with the *intention* of modelling an implement of more or less Chellean shape from a rostro-carinate by treating the carina as one of the lateral margins of a Chellean and flaking away the ventral plane to form the other lateral margin. That & all that you have to say about the necessity for a striking platform, (my dorsal plane or platform), and its retention in many completely symmetrical Chellean implements, is *enough* for one paper.[39]

Moir, though, could not be restrained for long. Once he had asserted a relationship between hand-axes and rostro-carinates, he turned his attention to more ancient ancestors. Despite Lankester's warning, Moir traced the origin of several Palaeolithic industries back to different Eolithic types from Kent, and published his views in 1919 in a book on *Pre-Palaeolithic Man*. He

---

the skate took place, and fashioned their artefacts on a similar plan. It would appear to be a mere chance that the Platessiform and Batiform implements were developed along analogous lines to the plaice and the skate.'

[37] Moir 1919, 36–37, 39.
[38] Lankester to Harrison, 15 April 1906, quoted in Harrison 1928, 271.
[39] Lankester to Moir, 3 September 1917: BLL: Add.Ms.44970/15–16.

provided his readers with considerable detail. One sequence led from simple 'Kentian' scrapers and points to Mousterian-like specimens in the Middle Glacial Gravel, moved on to Mousterian forms in the Chalky Boulder Clay above, and culminated in true Mousterian industries and the later industries of the Cave period. This was how Moir explained the origins of flake-dominated industries. Core-tools formed the basis of his other sequence, which led from Kentian points to Sub-Crag rostro-carinates, moved up though the twin hand-axe lines (Platessiform and Batiform) and on to the familiar Palaeolithic industries known as the Pre-Chellean, Chellean, and Acheulian.[40] Moir later added to this sequence, claiming that his 'Early Chellian' industry from the foreshore at Cromer represented the transition between the rostro-carinates and the Platessiform and Batiform hand-axes.[41] His influence on perceptions of the British Palaeolithic sequence is discussed in Chapter 8.

These approaches taken by Moir and Lankester to the Pre-Palaeolithic have offered a glimpse at expectations that also coloured Palaeolithic classifications. Their treatment of the Pre-Palaeolithic rostro-carinate, for example, recalled that of the Palaeolithic hand-axe: both were distinctive, attractive tool-types; both played a prominent part in the construction of classifications and sequences. Similarly, when Moir developed his parallel industrial lines running from the Eolithic to the Palaeolithic, he drew on much a broader assumption of progression; his division between industries dominated by flakes and flake-tools and those dominated by core-tools (such as hand-axes or rostro-carinates) was a division that would be familiar to many Palaeolithic researchers. It is disconcerting to see the ease with which Moir and Lankester identified progression and type-fossils in stones that Warren and Haward believed to be the products of natural rather than human forces, and the conviction carried by such identifications.

6.4 GETTING AHEAD IN SOCIETY

Lankester believed that arguments would have more weight if skilfully and strategically presented. This is evident even in the few excerpts from his letters to Moir given above. The tactics employed by Lankester and Moir, like the content of their arguments, offer another glimpse behind the scenes of everyday research on the British Stone Age. In many ways, the relationship between Moir and Lankester resembled that of Harrison and Prestwich, with

---

[40] Moir 1919, 48, Fig. 7.   [41] Moir 1921a, 429–430; 1921b, 385, 395.

Moir taking the part of the local researcher and Lankester playing the esteemed academic figure. In character, however, the two pairs were very different. Unlike the diffident Harrison, Moir could be confrontational and suspicious of the opposition, hot headed and, without Lankester's restraining hand, likely to rush heated words into print. But his impulsiveness has left valuable records: Lankester often found it necessary to explain the need for social strategy in his letters to Moir, more aware than his friend of the academic approval that could be gained by addressing powerful people with persuasive words and attractive exhibits in a supportive (or at least a prestigious) environment.

The Prehistoric Society of East Anglia provided a strong local support network for the pre-palaeoliths, a role that was mentioned in Chapter 5. Moir had published his first detailed paper on the pre-palaeoliths in their *Proceedings*, and a Special Committee of worthies from the PSEA had given his finds a favourable reception a few days before Christmas in 1910. The reputation of the pre-palaeoliths and their discoverer within the Society was thus decided.[42] Sturge, Underwood, Clarke, and other familiar figures in the PSEA then welcomed Moir, who soon rose to a position of prominence in the Society, assisted by his close association with the supposed local Pre-Palaeolithic industry.

By 1912, Moir was Vice-President of the PSEA. In his Presidency over the next two years he was accompanied by an equally supportive Vice-President: Lankester himself.[43] Moir and Lankester negotiated their positions between them, meanwhile securing a post at the Ipswich Museum for Clarke, supporter of the Pre-Palaeolithic, member of the Special Committee, and discoverer of the Norwich test specimen.[44] For a while, the reputations of Moir, the pre-palaeoliths, and the PSEA spiralled upwards together, and most of his papers on the subject appeared in the *PPSEA*. In this pro-Eolithic society, Haward and Warren were among the sceptical minority. At the local level, the pre-palaeoliths were doing well.

Lankester had also started to introduce the pre-palaeoliths into national societies: forums that had been established for longer than the PSEA, and which might add lustre to the East Anglian finds. Many metropolitan societies were welcoming contributions on Stone-Age subjects at this time. When the

---

[42] Underwood *et al.* 1911, 37.

[43] Lankester wrote to Moir, 'I shall be very glad to become Vice President of the Prehistoric Society of East Anglia, and if I feel able to do so, to become President' Lankester to Moir, 22 September 1913: BLL: Add.Ms.44969/76.

[44] 'I certainly would wish you to continue as President of the Prehistoric Soc. I could not attend to it *at all* at present—if ever. I think Clarke would be an excellent appoint. at the Ipswich Museum' Lankester to Moir, 11 May 1914: BLL: Add.Ms.44969/121.

PSEA was praised by Marcellin Boule for its focus on human palaeontology, the other society included in this tribute (which came from the prestigious pages of *L'Anthropologie*) was an older national society: the Anthropological Institute.[45] A number of early papers on the Palaeolithic had appeared under the auspices of the Anthropological Institute, which had welcomed Prestwich's paper on the Kent eoliths and Warren's paper on natural fracture, as well as contributions by Moir and Lankester.[46] The journal *Man*, published under the direction of the Anthropological Institute, also printed several papers on the pre-palaeoliths.[47] The Institute had similar interests to the PSEA and even hosted a joint meeting on flint chipping on 17 February 1914; Warren delivered one of the papers at this event.[48]

The pre-palaeoliths had soon swept their way into a more prestigious arena than even the Anthropological Institute in a development that alarmed their sceptics. Boule noted, '*La question n'était pas encore sortie du cercle fort restreint des spécialistes quand Sir Ray Lankester la prit sous son patronage et l'introduisit dans un milieu scientifique beaucoup plus étendu et plus officiel*' ('The question had not yet left the very restricted circle of specialists when Sir Ray Lankester took it under his patronage and introduced it into a much wider and more authoritative scientific circle').[49] It was through a memoir by Lankester that the pre-palaeoliths were guided into the Royal Society; Boule remarked: '*Naturellement, ce mémoire, magistralement composé et rédigé, fit une grande impression en Angleterre*' ('Naturally, this memoir, masterfully composed and written, made a great impression in England').[50]

The Royal Society, founded in 1660, stood foremost in age and respect among all the learned societies of Britain, and Lankester used his position in both the Royal Society and the PSEA to further the Pre-Palaeolithic cause. When Moir employed a collector and workman in 1912 and 1913, the work was paid for by the Committee of the Royal Society and was published in the *PPSEA* the following year.[51] Lankester used his standing in the PSEA to set up 'The Sub Crag Exploration Fund' to continue this funding for 1914. His draft proposal, which was to be sent to men of influence, was directed to Fellows of the Royal Society.[52]

In his first Royal Society paper on the pre-palaeoliths, Lankester highlighted the credentials of the geologists who had accepted the association of

[45] Boule 1915, 2–3.
[46] Prestwich 1892; Warren 1905; Lankester 1914; Moir 1916; 1921b.
[47] Warren 1913; 1922a; 1923b; Moir 1916; 1921b; 1922.
[48] Warren 1914a.   [49] Boule 1915, 6.   [50] Boule 1915, 8; Lankester 1912b.
[51] Moir 1913a.
[52] Lankester to Moir, 6 November 1913: BLL: Add.Ms.44969/95–97. The signatories to Lankester's 'Sub Crag Exploration Fund' document were Archibald Geikie, President of the Royal Society; Arthur Evans; Hercules Read, Keeper of Ethnology at the British Museum; John Marr; and Henry Balfour of the Pitt-Rivers Museum.

these specimens with the Red Crag: Whitaker and Marr were not only past Presidents of the Geological Society; they were also Fellows of the Royal Society.[53] In another paper, delivered to the Anthropological Institute, Lankester observed that although many researchers had accepted the human manufacture of the rostro-carinate, 'I do not intend to "count heads" nor to cite a list of the names of those well-known experts in the study of ancient flint implements'. He continued, nonetheless, with a list bloated with Fellows of the Royal Society: 'Such a list of names would include those of Sir Arthur Evans, F.R.S., Sir Hercules Read, F.R.S., Dr Flinders Petrie, F.R.S., and the late Lord Avebury'. John Lubbock (Lord Avebury to Lankester), another F.R.S., had given the Plateau eoliths and the pre-palaeoliths his approval in the seventh edition of *Prehistoric Times*, and was a prized addition to Lankester's list of respected names.[54]

Lankester's opponents, however, were concerned about this link between the pre-palaeoliths and the Royal Society. Dawkins begged Sollas to attack the rostro-carinates 'in the names of archaeology and geology' when they came up at the next Royal Society meeting; he complained that Moir and Lankester were 'lowering the character of the Phil. Trans' and 'misleading' those who assumed the Society gave the contents its stamp of approval.[55] Henry Howorth (1842–1923), ex-Member of Parliament for South Salford and a dedicated Palaeolithic researcher after his retirement, was no less dismayed that such material had infiltrated the Royal Society. He wrote to Dawkins:

it is scandalous that the first scientific society in Europe should publish such rubbish as Moir writes and that the Society should give it its *imprimatur*. The whole difficulty is the way in which the geologists are represented on the Council. The papers are sent I am told to the geological committee of the Society & rather than have a row with Lankester they pass anything.[56]

There were loose patterns to the choices made by supporters and sceptics about where to deliver their views on the Pre-Palaeolithic. Both were attracted to societies that had welcomed papers on the Palaeolithic, although critical papers that discussed flint fracture often appeared in geological arenas. There were, however, no firm rules. Moir, who viewed himself as a prehistoric archaeologist, directed many of his papers towards the *PPSEA*; but Haward delivered his counter-arguments on flint fracture to the same journal.[57] The Royal Anthropological Institute received papers from Moir and Lankester, but also published Warren's work on flint fracture.[58]

[53] Lankester 1912b, 285.   [54] Lubbock 1913, 421–423; Lankester 1914, 4.
[55] Dawkins to Sollas, 24 June (?1912): UMO: JP, Box 1.
[56] Howorth to Dawkins (?1920): BMD: WBD, 70493/1.
[57] Haward 1912; 1914; 1919; 1921.
[58] Lankester 1914; Moir 1916; Warren 1905.

Warren frequented geological societies more frequently than Moir or Lankester. He had a secure position within the Geologists' Association, where he served on the Council from 1913 until 1926, and he kept the members informed about his flint-fracture researches.[59] The paper Warren gave to the more prestigious Geological Society of London in 1920 also coincided with his term on their council (from 1917 until 1920).[60] Howorth contrasted the atmosphere in the Royal Society with that in the Geological Society, where Warren's paper was received with particular warmth: 'I was in good spirits and let fly vigorously & was much applauded. Reginald Smith was the only one on the other side and he made a hash of his case'.[61] It is noteworthy that Reginald Smith (1874–1940), who is not to be confused with Worthington Smith, worked at the British Museum and was not seen as a geologist. Moir, though, who did not regard himself primarily as a geologist, had nonetheless been elected to the Geologists' Association in 1908; although he preferred to launch his short defensive letters about Warren's conclusions from the pages of the *Geological Magazine*.[62]

The decision to present a case to a particular institution was taken carefully. Moir, Lankester, Haward, and Warren were influenced in their choices by personal habit, public position, and concern about the kind of reception they might receive. The prestige of a society would be reflected back on the papers it accepted, and personal standing within a society might assist a sympathetic reception; but opinions also had to survive the private dissections that took place after the paper had been read and published. Moir and Lankester wanted to share their reconstructions of a new industry with an interested audience; Warren, though, was more at home in geological circles, where his listeners understood his flint-fracture arguments.

6.5 WORDS AS WEAPONS

The care with which society rooms were selected for the presentation of papers was matched by wariness in the use of words. Lankester wanted to distance his East Anglian specimens from earlier finds made in Kent and Belgium, and emphasize their connection to accepted Palaeolithic industries. This desire was reflected in his terminology, and is revealed most clearly in his

---

[59] Anon 1913, 302; 1914, iii–iv; Warren 1923a; 1923c.
[60] Warren 1920.
[61] Howorth to Dawkins (?1920): BMD: WBD, 70493/1.
[62] Anon 1913, 301; Moir 1913c; 1915b; Moir and Barnes 1923.

reactions to the word 'eolith'.[63] The term 'Eolithic' seems first to have been used in *Musée préhistorique* in 1881, a volume co-authored by the father-and-son team of Gabriel and Adrien de Mortillet.[64] This new label was popularized in Britain after Brown used it to demarcate Harrison's rude Plateau specimens from the Palaeolithic implements of the river drifts: 'Eolithic; Roughly hewn pebbles and nodules and naturally broken stones, showing work with thick ochreous patina, found on the plateaux of the chalk and other districts in beds unconnected with the present valley drainage'.[65]

Lankester's early support for Harrison's Kent eoliths has already been mentioned. Later, however, when he received a box of early East Anglian discoveries from Moir, Lankester was swift to warn their finder about the dangers of the old term 'eolith'. He advised, 'I think the word Eolith & even Palaeolith should be avoided. "Worked flints of the so-&-so beds or horizon" is the best sort of term'.[66] Moir was soon defining his finds as 'pre-palaeoliths' that occupied 'a stage of culture mid-way between the Eoliths proper and the river-drift palaeoliths,' because they combined the characteristic surface-flaking of palaeoliths with the edge-chipping of eoliths.[67] By avoiding the word 'eolith', Lankester hoped to evade the negative associations of the Kent finds; his opposition to Moir's work on the Kentish ancestry of pre-palaeoliths, mentioned earlier, had been prompted by similar fears.[68]

Sloppy references to 'eoliths' were seized on by Lankester. In the first edition of his *Ancient Hunters*, Sollas explained how the use of this term 'has been extended to similar objects of any age earlier than the Palaeolithic'.[69] Lankester disagreed. Sollas was castigated in the *Saturday Review* of 16 March 1912 for this loose definition, for applying the term to Breuil's Belle-Assise flakings, and for ignoring the evidence of the rostro-carinates; Lankester asserted: 'It is not the part of a man of science to sweep away all such evidence with the exclamation "Eoliths!"'[70] He held a similar grievance against Charles Dawson (of Piltdown fame) who, at a meeting of the Royal College of Surgeons, 'used the silly word "Eolith" which is always done for the purpose

---

[63] Moir 1935, 93–95.

[64] De Mortillet and de Mortillet 1881. This origin of the term 'Eolithic' has also been noted by Grayson 1986, 84 and De Bont 2003, 607. There is also a misconception that John Allen Brown coined the term (see MacCurdy 1907, 546; Osborn 1915, 265).

[65] Brown 1893, 94.

[66] Lankester to Moir, 29 December 1910: BLL: Add.Ms.44968/16.

[67] Moir 1911, 22.

[68] Lankester 1912b, 311; 1912c, 250; Lankester to Moir, 8 June 1910, and 29 December 1910: BLL: Add.Ms.44968/12–13, and 44968/15–16.

[69] Sollas 1911, 55, footnote 1.

[70] Lankester 1912d, 333–334.

of making confusion and false suggestion'.[71] He also deplored as foolish its use 'by various continental people'—which was probably a dig at Aimé Rutot. Lankester applied the term 'Eolith' solely to the Kentish specimens, and instructed Moir, 'Keep the question of Eoliths on such flints *apart*. You have got *some* worked flints which are certainly Proto-Chellean'.[72]

That word 'Eolith' *must* be abandoned. Whenever one says a thing is an eolith, people think one is a fool, and accepts *all* that anyone has ever called eolithic. I think we must use 'prae-Chellean' instead, meaning 'prae river valley-gravel'. The Chellean is the oldest set recognised by the French, & we need not bother about the words Strepyan & Mesvinian of Rutot.[73]

The 'eolith' affair leaves no doubt about the power of words to raise or lower the tone of an argument, tones which might be missed now that their original strengths and meanings have dissipated. The judicious use of such terms as 'Pre-Chellean' or 'Pre-Palaeolithic' to describe the East Anglian industries helped Moir and Lankester to suggest that these were on the same respectable level as Palaeolithic industries, and were quite distinct from the less reputable 'eoliths'. It is interesting to see the concern that this strategy caused among supporters of the eoliths. Rutot wrote to Moir in 1912, accused him of suppressing the eoliths and argued that rostro-carinates *were* eoliths. But whereas Rutot believed that eoliths had not been intentionally shaped, but 'accommodated' (the edges blunted) to fit the hand, Lankester and Moir were adamant that their pre-palaeoliths had been shaped.[74] Lankester's response was indignant: he asserted that Rutot was 'a vain ticklish old fellow & can't bear to see any one in the same field with himself', and, with marvellous unconscious irony, added 'the old silly is merely anxious about a word'.[75]

6.6 STRATEGIES IN PRINT, IN PERSON, AND ARMED WITH SPECIMENS

Besides sending the right message through the careful use of individual words, the overall style of publications had to strike the right tone if the writer wished to gain the reader's confidence. Despite variation in styles between different

[71] Lankester to Moir, 27 February 1911: BLL: Add.Ms.44968/25–26.
[72] Lankester to Moir, 22 April 1911: BLL: Add.Ms.44968/31–32.
[73] Lankester to Moir, 6 March 1912: BLL: Add.Ms.44968/124–125.
[74] Rutot 1900, 712, 714; De Bont 2003, 608.
[75] Lankester to Moir, 23 May 1912: BLL: Add.Ms.44968/161–162.

specialist subjects, societies (local or national), and individuals, there were also general rules to be followed in constructing a convincing scientific case. Lankester became concerned that Moir's 'pugnacious & suspicious' temperament and 'baseless suggestions' about opponents was 'not the way to learn all the facts and to become a real leader in the subject'.[76] He often wrote to remind Moir of the benefits ensuing from appearing detached, methodical, 'scientific', and therefore 'expert' in print.

Lankester taught Moir the subtleties of writing slippery articles: 'Do not mind my telling you that you should not say as you often do "It is quite certain that" or "there is no doubt possible etc", but rather "It seems probable that" or "there is much reason to suppose" '.[77] When McKenny Hughes, the elderly Professor of Geology at the University of Cambridge, attacked the rostro-carinates in a communication to the Cambridge Antiquarian Society, it was Lankester who advised Moir on the style of his response: 'be *very careful* to treat him as a poor deluded nice old gentleman, who means no harm. Say it is to be regretted. Point out that he has written without knowledge, but speak of him in a forgiving spirit'.[78] Lankester also had to restrain Moir from personal confrontation; he explained the advantages of a precise and clinical published account: 'I don't see how Mr. Hazzledine Warren can do or say anything about your sub-Crag implements, and I should leave him alone. [...] They must *wait* now if they wish to talk about your things, until my account of them is out. There is no hurry!'[79]

A published article carried authority, particularly if it appeared under the aegis of a prestigious society; the personal touch could also be useful if handled with delicacy. Moir recalled a steely moment after he had read his paper to the Royal Society in 1917 on the passage from rostro-carinates to hand-axes, when Lankester 'looked round in a threatening way at the audience and said: "If there are any present who do not accept these conclusions, Mr. Moir and I will no doubt be able to deal with them" '.[80] Fierce disagreement in society rooms might, however, make the hopeful persuader look foolish, and on another occasion Lankester cautioned Moir: 'I should strongly advise you to let Sollas alone. No one cares about these wrangles. Sollas will have to *eat his words*, as soon as my paper on the Wittlingham flint is out. It is far best to let people like Sollas jaw. They know *nothing* about the subject. The more you argue with them, the more importance they get. *Leave him alone*'.[81]

---

[76] Lankester to Moir, 23 January 1911: BLL: Add.Ms.44968/21–22.
[77] Lankester to Moir, 18 January 1911: BLL: Add.Ms.44968/17–18.
[78] Lankester to Moir, 16 March (bound with letters from 1911, but must date to 1915): BLL: Add.Ms.44968/27–28. See also Hughes 1913, 58–64 and Moir 1915c.
[79] Lankester to Moir, 4 January 1912: BLL: Add.Ms.44968/106–107.
[80] Moir 1935, 78.
[81] Lankester to Moir, 31 July 1913: BLL: Add.Ms.44969/71–72.

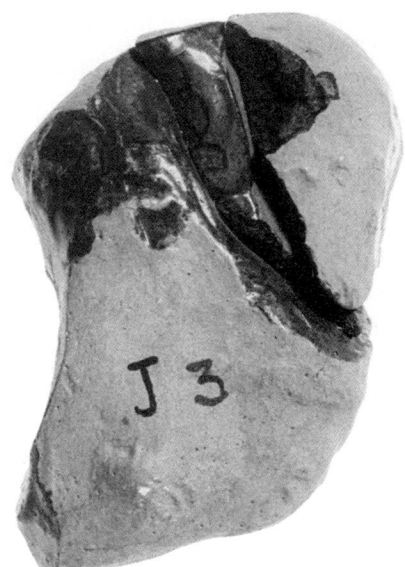

Figure 6.5. One of the Eolithic casts that Warren displayed at meetings to convince his viewers that eoliths had been fractured by natural processes: 'Fractured flint with 2 flakes replaced as found' (PRM: Accessions Book No. X, p. 454; PRM: 1940.12.161.3).

Lankester also believed in the value of good published illustrations, and frequently warned Moir: 'I don't think you realize how much *weight* of conviction is carried by a first-rate set of drawings'.[82]

Another useful strategy was to confront wavering sceptics directly with the physical evidence. When leafing through old society journals, brief references to past exhibits can occasionally be found interleaved between lengthy published papers, an indication of an elusive aspect of debate that had once provided the focus for much energy. The exhibition of specimens in society rooms or museums, to illustrate papers or to promote general discussion, was standard practice, as will be recalled from the scene at the start of Chapter 5 when Lankester confronted Dawkins with his flints at a Royal Society *soirée*. Warren countered with casts and flints that illustrated the role of natural pressure in chipping flints, which he displayed at the Geologists' Association and the Geological Society (see Figs. 6.5 and 6.6).[83] At the Geologists' Association, garrulous pro-Eolithic and anti-Eolithic exhibitors tended to block the narrow passage at the end of the tables which ran down the centre

[82] Lankester to Moir, 2 March 1922: BLL: Add.Ms.44970/207. See also Moir 1935, 58–64.
[83] Anon 1900a, 60; 1909, cxxv; 1913, 302; 1914, iii–iv.

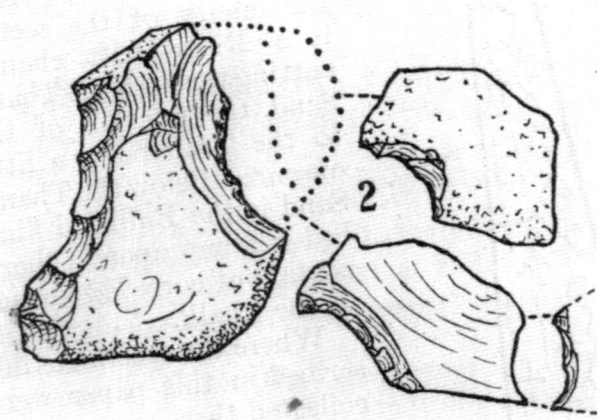

Figure 6.6. Warren also published illustrations to convince his readers of his arguments. This is his drawing of the eolith from which the cast in Figure 6.5 was made, demonstrating how the two larger chips refit (Warren 1920, 240, Fig. 2). By permission of the Geological Society of London.

of the main library of University College London, to the irritation of members wishing to pass through to see other, less controversial, displays.[84]

Moir's sub-Crag specimens made their first substantial debut in the Geological Society on 7 November 1910, and thereafter enjoyed a varied career.[85] In 1912, Lankester informed readers of *Nature* that Moir's specimens could be seen at the British Museum: Moir later suggested that they might view the results of his experiments at the same prestigious location.[86] It must have helped that Reginald Smith, an Assistant Keeper at the British Museum, was sympathetic to the pre-palaeoliths; but Moir could not prevent Warren from visiting the exhibit and publishing critical remarks.[87] When rostro-carinates were later on display at the Museum, Moir was very particular about *how* they should be displayed (Fig. 6.7). He wrote to Sir Hercules Read, the Keeper and Smith's superior, 'May I suggest that when exhibited, the specimens should be posed in the rostro-carinate manner [sketch] not as most palaeoliths are shown [sketch], as then their significance is more easily recognized'.[88]

---

[84] Brown 1958a, 35.
[85] Lankester 1912b, 285.
[86] Lankester 1912c; Moir 1912b, 463.
[87] Warren 1913, 37–38.
[88] Moir to Read, 22 November 1919: BM(F): Misc. Doc. Files, Moir.

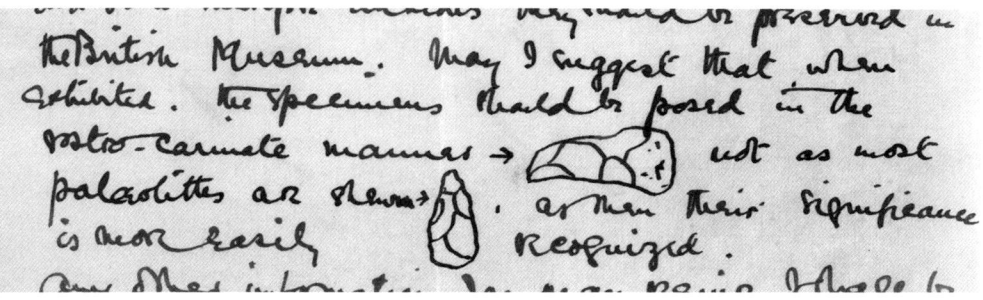

Figure 6.7. Moir's suggestion to Read about how his rostro-carinates ought to be positioned in a forthcoming exhibition at the British Museum (Moir to Read, 22 November 1919: BM(F): Misc. Doc. Files, Moir).

The choice of where to display exhibitions followed similar reasoning to the choice of where to present papers. Lankester suggested to Moir where he might best exhibit his finds. There was the Geological Society, 'the place where you have most *claim* to exhibit since you are a fellow'. Then there was the Royal Society, where Lankester had 'obtained a promise from the Secretary [...] that I shall have a large table'. Lankester also outlined plans for cases in the Jermyn Street (Geological Survey), British, and Natural History Museums, the latter requiring 'careful coaxing of Harmer & also Smith Woodward'. He continued, 'Once you can *place* the things on view, you can *send* people (through a special note in "Nature") and get one of the officials to keep the key & show the things to visitors'.[89] A simple display might have been eased into existence through complicated negotiations that relied on personal standing in societies, knowing the right people and approaching them appropriately. The display then had to be advertised, although *Nature* was used by Moir and Lankester for attacks as well as announcements.[90]

The presentation of the pre-palaeoliths was, then, carefully managed, and Lankester kept a watchful eye on Moir's activities. Moir had a firm foothold in his local society, the PSEA; Lankester approached the national societies with a combination of cautiously written articles and a combative personal presence. They favoured arenas that had welcomed Stone-Age research in the past. The sceptics, however, who did not believe in the archaeological nature of the pre-palaeoliths, often delivered their arguments to geological societies. Both sides set up exhibitions: some displayed pre-palaeoliths; others, the results of flint-fracture experiments. Each gathered together a network of supporters,

[89] Lankester to Moir, 7 February 1921: BLL: Add.Ms.44970/143–146.
[90] Lankester 1912c; 1912d; 1921, Moir 1912b; 1912c; 1916; 1921c.

winning them over in person, through papers and correspondence, and by direct confrontation with exhibits.

6.7 'THOSE TRICKY FRENCHMEN': 'INTELLIGENT MEN & EXPERTS'[91]

Personal confrontation was a common strategy in society-room debates, and considerable efforts might be made to gain the good opinion of learned society lions. The benefits of converting a prominent, respected opponent were immense: such individuals could initiate an extensive shift in opinion. Many of the most venerable British Palaeolithic researchers, though, were no longer to be seen in society. Evans had died in 1908 and Prestwich in 1896, his passing much regretted by Harrison. Dawkins, fierce in his attacks on the eoliths, was alive but not particularly lionized. But powerful opinions on the eoliths had also been expressed in France, and in 1913 Lankester complained to Moir, 'Those tricky Frenchmen have not yet published any notice of my paper or of their visit to England'.[92] Lankester was referring to a visit by Boule and Breuil to Ipswich in September the previous year, when he and Moir had tried to convert these 'tricky', but influential, Frenchmen who had published such devastating attacks on the eoliths in the past.

Around the time of this visit, some British researchers complained that the French were taking the lead in Palaeolithic questions.[93] They had the rich Palaeolithic deposits of the Somme Valley and the caves of the Dordogne, and the papers of Boule and Breuil had given them a prominent place in the Pre-Palaeolithic question of natural *versus* human flaking (see Chapter 5).[94] In addition, Prince Albert I of Monaco (1848–1922) had set up a prestigious Institute of Human Palaeontology in Paris in 1910. Boule directed this new Institute; Breuil was his Professor of Prehistoric Ethnography.[95] The opinions of these two individuals were much respected in Britain; if they converted to the Pre-Palaeolithic cause, British sceptics might follow their lead.

The visit of Boule and Breuil to Ipswich in September 1912 was remarkable for three reasons. First, the manoeuvrings of Lankester and Moir as they tried to encourage a favourable impression of their pre-palaeoliths. Second, the

---

[91] Lankester to Moir, 18 March 1913, and 19 September 1912: BLL: Add.Ms.44969/59–60, and 44968/193.
[92] Lankester to Moir, 18 March 1913: BLL: Add.Ms.44969/59–60.
[93] Sturge 1908, 9; Boule 1915, 1–2.
[94] Boule 1905; Breuil 1910.
[95] Boule 1912b; Burkitt 1925, 14; Vallois 1941–46, 208; Brodrick 1963, 65–68.

conflict between the Englishmen's belief in their 'artefacts' and the Frenchmen's grounding in flint-fracture research. Third, the differences between Breuil's response to the pre-palaeoliths in 1912 and in 1920, on his second visit to Ipswich—of which more below.

Lankester had been trying to gain support for the pre-palaeoliths on the Continent for months before the Frenchmen arrived. He had urged Moir to travel to Paris and Brussels, armed with Lankester's letters of introduction, so that he could show his best pre-palaeoliths and illustrations to people like Boule and Rutot.⁹⁶ He had sent casts to the Museum of Saint-Germain-en-Laye via Boule, who was invited to come to England to see the originals and the sediments in which they had lain.⁹⁷ He had suggested to Moir that if Breuil could not come to Ipswich, Moir might instead send casts of his '*best & most convincing* specimens'.⁹⁸ Boule and Breuil, however, both confirmed that they would arrive in September 1912. Lankester had hoped to welcome their patron as well, but Prince Albert of Monaco did not choose to attend.⁹⁹

News travelled fast. Several British researchers became excited at the prospect of such distinguished visitors, but Lankester was determined to keep the event under his thumb. He informed Moir, 'there must be no fuss & intrusion of strangers or even local geologists etc. They want to go over the thing with you and me'.¹⁰⁰ Sollas, who at this time was still critical of the rostro-carinates, was still determined to join the September party; Lankester's efforts to stop him reveal the value he attached to orchestrating this visit. He began by complaining to Moir: 'Sollas wants to come. He will only be a nuisance, but I shall try to manage him. He will inevitably try to draw their attention off, on to *other* & irrelative problems'.¹⁰¹ Lankester then wrote to Sollas, forbidding him to attend:

with regards to Boule & Breuil. They are as you know coming over expressly to see for themselves the sub-Crag implements and the actual conditions in which they are found. I have given up my holiday & have gone & shall go to expense & trouble in various ways to ensure their seeing what I wish them to see, and I do not wish to have any chance of their attention being turned from the evidence.

If they should be able to go to Oxford *after they have done with me*—well and good. But *this* (here) is my special and, as I think, very urgent affair and I must beg

⁹⁶ Lankester to Moir, 4 January 1912: BLL: Add.Ms.44968/106–107.
⁹⁷ Boule 1915, 9.
⁹⁸ Lankester to Moir, 31 May 1911: BLL: Add.Ms.44968/39–40. See also Lankester to Moir, 29 May 1911: BLL: Add.Ms.44968/37.
⁹⁹ Lankester to Moir, 5 September 1912: BLL: Add.Ms.44968/188–189.
¹⁰⁰ Lankester to Moir, 2 July 1912: BLL: Add.Ms.44968/172.
¹⁰¹ Lankester to Moir, 5 September 1912: BLL: Add.Ms.44968/189.

you to leave me un-embarrassed and to leave them also free to give their attention to the matter for which they have undertaken to come over. I do not want any assistance in explaining to them the geology of this district—nor do I wish to have other audience & witnesses to distract my attention & theirs. I do not know how long it may take or how long they may be able to give, & I desire that no unnecessary limit shall be put to their enquiry here, or to my guidance of their movements.[102]

Even Marr was not allowed to come, despite his role in confirming the ancient geological context of the pre-palaeoliths.[103] When Moir asked Lankester if Marr could attend, the response was apoplectic:

I most emphatically object to *any one* coming to take part in our inspection or to 'look up' the French geologists whilst they are with me. No one ought to know when or where they are coming, excepting you and me. This is a serious job and not a pic-nic: and if any one tries to join us or take part in our visitation, I shall decline to let them do so. It would be too fatiguing, full of delays & jaws & altogether *a failure*. [. . .] What we have to do is to bring Boule & Breuil face to face with *the facts* and, since they are intelligent men & experts, let them draw their own conclusions. Whatever information they want, you & I can give them and no one so well.[104]

In the same letter, Lankester set out the subjects that were permitted within the hearing of the Frenchmen, and those that were forbidden. He told Moir:

Pray do not (unless there is time after we have seen everything) raise the question of fracture by sand-pressure, or any other story than that of these sub-Crag flints. Once that is dealt with, all other hares may be let loose for them to hunt. But not before. [. . .] I do not wish any one to be told what day we expect B & B, nor where they will stop or go or when they will leave.[105]

This mention of sand-pressure was a reference to Breuil's paper on natural fracture at Belle Assise, published two years before, which (together with Boule's earlier paper on Mantes) had raised doubts in Britain about the human workmanship of eoliths. Lankester valued the good opinion of the Frenchmen, both 'intelligent men & experts', too much to leave it to chance; he tried hard to control their environment and influence their judgment. Even Moir had to be warned against attacking them or distracting them from the main point: the sub-Crag flints.

---

[102] Lankester to Sollas, 19 September (undated, but must be 1912): BLO: MS.Eng.lett.d.329, fol.14.

[103] Whitaker and Marr 1911.

[104] Lankester to Moir, 19 September 1912: BLL: Add.Ms.44968/192–193.

[105] Lankester to Moir, 19 September 1912: BLL: Add.Ms.44968/193.

6.8 'THOSE TRICKY FRENCHMEN':
'TWO HASTY FOREIGNERS' [106]

As events turned out, the Frenchmen were unimpressed by the pre-palaeoliths of East Anglia. They left Ipswich unconvinced. Moir was furious. Lankester never forgave Breuil, though he wrote to calm Moir's wrath and to prevent him from aggravating the problem:

My dear Mr. Moir,

Pray keep a calm attitude about Boule & Breuil. A matter of inference based on tangible evidence, can not be dealt with by the mere opinion of two hasty foreigners who deliberately chose not to consider the specimens & other facts which are opposed to their fanciful notions.

Who cares if Sollas or any one else is rejoiced? Don't allude to it, or take any notice of it. By & bye I shall have to write more on the subject, but *note this*—neither Breuil nor Boule made any close observation of the really best and convincing specimens: Breuil talked like a conceited school-boy who had just learnt some terms about flints & wanted to show it. His manner was impudent & arrogant (that of the priest) and he absurdly said, abruptly of the Lakenheath specimen of Sir John Evans 'That is simply a chisel.' Of course it is not. But he kept trying to make 'assertions' showing his shrewdness. I don't consider that either he or Boule have any adequate knowledge of flints, especially older palaeolithic ones—Remember Breuil's absurd notion that the weight of sand could fracture the flints of Belle Assize. Boule was very anxious to *assert* that the striae are not due to glacier action. Who cares whether they are or not? Breuil wanted to trace early fracture to one of the scratches. Silly!

Yours sincerely,

E. Ray Lankester.[107]

So the 'intelligent men & experts' became 'two hasty foreigners' who had dismissed Lankester's selection of convincing specimens too rapidly.[108] Their rejection of the Lakenheath specimen (Fig. 6.8), a 'rostro-carinate' that had been discovered in the collection of the late Sir John Evans himself, was particularly hurtful.[109] Evans would, no doubt, have been indignant at the suggestion that his implement was a rostro-carinate; the Frenchmen were

---

[106] Lankester to Moir, 18 March 1913, and 15 October 1912: BLL: Add.Ms.44969/59–60, and 44969/7.

[107] Lankester to Moir, 15 October 1912: BLL: Add.Ms.44969/7–8. 'Striae' were scratches on the flints: some thought these had been produced by glacial action; others, by the pressure of sediments above.

[108] Lankester to Moir, 19 September 1912, and 15 October 1912: BLL: Add.Ms.44968/193, and 44969/7.

[109] Lankester 1912b, 295–299.

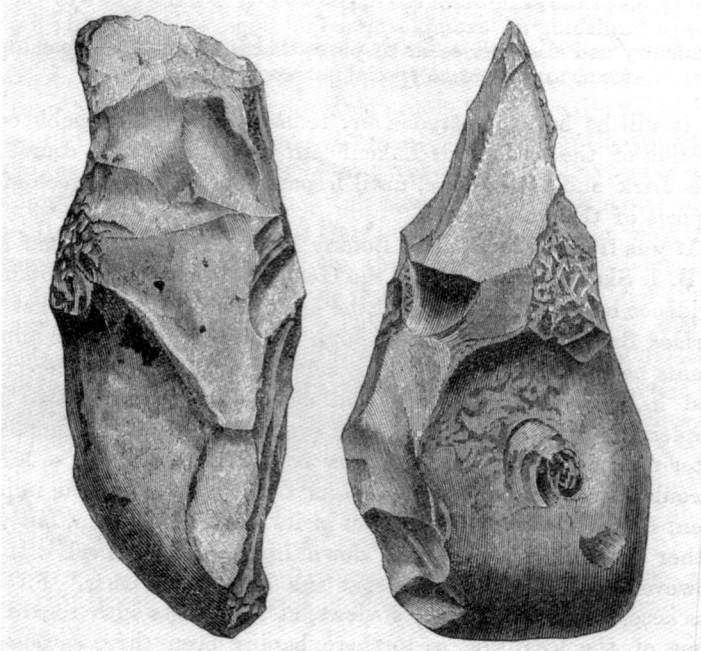

Figure 6.8. The Lakenheath implement, claimed as a rostro-carinate by Lankester (Evans 1897b, 567, Fig. 444).

equally unimpressed by this attempt to deploy the name of a hero from the past to fight a present cause.

Far from offering themselves as willing converts, Boule and Breuil had tried to convert Moir and Lankester. In a move that seems to have surprised their hosts, they had used this field-trip to explain and illustrate their own work on natural flaking. Lankester recalled:

The more I think it over the more I am disgusted by Boule & Breuil. They were and behaved as absolutely prejudiced persons—not fair-minded enquirers. Boule with his 'histoire de Mantes' & Breuil with his 'affaire de Bel assize' wish in the most arrogant & childish spirit to bring everything concerning such flints, as are not of the most obvious & orthodox character & history, under one or other of these two explanations.

From the first they started on this line and never *even looked* carefully at the best specimens which are so definite in shape & in chipping as to render their 'Mantes and Belle assize' theory absurd.[110]

---

[110] Lankester to Moir, 12 October 1913: BLL: Add.Ms.44969/79–80.

A few months after leaving Ipswich, Breuil echoed Lankester's opinion that the other side was absurdly prejudiced. He remarked to Haward, 'as with all their "Eolithic-loving" confreres, it is difficult to discuss [matters] with these gentlemen. They affirm their opinions with too much enthusiastic conviction which prevents them from appreciating the rights of others to doubt'. Just as the English pair had shown him their most convincing specimens, Breuil had shown Moir and Lankester 'the best flints from *Belle Assise*', but 'M. Reid Moir would not say that these were not made by Man. Sir Ray Lankester was more prudent: he said "that they were not due to Pressure"'.[111]

Boule published a considered opinion of the pre-palaeoliths of East Anglia in *L'Anthropologie* three years after his visit to Ipswich. He described how these flints had been carefully selected from hundreds of less attractive neighbours and criticized the reasoning that had led Lankester and Moir to think that they might have been made by humans. Moir informed the readers of the *Geological Magazine* that this article was 'the most extraordinarily biased statement it has ever been my ill-fortune to read'.[112]

The observation of a site in the field or a specimen in the hand could often be more persuasive than printed argument, but not in this case. All four participants saw themselves as experts in their field; all had published their opinions in the past; none were willing to change their minds. The Frenchmen showed the Englishmen evidence of natural fracture and made them angry; the Englishmen showed the Frenchmen their best specimens and were surprised at their failure to see past the detail of the flaking to the artefact beneath. The clash of approaches was matched by a clash of personalities. Lankester was annoyed that Breuil had not been more deferential, and described him as a 'conceited little Jesuit'; Moir recorded that 'his feelings on the matter were not by any means concealed during our tour'.[113] Moir, Lankester, and Breuil have all been described by their peers as impatient, and the meeting of three hot-headed individuals must have been tense.[114]

---

[111] Breuil to Haward, 27 February 1913 (translated typescript): BM(F): 'Cromer, Norfolk', F. N. Haward coll.

[112] Boule 1915, 16–31; Moir 1915d, 476.

[113] Lankester to Moir, 18 April 1915: BLL: Add.Ms.44969/159–160; Moir 1935, 143.

[114] Burkitt recollected that Breuil 'had an electric, not to say impatient, temperament' (Burkitt, transcript of radio broadcast, transmitted on 5 March 1962: ULC: 7959, Box 3). Lankester was described in his obituary as 'a man of strong feelings' with an 'impetuous temperament' (Goodrich 1930, xiv). Burkitt 1944, 369 remarked on Moir's 'characteristic and ferocious brand of language'; another obituarist reported, ' "sweet reasonableness" did not enter into Moir's method of controversy' (Keith 1944, 741).

6.9 THE CONVERSION OF BREUIL AT FOXHALL HALL

Although the arguments used by Lankester failed to strike the right chord with Breuil in 1912, later events showed that he was right to value Breuil's conversion. In 1920, Breuil visited Ipswich again. This time he viewed a series of Pre-Palaeolithic flakes, found by Moir the previous year in an old Crag pit at Foxhall Hall, east of Ipswich. Here, Moir had observed two layers of finely stratified flakes (Fig. 6.9) *within* the Red Crag deposits (Fig. 6.10). The flakes were found in sand, they were not in contact with each other, and seemed to represent convincing evidence of an ancient working floor lying *in situ* in a context where natural fracturing of flint was impossible. Moir had also taken care when excavating in this pit to stop post-Crag specimens falling down into these lower horizons.[115]

Miles Burkitt, a *protégé* of Breuil, told the tale of how Breuil came to view these finds. Burkitt had lectured on prehistoric archaeology at the University of Cambridge since the end of the war. At this time, his lectures were voluntary—it would be a few years more before an official lectureship was created. In 1920, Burkitt was called in to examine some of the latest

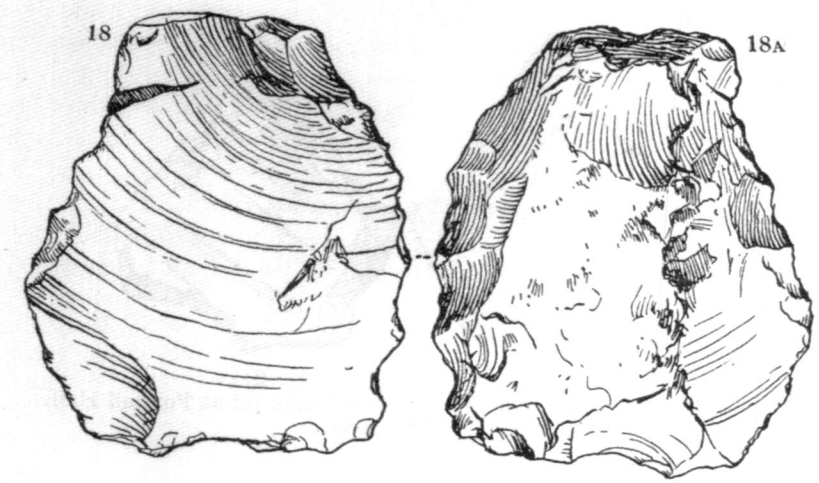

Figure 6.9. 'Two views of *racloir* [scraper] from pit at Foxhall Hall' (Moir 1921a, 409, Figs. 18 and 18a).

[115] Moir 1921a, 398–401; 1924a, 647.

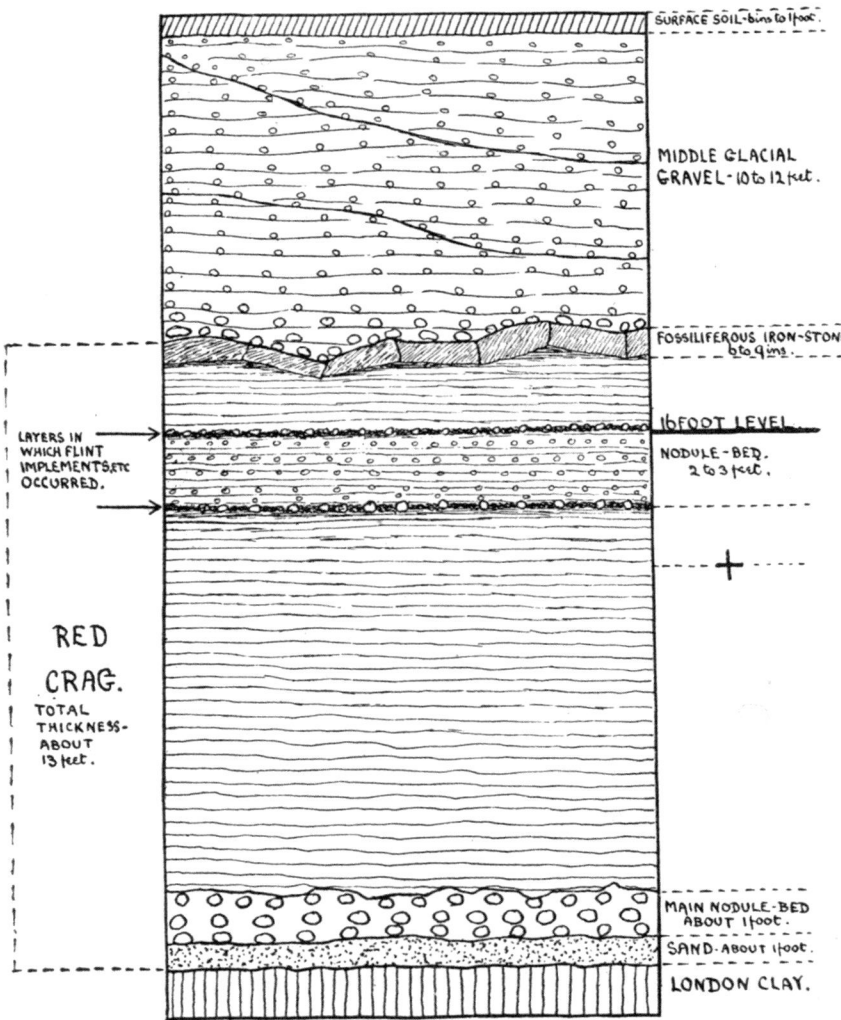

Figure 6.10. 'Diagrammatic section of western face of pit just south of Foxhall Hall' (Moir 1921a, 400, Fig. 3).

discoveries that Moir had sent to the Sedgwick Museum of Geology at Cambridge.[116] In turn, Burkitt invited Breuil to see these finds. After looking over Moir's flints at his home in Ipswich, Breuil accepted Moir's assessment of their origin. The situation in which they had been found seemed to preclude the possibility of natural fracture, and Breuil saw evidence of human workmanship in certain characteristics of fracture. Driving home with Burkitt, Breuil quietly observed, '*aujourd'hui a beaucoup vieilli l'humanité*' ('Today has greatly aged humanity').[117]

The news spread rapidly. Sollas, whom Moir had angrily described rejoicing at Breuil's earlier rejection of the pre-palaeoliths, now added a postscript to his 1920 paper in the *PPSEA*, congratulating Moir on a 'well-deserved triumph'.[118] Sollas also wrote to Dawkins in Manchester to explain why this event had catalyzed his previous doubts into a full turnabout. His letter was full of 'Breuil':

Dear Sir Boyd Dawkins,
    Things have happened since the Abbé was working in your laboratory! He has been to Ipswich & admitted many of Moir's flints from the Crag to be human implements. With this recognition on the part of Breuil the one objection which I made to the artefact nature of the specimens I described before the Geol Soc falls to the ground. So I am joined to Breuil, Marr, Birket [Burkitt] and the rest over this momentous issue. Nevertheless I went to London to listen to what Ray [Lankester] might have to say, this more especially as the rostro-carinates are not among the forms which Breuil accepts! But Ray was not there. Paper taken as read.
    Nothing has done more to retard the acceptation of Moir's implements than Lankester's insistence on the merits of this one type; it is this which has diverted attention from much more impressive documents.
    I may add that Moir has found a true working floor in the Norwich Crag with hearths I believe & goodness knows what. Breuil went to see it.
    Nothing is published yet. Breuil thinks the Piltdown skull is probably the same age as the Crag things.
    Ever truly yours
    W. J. Sollas.[119]

Sollas's letter to Dawkins reveals why these flints from Foxhall Hall were so much more convincing to Breuil than the specimens he had seen in 1912. On

[116] Burkitt, typescript notes on 'The Foundation of the Museum of Archaeology and Ethnology in Downing Street and of the teaching of Anthropology in the University of Cambridge' (undated), p. 17: ULC: 7959, Box 3.
    [117] Burkitt 1944, 369.
    [118] Sollas 1920, 267. This paper entered the *PPSEA* after it was rejected by the Geological Society of London, and described a particularly convincing sub-Crag specimen of Moir's that Sollas seemed almost to accept before concluding that he suspended his judgement.
    [119] Sollas to Dawkins, 27 June 1920: BMD: WBD, 70494. The connections between the Piltdown find and the eolith debates have been discussed by Spencer 1988, 1990.

that occasion, Lankester showed Breuil selected types; but the sceptical Breuil, steeped in flint-fracture research, was more interested in the evidence of human workmanship. In 1920, Breuil saw what Lankester might have viewed as insignificant flakes; but these came from a geological context where natural fracture seemed unlikely, and were therefore more persuasive to Breuil than any number of rostro-carinates. Indeed, Breuil still refused to accept the rostro-carinates as artefacts.[120] Sollas received a description from Burkitt of the features of the Foxhall flakes taken by Breuil as evidence for human workmanship:

Breuil at Ipswich paid no attention to a trimmed edge alone nor a wavy flake, even if there was a bulb. These latter are common in the boulder clay and Breuil considered them probably natural. What he considered most was the presence of fairly long flat flakes with bulb etc. and more so if these had a trimmed edge, the face of the flake showing no striae which could be associated with the trimmed edge.

He pointed out to me that the trimmed edge alone; or the flake alone; even the presence of striae did not preclude the possibility of human workmanship. But the association of (I) flake (II) trimmed edge (III) absence of striae made the human agency almost certain. I always feel too that the presence of small chipped awl-like forms is significant for surely nature would have broken anything so fragile![121]

Whether or not these were true artefacts fashioned by humans is beside the point (Palaeolithic archaeologists of today assert that they were not). What is at issue here is the information that Breuil considered relevant to the case. His attention was drawn to the character of the flaking (he associated bulbs of percussion and retouch with human activity) but his conclusions about their human origin were given further support by the absence of striae (scratches he believed to be the marks of pressure, although others had suggested that striae were made by glacial action). Breuil believed that these features, when taken together, were not explicable by natural processes. He was also impressed by the geological context of the finds, which not only supported a great age but also seemed to have preserved the flakes in the positions where they had been dropped.[122] In contrast, Moir's earlier discoveries, scattered far from their original positions, lacked this primary context that hinted at human activity. They had been selected from unstratified deposits of flints with no discernible layers, all mixed together and separated by millennia, not by minutes. Moir had been criticized for trying to arrange some of these mixtures into groups of different ages merely on the basis of patina, staining, form, and flaking.

---

[120] Breuil 1921b, 357.
[121] Burkitt to Sollas, 10 October (no year, but must be 1920): BGS: GSM1/445.
[122] Breuil 1921b, 357.

Breuil's conversion cut a swathe through the British opposition, just as Moir and Lankester had hoped it might years earlier. Indeed, the quantity of prestigious reversals even alarmed Warren, who delayed giving his opinion on the Foxhall Hall flints, in part because 'the change of opinion of so many of the highest authorities has been a severe shock, which gave one pause to think twice'.[123] In France, Breuil seems to have persuaded almost everyone to accept the human origin of the pre-palaeoliths in a series of meetings of the Institut International d'Anthropologie in 1921 and 1922; Boule, though, was still not convinced.[124] A commission selected by this Institute set off for Britain in September 1922. They visited Moir's sites (including Foxhall Hall) and saw several other collections, including Warren's pressure-flakings from the sub-Tertiary Bullhead Bed at Grays, Essex. This was Warren's finest discovery of a natural flaking site: it had yielded a series that Breuil thought approached the Belle Assise flakings, and included rostro-carinate forms. The members of the commission, however, concluded that Warren's natural flakings from Grays were different in character from Moir's sub-Crag specimens.[125] The tables had turned since Breuil's dismissal of Lankester's Lakenheath rostro-carinate in 1912.

The Foxhall and the older sub-Crag finds also inspired the American palaeontologist Henry Fairfield Osborn (1857–1935) to change the well-known correlation table he had published in 1916, in his book on *Men of the Old Stone Age*.[126] Osborn was Research Professor of Zoology at Columbia University and President of the Board of Trustees of the American Museum of Natural History. He had been invited to join both institutions in 1891. At Columbia, he had founded the Department of Biology (later Zoology); at the Museum, he had promoted his particular field of interest: mammalian palaeontology. Osborn applied his research on the evolution of mammals to the problem of correlating the geological deposits of America and Europe, and he believed that Moir's finds at Foxhall had finally added humans to the Tertiary fauna.[127] Remarking on Osborn's paper in the *American Naturalist*, Lankester observed, 'One would think from what Osborn writes that the Foxhall flint really *started* your discoveries or made a great difference in their value—That seems to me to be an erroneous view of the case'.[128]

But how many would have agreed with Lankester that this was an erroneous view of the case? It seems that Lankester, once again, had not fully appreciated the perspectives held by Breuil and his other opponents. The

[123] Warren 1922a, 87.
[124] Burkitt 1921, 457; Breuil 1922, 226–229; Capitan 1923, 67; Lohest and Fourmarier 1923, 56.
[125] Warren 1920, 248; Hamal-Nandrin and Fraipont 1923, 58.
[126] Osborn 1916; Osborn and Reeds 1922, 471.
[127] Osborn 1922, 440. Osborn's life and work has been examined by Gregory (1938), Rainger (1991) and Regal (2002).
[128] Lankester to Moir, 6 March 1922: BLL: Add.Ms.44970/212.

eventual collapse of Breuil's opposition was indeed triggered by Moir's flints from Foxhall Hall, and it was this conversion that led so many others to reconsider their old opinions of the pre-palaeoliths. The Foxhall flints had relatively little merit in themselves; it was the response of a powerful figure-head that gave them value. Although Lankester recognized the importance of presenting his pre-palaeoliths in a strategic manner to prominent peers, the final conversion of Breuil took place on his own terms, and was based on an approach that was alien to the perspectives of Moir and Lankester. Breuil had remained unconvinced by Lankester's rostro-carinate, and he had not been swayed in the past by Moir's sequences of Pre-Palaeolithic industries. The corpus of what passed for accepted knowledge amongst fellow workers might have appeared self-evident on the surface but, digging deeper, many individ-uals could not see past their own approaches and expectations, and were persuaded by different aspects of the case.

Before moving into the Palaeolithic timescales of the next chapter, it will be helpful to take a final glance at the classifications and sequences of the Pre-Palaeolithic that were discussed earlier. The pre-palaeoliths were treated in a similar way to the palaeoliths, but the interest of these similarities is enhanced by the fact that most of the stones described in the past two chapters are now regarded as natural products. Their classification was not very remarkable: chipped stones were organised into sets, or 'industries', according to their date and character. But the desire to see a progression in skill through time must have been extremely strong for such patterns to be seen in sequences of Eolithic and Pre-Palaeolithic industries.

The treatment of variety within a single Eolithic or Pre-Palaeolithic indus-try is also revealing. Industrial definitions were weakened by too much variation in the character of tools that had been grouped together as a single industry, dating to a discrete period of time. Prestwich had been perplexed by the presence of river-drift forms in the collections of Kent eoliths; Harrison tried to suppress them (see Chapter 5). Moir, however, dealt with variety in a different way. He peered down the single line of Eolithic industries and divided them between a number of different, parallel sequences, separating the ancestors of the flake-dominated Mousterian from the precursors of the Chellean hand-axes.[129] This solution to the problem of variety would also prove attractive to Palaeolithic researchers faced with growing numbers of industries that overlapped in time and space. The next three chapters will examine the emergence of a standard sequence of British Palaeolithic indus-tries and the different ways of dealing with anomalous industries that fell outside their expected time.

[129] Moir 1919, 48, Fig. 7.

7

Chronologies of the Early
Twentieth Century

Dr Allen Sturge spelled out some of the problems facing researchers who worked on the British Palaeolithic in his first Presidential Address to the Prehistoric Society of East Anglia. For the Drift period, the main task was 'to ascertain the relative ages of the humanly-worked stones, and the number of the periods concerned'.[1] In 1908, when Sturge made this suggestion, he was troubled by suspicions that the French divisions—Chellean, Acheulian, and Mousterian—were too broad to encompass the variety of British tools. He called on the younger school of geologists to help solve the difficulty.

A few elderly figures, familiar from previous chapters, would accompany Sturge's younger school of geologists as they worked on various sequences that could give a date to the stone tools of Britain. James Geikie published *The Antiquity of Man in Europe* in 1914, still defending his sequence of interglacials.[2] Harmer was inspired to take up the glacial researches of his old friend Wood as the twentieth century dawned. The reflection left by the Glacial epoch in East Anglia also led the young geologist Percy Boswell (1886–1960) to consider the connections between boulder clays and Palaeolithic industries.

Dawkins continued to promote his classification of Quaternary mammals and to attack Geikie's views, but two newcomers took a different approach to the palaeontological sequence. Martin A. C. Hinton and Alfred Santer Kennard used the bones of smaller mammals and the shells of molluscs to reconstruct the geological history of the Thames Valley. They maintained the traditional antagonism of palaeontologists towards the theories of glacial geologists by suggesting controversial links between the river drifts and the glacial sequence.[3] Meanwhile, Warren, the eolith sceptic, developed his own opinions about the British Palaeolithic sequence and its place in geological time as he worked on Palaeolithic sites around Essex. Several different answers to Sturge's question about the relative ages of stone tools would be extracted with the help of these sequences of glacial deposits, bones and shells, and river sediments.

---

[1] Sturge 1908, 13.   [2] Geikie 1914, 113–114, 218–219.
[3] Hinton and Kennard 1905; Dawkins 1910, 246–247, 254; Hinton 1910; Kennard 1916.

7.1 THE GLACIAL DEBATES CONTINUE

The glacial sequences of the early twentieth century owed much to Wood, who had worked on the drifts of East Anglia in the 1860s and 1870s (see Chapter 2). That his tripartite framework (see Table 7.1) continued to structure interpretations decades after his death was due in part to the work of his old helpmate, Harmer (Fig. 7.1). The two geologists had forged a friendship from a shared interest in geology. After they met in 1864, Wood had drawn Harmer into his enormous, self-appointed task of mapping the glacial deposits of eastern England on the scale of one inch to the mile. He had assigned Harmer the glacial deposits of Norfolk and northern Suffolk to work on. But when Wood died in December 1884, Harmer was distraught: he abandoned geology for politics and became Lord Mayor of Norwich.[4]

Then, in the last years of the nineteenth century, Harmer suddenly reappeared at the meetings of the British Association. He had returned to geology and wanted to share his thoughts on the Pliocene Crags.[5] In 1899, James Geikie received a letter from Harmer indicating that his interest in glacial drift had also revived: 'I don't know whether you have ever been over the chalky boulder clay country of Norfolk & Suffolk. That deposit has not received much attention from geologists since Searles Wood's death but it sadly wants working out'.[6]

Harmer set to working out the boulder clays himself, but he now had a valuable assistant. In 1910, he reported that his recent investigations around East Anglia had been 'rendered possible to an old man by the fortunate invention of the motor-car, which literally gave me a new lease of geological life for field work'.[7] Harmer's distinction as one of the early motorized geologists was also convenient for his geologist visitors. They were used to travelling between pits, cuttings, and other exposures scattered around the countryside by train, dog-cart, horse, trap, bicycle, or on foot; but by motor car, they could see far more geology. The increasing use of cars by wealthier geologists was reflected in a minor episode of Geological Survey bureaucracy: the appearance in 1904 of Survey Circular No. 294c on the 'Use of Motor Cars and Motor Cycles on the Public Service'.[8]

---

[4] Harmer 1910a, 1–2; Harmer to J. Geikie, 19 March 1900: BGS: GSM1/512; Kendall 1923, 779; Boswell, autobiography, 1942–48, p. 23: ULL: D4/1.

[5] Harmer 1895a; 1895b; 1899.

[6] Harmer to J. Geikie, 25 August 1899: BGS: GSM1/512.

[7] Harmer 1910a, 3.

[8] Circular 294c, 1904: BGS: GSM2/276. The benefits of Harmer's car were remarked on by Bonney in a diary entry for 20 July 1908 (SMC: TGB, Diary of Tours 1907–1909). On the influence of motorized transport on Survey work, see Wilson 1985, 110–111.

Figure 7.1. Frederic Harmer
(1835–1923), glacial geologist
(Boswell, autobiography,
1942–48, p. 23: ULL: D4/1).

Assisted by his car, Harmer gathered material for several publications on the boulder clays of East Anglia, culminating in his memoir on *The Glacial Geology of Norfolk and Suffolk* (1910).[9] He suggested that the area had been traversed by two ice-sheets: one had travelled across the North Sea from Scandinavia, leaving the North Sea Drift (of Wood's Lower Glacial series); the other ice-stream had advanced later from an inland source in the north west, leaving the Chalky Boulder Clay (of Wood's Upper Glacial series). Harmer suggested that the period between the retreat of the first ice-sheet and the advance of the second (when Wood's Middle Glacial Sands and Gravels had been deposited) might represent one of the interglacial periods proposed by Geikie and Penck (see Table 7.1).[10]

Few geologists had adopted Geikie's multiple interglacials, but two glacial periods had long been included in the British glacial sequence. Harmer was

[9] Harmer 1902; 1904; 1910a; 1910b.   [10] Harmer 1910a, 3, 25.

not unusual in suspecting the existence of an interglacial period between the two. The Palaeolithic, of course, was generally supposed to post-date *both* ice-sheets, for reasons outlined in Chapter 2. Clement Reid had suggested in the late 1890s, in his report on the Hoxne sequence, that alternations of temperature might have occurred after the deposition of the Chalky Boulder Clay.[11] But as the twentieth century arrived, most English geologists still grumbled at Geikie for classing implementiferous deposits as 'interglacial' merely because glacial conditions had continued further north. They argued that the influence on southern England of these distant and minor glaciations was questionable—their tool-bearing deposits post-dated their local Chalky Boulder Clay, and that was sufficient for them to label the Palaeolithic 'post-glacial'.

Meanwhile, Harmer had become acquainted with a young local geologist who, a while later, would promote a different view of Palaeolithic times. This was Boswell, whom Harmer came to regard as his geological son. Boswell's youthful interest in science and geology had led him to disagree with his family. His younger brother entered the family printing business in Ipswich; Boswell decided to become a schoolteacher. He found the job dispiriting but it provided him with some independence, and he was later able to study geology at university. Courses in geology had been multiplying steadily and Boswell was one of a growing number of geologists who gained university degrees in the subject.[12]

Table 7.1. The East Anglian glacial sequences of Wood (1870b), Harmer (1910b), and Boswell (1914).

Wood, 1870	Harmer, 1910	Boswell, 1914
Upper Glacial series (Chalky Boulder Clay)	Chalky Boulder Clay left by the Great Eastern Glacier	Chalky Boulder Clay
Middle Glacial series (Sands and Gravels)	Middle Glacial series: possibly an interglacial	Middle Glacial series: a brief amelioration of climate, not an interglacial
Lower Glacial series (Lower Boulder Clay)	North Sea Drift (Cromer Till; Contorted Drift; Norwich Brickearth) left by the North Sea Ice	North Sea Drift (Cromer Till; Contorted Drift; Norwich Brickearth)

[11] Reid 1896, 411.
[12] Boswell, autobiography, 1942–48, p. 23: ULL: D4/1; Mitchell 1961, 17–18; Williams 2004. O'Connor and Meadowes (1976) discuss the expansion of training opportunities for geologists in the late nineteenth and early twentieth centuries, and the relationship between professionally employed geologists and other members of geological societies.

When he was a research student in geology at Imperial College, London, Boswell gave Harmer enthusiastic acknowledgement in the paper he published on glacial deposits in the *Proceedings of the Geologists' Association*: 'If the study of glaciology in East Anglia ever becomes an exact science, it is he who had laid the foundation stone'.[13] Nonetheless, Boswell preferred Whitaker to Harmer as his geological father. Whitaker's interests were broader than those of Harmer, and he was not disturbed if Boswell strayed from his own views. Whitaker was also, by many accounts, a patient, gentle, good-humoured man and one of the most popular members of the Geologists' Association.[14]

Boswell had joined the Geologists' Association in 1906 at the age of nineteen. He was elected to the Geological Society the following year, but had to borrow the eight pounds for entrance and subscription from a friendly businessman; his own father would not help.[15] The Geologists' Association, based in London, was an important centre for Quaternary research in the early twentieth century. Subscription rates were relatively cheap, the society was friendly, and the environment nurtured the interests of geologists who felt unable to meet the financial, scientific, or social requirements of the Geological Society. Although Quaternary geology was shedding its unfashionable reputation, they might have felt that their research was marginalized at the larger Society. The Geologists' Association prided itself on a cosmopolitan character and regular field excursions; newcomers and established figures might meet over a geological field-section or in lively meetings at University College London.[16]

When Boswell gave his paper to the Geologists' Association in 1914, he followed Harmer relatively closely in his assessment of the East Anglian glacial sequence—apart from the interglacial episode. Boswell disagreed with Harmer's interpretation of the Middle Glacial Sands and proposed instead that these had been left during a slight amelioration of climate within a single, long glaciation.[17] George Lamplugh of the Geological Survey had been making similar attacks on the interglacial theory, but with more vigour. Lamplugh believed that southern England had witnessed only one glacial period, albeit with minor climatic oscillations. He identified several different centres from which ice had expanded, agreed that individual ice-sheets might have advanced and contracted before they all disappeared completely, but saw no

---

[13] Boswell 1914, 151.
[14] Boswell, autobiography, 1942–48, p. 41: ULL: D4/1. On Whitaker, see Strahan 1925, xi–xii, Dewey 1926, 231, and George 2004, 59, 61–63.
[15] Boswell, autobiography, 1942–48, p. 34: ULL: D4/1.
[16] Sweeting 1958, 14–15.   [17] Boswell 1914, 129.

evidence for even a single interglacial episode during this long period of glaciation.[18]

Lamplugh's single glacial period was an extreme vision on the spectrum of views about the number and magnitude of climatic shifts during the Glacial epoch. At the other end of this spectrum lay the observations of James Geikie, who envisaged four major (and two minor) glaciations. Arguments over Geikie's multiple interglacials had been reinvigorated by the appearance of the final, authoritative report by Penck and Brückner on their work in the Alps. Their book, *Die Alpen im Eiszeitalter*, appeared in successive parts between 1901 and 1909 and described a sequence of four Alpine glaciations. It would not be long before other geologists joined Geikie in aligning the sequence of British Palaeolithic industries to this Alpine glacial sequence.[19]

As yet, most researchers in Britain still assumed stone tools to be post-glacial, so a glacial–interglacial sequence for stone tools was irrelevant.[20] Two avenues might lead to wider acceptance of an interglacial Palaeolithic: further glacial episodes would have to be identified in deposits *later* than the Chalky Boulder Clay in southern England, thus demonstrating that this boulder clay was not the last representation of glaciation in the area; or industries would have to be found *below* the Chalky Boulder Clay, thus bringing Harmer's interglacial into Palaeolithic times. Either route encountered the Chalky Boulder Clay: the time-marker that defined the post-glacial period for most Palaeolithic researchers. The first avenue is explored here; the second is left to Chapter 8. It would take another decade or so before the retreat of the Palaeolithic into pre-Chalky Boulder Clay times met with any approval.

7.2 AN INTERGLACIAL PALAEOLITHIC IN THE THAMES VALLEY

The stone tools of the Thames Valley had seemed secure in their post-glacial position ever since 1892, when Holmes had described the oldest valley-gravels lying above the Chalky Boulder Clay in the Hornchurch section (see Chapter 2). Many of his fellow geologists had been hoping to find such a section since the early 1860s.[21] But they did not necessarily assume that the climate had stayed clement after the last glaciation had passed. The drift that Osmond Fisher

[18] Lamplugh 1906, 533, 557; Sturge 1911, 101.
[19] Penck and Brückner 1901–09; Lamplugh 1906, 537; Boule 1908b, 6; Dawkins 1910, 254; Read 1911, 10; Geikie 1914, 191.
[20] Chandler 1914, 62; Kennard 1916, 259; Halls and Sainty 1926, 89; Boswell 1936, 149.
[21] Holmes 1892, 370.

called 'Trail'—which lay festooned over the Palaeolithic Floors of the Thames Valley—was often linked to a time of great cold. Frozen soils were thought to have become soggy in the summer sun and slumped down the slopes, covering stone tools in twisted masses of Trail on their way downhill. Although associated with a cold climate, the Trail was rarely linked to a glaciation. When Dawkins had suggested in 1867 that the Trail might be correlated to the same glaciation that deposited the Chalky Boulder Clay, he had gained little support.[22]

In the early twentieth century the Trail was again linked to a glaciation, only this time to one *later* than the Chalky Boulder Clay. If this link held, then most of the Palaeolithic river drifts of the Thames Valley would have to be described as interglacial. This might have been on Warren's mind as he worked through the winter of 1910 at a ballast pit known as 'Pickett's Lock' on the northern outskirts of London. During his usual round of observations in pits, trenches, and other excavations near his home, Warren noticed that the Great Eastern Railway Company had exposed the low-level river drift in their pit at Ponder's End in the Lea Valley, Middlesex. That winter, he was busy at the cold task of washing and collecting the remains of plants and shells from an interesting dark bed of vegetable matter.

In December 1910, Warren reported to *Nature* his discovery of arctic plant remains in this bed: evidence of a cold stage that post-dated the Palaeolithic. Warren linked this cold stage, which was later than the Mousterian epoch, to the 'Trail' of Osmond Fisher. In a longer article presented to the Geological Society, Warren suggested that it could be called the 'Ponder's End Stage.' He agreed that tool-makers had arrived in Britain after the maximum glaciation and the Chalky Boulder Clay, but concluded that his evidence for a *return* of glacial conditions meant that the Palaeolithic had to be interglacial in date. He recalled that Reid had seen similar climatic oscillations reflected in the plant remains from Hoxne at the end of the nineteenth century.[23]

Warren thought that the deposition of the Chalky Boulder Clay had been followed by two arctic stages: one represented by Reid's arctic bed at Hoxne; the other, by his own Ponder's End Stage. He linked these climatic oscillations to those on the Continent by using the names of Alpine glaciations.[24] The Alpine glacial sequence would appear frequently in British publications as researchers recognized more glaciations in their drifts. Penck and Brückner had named the first, second, third, and fourth glaciations in the Alps after four rivers where river-gravel deposits from each glaciation were clearly displayed: the 'Günz,' 'Mindel,' 'Riss,' and 'Würm.' (This alphabetical order

---

[22] Dawkins 1867, 109.
[23] Warren 1910; 1912a, 219, 225; Reid 1896, 411.   [24] Warren 1912a, 223–224.

Table 7.2. The glacial and interglacial terminology of Penck and Brückner, with the equivalent numbered stages that were sometimes used in Britain.

Glaciations	Interglacials	Numbered stages
Würm		Fourth glacial
	Riss-Würm	Third interglacial
Riss		Third glacial
	Mindel-Riss	Second interglacial
Mindel		Second glacial
	Günz-Mindel	First interglacial
Günz		First glacial

made them easy to remember.) The intervening interglacials were described by linking the glacial labels on either side. Thus the Third interglacial, which fell between the Riss and the Würm glacial episodes, was known as the 'Riss-Würm' (see Table 7.2 for a summary of Alpine glacial terminology).

Warren, like a few other British researchers, was starting to consider the connections between the glacial deposits and Palaeolithic industries of Britain and the supposed sequence of Alpine glaciations. He was aware that two great opposing industrial–glacial schemes had been developed on the Continent: Penck and Brückner saw four glaciations; Boule saw only three. Penck, along with much of the German school, supported a long glacial chronology for the Palaeolithic stretching back to the Second (Mindel-Riss) interglacial; Boule and the French school defended a short chronology: they thought that the Chellean had arrived in Europe during Penck's Third (Riss-Würm) interglacial, and associated the Mousterian with the Fourth (Würm) glacial episode.[25]

It seemed to Warren that neither Penck's nor Boule's version accorded exactly with the English evidence; neither had room for his Ponder's End stage, which had occurred after the time of the Mousterian tool-makers. Lubbock, in his seventh and last edition of *Prehistoric Times* (1913), also remarked on the wide range of correlations that could be drawn between glacial and industrial sequences. But both the long and the short glacial chronology tipped British stone tools into interglacial times and caught them in the mesh of a glacial sequence; tipped back further, they would be caught in a finer mesh of greater value as a Palaeolithic timescale.[26]

Unfettered by belief in a post-glacial Palaeolithic, Skertchly and Geikie had managed to construct industrial–glacial sequences in the 1870s (see Chapter 2

---

[25] Boule 1888, 672–680; 1908b, 7–9; Obermaier 1906a, 374; Breuil 1913a, 163–164; Geikie 1914, x; Osborn 1915, 236, 287.
[26] Warren 1912a, 224; Lubbock 1913, 420; Smith 1915a, 7.

and Table 2.7). Geikie, fond of interglacials, favoured Penck's long chronology. He was sure that Chellean and Acheulian tool-makers had lived in Britain between the Second (Mindel) and Third (Riss) glaciations.[27] When he gave the Munro Lectures at the University in Edinburgh in 1913 (a Lectureship in Anthropology, founded in 1910), Geikie thanked the growth of research into interglacial deposits, particularly those of Penck and Brückner in the Alps, which allowed him to discuss the geological dates of later culture-stages in more detail.

Geikie's lectures were published in 1914 as *The Antiquity of Man in Europe*. This book updated his third edition of *The Great Ice Age* (1894) and included more information on Palaeolithic matters. It would still take another decade or two for the glacial sequence to gain a powerful grip on ideas about the British Palaeolithic. Then, industrial–glacial correlations like those promoted by Geikie (see Table 7.3) would shape Palaeolithic sequences and direct the relationships that were reconstructed for their makers. The reasoning behind such industrial–glacial correlations is clear from Geikie's account in *The Antiquity of Man*. The details, though turgid, introduce a refrain that would become louder as correlations to the Alpine sequence of glaciations grew more popular.

Table 7.3. The first four of James Geikie's six glacial episodes, with associated sediments, bones, and tools (after Geikie 1914, 246–269).

Alpine glacial sequence	British deposits	Palaeolithic industries
Würm (Fourth glacial)	Warren's arctic plant bed in the Lea Valley; Trail covers Palaeolithic Floors. Severe climate; Magdalenian reindeer hunters retreat to France	Aurignacian; Solutrean; Magdalenian
Riss-Würm (Third interglacial)	Later river drifts; temperate conditions; warm fauna at the start of the interglacial episode	Mousterian
Riss (Third glacial)	Later river drifts; fauna includes mammoth, woolly rhino and reindeer	Mousterian
Mindel-Riss (Second interglacial)	Older river drifts; southern fauna (straight-tusked elephant), which were then succeeded by northern species (woolly rhinoceros, mammoth)	Acheulian; Chellean
Mindel (Second glacial)	Chalky Boulder Clay	
Günz-Mindel (First interglacial)	Cromer Forest Bed	
Günz (First glacial)		

[27] Geikie 1894, 689. Miller and Skertchly 1878, 551 had come to similar conclusions in 1878, but wrote about 'Ancient Valley Palaeoliths' rather than 'Chellean' or 'Acheulian' industries.

Geikie reflected that the borings carried out at Hoxne in the late nineteenth century had placed the Palaeolithic after the Chalky Boulder Clay, which he associated with the Second (Mindel) of four great ice-sheets. With the arrival of temperate conditions came the Chellean and Acheulian of the High (100 ft) river terrace of the Thames. The Mousterian of the Middle (50 ft) terrace was linked to the arctic conditions of the Third (Riss) ice-sheet, and Geikie believed that this industry continued through to the Third (Riss-Würm) interglacial.[28] He was delighted with Warren's recent discovery of the arctic plant-bed at Ponder's End, which he connected to the Trail and associated both deposits with the Fourth (Würm) ice-sheet. This Fourth glaciation, he connected to the Reindeer period, when Britain was occupied by Aurignacian and Solutrean tool-makers; at its climax, the Magdalenian reindeer-hunters retreated south to France and Spain. Geikie's Fifth and Sixth glacial episodes post-dated the Palaeolithic; they were relatively minor and need not concern us.[29] This was one of Geikie's last contributions to glacial geology; he died in 1915, a year after the publication of his book.

The elaborate scheme developed by James Geikie was founded on several controversial sources: glacial deposits, the remains of animals and plants, river drifts, and Palaeolithic industries. Each presented Geikie with a chronological sequence, and he used the four marker points supplied by Alpine glaciations to connect these sequences together. But few of Geikie's peers agreed with his version of events. The value of the Alpine scheme was questioned; the number of glaciations in Britain was contentious. Each of the sequences forged by Geikie from glacial geology, palaeontology, stratigraphy, and archaeology could be constructed differently. These sequences could also be connected or aligned differently. And while Geikie settled on glaciations as a chronological anchor, some of his colleagues made a different selection, believing that one of the other sequences offered a more reliable basis on which to build their own chronological schemes. With all these clocks set so differently, the sequence of Palaeolithic industries might occupy a variety of places in time.

7.3 FROM ELEPHANTS TO VOLES

Palaeontological clocks often struck a different time to glacial clocks. Dawkins had skirmished with Geikie in the nineteenth century, believing that his own sequence of mammals was more trustworthy than any time-scale based on

---

[28] Geikie 1914, x, 119, 228–230, 259, 264.
[29] Geikie 1914, 266–268, 278–281.

glaciations. Falconer had not always agreed with the patterns drawn by Prestwich between river drifts and the glacial epoch, suggesting that his sequence of fossil elephants revealed different connections. Palaeontologists of the twentieth century continued the conflict as they refined their bony frameworks of time. They constructed their sequences on traditional sources: fossil elephants, for example, were to provide powerful time-markers for many decades to come.[30] But the small bones of ancient voles and lemmings also crept into faunal sequences during the twentieth century. These members of the family *Microtinae* had evolved rapidly and were sensitive to changes in climate. As they had arrived, evolved, and disappeared from Britain, they had left a very fine trail of time behind them in their bones.

Hinton saw great stratigraphic potential in voles and lemmings. He had studied fossils in the museums of London since his early teens, and collected specimens from the gravel pits that sliced down into London's geological past. As he grew more interested in the teeth and skulls of the smaller mammals, Hinton developed his own opinions about climate changes in the glacial epoch and their connection to the river drifts and implements of the Thames Valley. But he also had to spend much of his time working as a junior clerk in barristers' chambers to ease the financial difficulties of his family.[31]

In 1900 Hinton heard Geikie's brother, Archibald, speak at the Geologists' Association. Moved by a comment in the paper, he wrote soon afterwards to Archibald Geikie, Director-General of the Geological Survey, and asked about the likelihood of gaining financial support for his studies in Quaternary stratigraphy and palaeontology. Paid employment in the subject was, however, difficult to find: it was not until 1921 that Hinton managed to obtain a post related to his interests, in the Zoology Department of the British Museum (Natural History), London. His stocky, heavy-booted figure became a familiar sight around the museum, pockets bulging with notebooks and bone-full tobacco tins.[32]

The Geologists' Association encouraged and supported Hinton's early interest in Quaternary sediments and bones. He read his first paper there in 1899, aged sixteen: the Association placed no limit on the age of members. In this early work on the stratigraphy and fauna of the Thames (summarized in Table 7.4) Hinton declared that the conclusions of James Geikie were 'incontestable, in so far as they related to the Palaeolithic era'.[33]

Hinton would reflect that this early work was carried out with 'implicit trust in my text book and in the teachings of my elders'.[34] Later, he placed more trust in bones. The glacial teachings of Geikie would be rejected for a

---

[30] Oakley 1964b, 15, 35.   [31] Savage 1963, 155–156.
[32] Hinton to A. Geikie, 3 June 1900: ULE: Gen 521/3; Anon 1900b; Savage 1963, 156, 165.
[33] Hinton 1900, 271.   [34] Hinton 1926a, 325.

Table 7.4. Hinton's opinions in 1900 about climate change in the Thames Valley, before he turned against interglacial schemes (after Hinton 1900, 281).

Geology of the Lower Thames Valley	Climate change and the Palaeolithic sequence
Trail	End of the Palaeolithic period. Cold
Middle (50-ft) terrace gravels (in part) and brickearths	Newer Palaeolithic. Genial (interglacial)
Middle (50-ft) terrace gravels (in part)	Older Palaeolithic. Cold
High (100-ft) terrace drift	

different climatic pattern. In 1910, half-way through research for his *magnum opus*—the *Monograph of the Voles and Lemmings (Microtinae) Living and Extinct* (1926)—Hinton delivered a report of his findings to the Geologists' Association. Some members of his audience might have expected recent palaeontological research to support the usual climatic sequence: travelling from the early warm period of the Cromer Forest Bed, up through the glaciation that left the Chalky Boulder Clay, and into post-glacial times when the 100-ft river drifts were left in the Thames Valley. They would have been surprised by Hinton's opinions.[35]

Hinton argued that the bones of voles and lemmings from 'Cromerian' (or Cromer Forest Bed) times and those from the High (100-ft) terrace of the Thames Valley were so similar that it seemed impossible for any major, disruptive climate change to have occurred in the interim. Indeed, the river-drift species looked positively archaic when compared to their cousins from Continental river drifts. Hinton could see no evidence for an influx of the arctic species that might be expected from sediments dating to early post-glacial times; he saw only relics of a warm pre-glacial fauna in these 100-ft deposits.

The character of the fauna did not change until part-way through the next (Middle or 50-ft) terrace. Still Hinton did not link this change to a glaciation. He explained that the early-Middle-Terrace fauna from Grays Thurrock and Ilford was very similar to the fauna from the High Terrace, whilst the later-Middle-Terrace fauna from Crayford and Erith included many newcomers: an influx that Hinton connected to the re-establishment of a land-bridge between England and the Continent. The major glaciation, he declared, had occurred later still: long after the Acheulian and Mousterian peoples had left their tools in the High and Middle terraces of the Thames.[36] The Palaeolithic, in Hinton's view, was neither post-glacial nor interglacial, but *pre*-glacial. Hinton, like Dawkins and Falconer before him, saw a complicated picture in the bones of the Thames Valley that conflicted with the views held by many of his peers.

---

[35] Hinton 1926b. See Hinton 1926a, 326–328 for his criticism of glacial geologists.
[36] Hinton 1910, 493–494, 498–507.

Table 7.5. Hinton's connections between changes in microtine fauna (voles, lemmings) and the sequence of river terraces in the Thames Valley (after Hinton 1910, 491–494). Later, he correlated the Ponder's End Stage to the period of maximum glaciation.

Terrace stages	Sites	Microtine fauna (voles, lemmings)
	Ightham Fissures; Langwith Cave	Period of maximum glaciation
Later Middle (50-ft) terrace	Crayford; Erith	Appearance of new species from the Continent; northern mammals appear and begin to replace southern types
Early Middle (50-ft) terrace	Ilford; Grays Thurrock	Little modified survivals of the later Pliocene (southern) fauna of Cromer
High (100-ft) terrace	Ingress Vale (Swanscombe)	Forest Bed times (Cromerian)

Kennard was one of the few geologists who welcomed Hinton's interpretation. While Hinton was hunting for the bones of voles and lemmings, Kennard had been picking shells out of similar sediments. These shells led him to the same conclusion: Britain's major glacial episode had occurred very late in the sequence of Thames Valley river-terraces. Hinton and Kennard shared similarities in background as well as opinion and built a strong collaboration against their many opponents. Back in 1899, Kennard had identified Hinton's shells from the Thames gravel pits, and they often met at the Geologists' Association.[37] Like Hinton, Kennard carried out his research in time spared from unrelated employment (in the warehouse of a London firm); it was only after retirement that he became a Research Specialist at the Geological Survey Museum. Although it might be assumed that these two would be disciples of Dawkins, who was renowned for his attacks on an interglacial Palaeolithic, Kennard had more respect for Falconer whom he regarded as one of Britain's finest palaeontologists. He treated Dawkins scathingly and hinted that, after Falconer's early death, Dawkins had appropriated his work.[38]

Kennard began his research on the fossil shells from Britain's ancient lakes and rivers (non-marine mollusca) in the latter years of the nineteenth century. Initially, he studied them in collaboration with Bernard Barham Woodward (1853–1930), one of the early researchers in this field. Woodward, Librarian in the British Museum (Natural History), belonged to the great clan of geological Woodwards: grandson of Samuel Woodward and nephew of Henry Woodward, editor of the *Geological Magazine*.[39] Around this time,

---

[37] Hinton 1900, 281; Kennard 1916, 250.   [38] Kennard 1947, 275; Warren 1949.
[39] Kennard and Woodward 1897; Kennard 1931, 72.

Kennard was described by Harrison, the eolith hunter, as 'young, strong, full of go, and a demon for work'.[40] Warren considered his friend to be 'a good conversationalist and gifted with a charming personality.' Another biographer observed that Kennard was old-fashioned in his opinions: he distrusted the French and Roman Catholics.[41] Though respected for his skill and frequently asked by his peers to identify their shells, Kennard's name, like Hinton's, was not as prominent as might be expected. The specialist nature of their research directed much of their work towards short reports, hidden at the end of papers published by their fellows.

When Warren sent his Ponder's End shells to Kennard for identification, Kennard was pleased to see arctic varieties in the sample. Hinton also welcomed Warren's discovery of this glacial phase, which fell right where they both expected it: late in Quaternary time.[42] Warren also remarked on the match between his discovery of a cold climate in the low-level river drift and Hinton's climatic observations in the bones, but differed from his conclusion: Warren saw a *late* glaciation; he did not connect this to the *major* glaciation of Britain.[43] Nonetheless, Hinton and Kennard remained convinced that the period of most severe cold had occurred late in the Quaternary history of southern Britain. They never became reconciled to the kinds of glacial chronologies promoted by Geikie and Penck.

7.4 HINTON, KENNARD, AND THE THAMES-VALLEY SEQUENCE

The interests of Hinton and Kennard were not restricted to bones, shells, and the glacial sequence; they also worked on the Palaeolithic sequence. In 1905 they published a paper on the subject in the *Proceedings of the Geologists' Association*: 'The Relative Ages of the Stone Implements of the Lower Thames Valley.' Hinton and Kennard connected the tools from this definitive Palaeolithic district to the spreads of sediment that lay in and around the river valley. They offered their results as a basis for future research. This suggestion was welcomed: an outline of their scheme can be glimpsed in the elaborations of later authors, although their paper was rarely cited and was soon forgotten.[44] Hinton and Kennard gave a clear, but detailed explanation for their sequence

---

[40] Harrison to E. Harrison (?1896), quoted in Harrison 1928, 204.
[41] Warren 1949, lviii; Wrigley 1949–51, 5.
[42] Kennard 1916, 250–251; Hinton 1926a, 328.
[43] Warren 1911, 168–169; Warren 1912a, 218, footnote 4.
[44] Hinton and Kennard 1905, 77; Chandler 1916, 240.

of industries and sediments. These details introduce a way of thinking that was central to interpretations of the date of British tools, the relationships between industries, and the peoples who made them.

Hinton and Kennard divided the history of the Lower Thames Valley into eight stages, founded on differences in the height of deposits above the current river. Height was significant because the ancient river had once flowed at a far higher level before cutting into the valley: carving down during spurts of erosion that coincided with periods when the land was rising. Cycles of erosion (down-cutting)—and aggradation (when the river channel grew thick with deposits)—formed the terraces that marched up the side of the valley like a series of steps. It was these terraces, and other reflections of ancient erosion and aggradation processes, that enabled Hinton and Kennard to group the spreads of gravels and brickearths into different stages. They observed that each stage was characterized by a different group of implements; although, of course, sediments might also hold older tools derived from earlier deposits.[45] All eight stages are given in Table 7.6: the three most recent are less relevant for the Palaeolithic period.

Table 7.6. The Palaeolithic sequence developed by Hinton and Kennard (1905), showing the connections they drew to stratigraphical stages in the Thames Valley and to Palaeolithic sequences developed on the Continent.

Thames-Valley stages	Thames-Valley tools	Continental industries
Holocene Alluvium		
Buried Channel		Magdalenian
Fourth Terrace		Solutrean
Third Terrace: Crayford Brickearths (50-ft or Middle Terrace)	Well-made; fine chipping; many long thin flakes	Mousterian
Second Terrace: Swanscombe Gravel (100-ft or High Terrace)	Pointed implements common; ovoids rare	Acheulian
First Terrace: Dartford Heath Gravel	Implements very rare	
Hill Gravels	Hill group palaeoliths	Chellean; Mesvinian of Rutot
	Transitionals (between the eoliths and Hill palaeoliths)	Reutelo-Mesvinian of Rutot
Plateau gravels (Middle Pliocene?)	Eoliths	Reutelian of Rutot

[45] Hinton and Kennard 1905, 86. On the complicated behaviour of rivers, see Dury 1986, 82–107.

The Palaeolithic sequence presented by Hinton and Kennard began with the eoliths found by Harrison on the Kent Plateau, far above the river drifts of the Thames Valley.[46] Kennard's opinion on the eoliths would fluctuate; but, when he published this paper, he agreed with many of his peers that eoliths were the earliest tools in Britain. Next in their sequence came the hill-gravels of Prestwich and his Hill group of implements: Kennard dated these to a time between the Kent eoliths (above) and the river-drift palaeoliths (below).[47] Even Warren, who did not believe in the eoliths, saw evidence of a Palaeolithic stage earlier than de Mortillet's Chellean in the implements lying at high levels above the river drifts in Kent.[48]

As they travelled down from the high ground into the main valley, Hinton and Kennard differed slightly from their predecessors in their division of the Thames river-terraces. Three terraces were still the mainstay of most geologists. The Geological Survey had carried out a partial re-survey of the gravels and brickearths of the Thames Valley on the larger six-inch scale during the early 1900s, refining the suggestions made by Whitaker a decade or two before.[49] In 1902 Theodore Pocock, one of the Survey officers working in the area, had established that the gravels formed three terraces—upper, middle, and lower— separated by two great periods of valley erosion. The upper terrace lay at around 100-ft above sea level, the middle terrace fell at the 50-ft level, then came the brickearths, and finally the lower terrace of gravel. Pocock had also confirmed that the boulder clay was indeed the oldest of the Thames Valley drifts.[50]

Hinton and Kennard added yet another terrace to this sequence: older, higher, and represented by the Dartford Heath deposits at 136-ft O.D. This meant that the 100-ft terrace of gravel (the High or upper terrace) became their Second terrace, and the 50-ft terrace of gravels and brickearths (the Middle terrace) became their Third terrace.[51] To avoid confusion, the terms '100-ft' and '50-ft' will be used here for these two terraces. Erosion then set in again, and left their Fourth (or Low Terrace) lying a little above the present river.

The 100-ft and 50-ft terraces were both rich in Palaeolithic tools. The 100-ft terrace included the productive pits of the Swanscombe area: these implements were associated with the Acheulian stage of Continental researchers. Hinton and Kennard identified two groups of implements in the 50-ft terrace: a mixed group

---

[46] Hinton and Kennard 1905, 78–79.
[47] Kennard to Buckman, 11 October 1901: BGS: IGS1/1151.
[48] Warren 1902, 97.
[49] Whitaker 1864, 82–83; 1875, 61; 1889 (i), 389–391.
[50] Pocock 1903, 200, 202–207.
[51] Hinton and Kennard 1905, 77, 80, 81–82. Brown 1886, 192 had also extended the highest point of Whitaker's High Terrace from 100 ft to 130 ft. On the lengthy arguments over the Dartford Heath deposits, see also Bridgland 1994, 187–189, 191–193.

from the gravels; and a more homogeneous group from the brickearths, which included the Palaeolithic Floor specimens from Crayford, Grays, and Stoke Newington. This brickearth group was connected to the Mousterian of the Continental scheme by the two authors, who noted that Worthington Smith had been the first to recognize their similarity to implements from Le Moustier.[52] Hinton and Kennard emphasized that the Palaeolithic industries from each of their stratigraphic stages exhibited no abrupt breaks, but merged into each other: 'New types occur at various horizons, but always accompanied by the older types so as to conclusively prove that we are dealing with an ascending scale of progress'.[53]

Behind the clear sequence of terraces presented by Hinton and Kennard in 1905 lay controversy. The distinction between a First (Dartford Heath) and a Second (100-ft) terrace was questioned by R. H. Chandler and Arthur Leonard Leach (1869–1957), who argued that these should be classed as a *single* terrace since they both shared a similar base-level. Chandler and Leach traced uncertainty about terrace classifications to the variety of methods used to record their height. They complained to the Geologists' Association that while Hinton and Kennard had used the summit-level to determine the height of the Dartford Heath spread, others might use the average level or the base-level of the deposit.[54] There was no systematic practice. Leach, a London schoolteacher, had met Chandler at a series of geological lectures given at the Woolwich Polytechnic sometime before 1905. They would keep a watchful eye on pits around London for exposures of the old river drifts.[55]

The distinction between the 100-ft and 50-ft terraces was not clear either. Various intermediate terraces had been reported lying between them. Llewellyn Treacher (1859–1943) argued that the 100-ft terrace ought to be split into two because he had seen an additional terrace between the 50-ft and 100-ft terraces near Maidenhead: the district where Whitaker had once constructed his terrace divisions.[56] Treacher was yet another member of the Geologists' Association. He had developed his knowledge of the Thames Valley sediments in the late nineteenth century as he helped his father with the family's market-gardening and fruit-growing business.[57]

---

[52] Hinton and Kennard 1905, 77, 81–82, 94, 99.

[53] Hinton and Kennard 1905, 100.

[54] Chandler and Leach 1911, 172; 1912, 104–105. The lack of a standard way to take the level of a deposit was still a problem in the 1930s (see Chandler's comment in Boswell 1931, 110).

[55] Bull 1942, 43; Brown 1958b, 67–68.

[56] Treacher 1910, 198. See also Warren 1926 43, Burchell 1934, 37, and Oakley 1937a, 278. Cranshaw 1983, 12–27 summarises the different opinions about the number and order of terraces and sub-stages in the Thames Valley.

[57] Dewey 1944, 42–43. On Llewellyn Treacher and his collection of implements, see Cranshaw 1983, 1–10.

Few geologists working in the early twentieth century assumed the ancient behaviour of rivers—or seas—to have been simple or uniform. Different regions, such as the valleys of the Somme and the Thames, were known to have different histories—of elevation and depression, tectonic movement, and denudation—which influenced the cycles of erosion and aggradation that produced river-terraces.[58] Ambiguity about the number of river terraces in the Thames Valley and the complex history of these sediments gave researchers considerable flexibility in their interpretations. The positions occupied by tool-bearing and bone-bearing deposits in the river-drift sequence could be stretched to match expectations. This point must be remembered.

7.5 PALAEOLITHIC SEQUENCES IN BRITAIN AND ABROAD

Another source of time was starting slowly to tick alongside the detailed sequences of glaciations, bones, shells, and river terraces. Sturge suggested in 1908 that Palaeolithic implements 'are the true fossils of the gravels, and will, I feel sure, be of as much assistance in forming "zones" of gravels as ammonites are in forming "zones" of the secondary formations'.[59] With these words he exhorted the younger school of geologists to try and unpick the British Palaeolithic assemblages that were sometimes lumped in the Chellean and Acheulian industries of the French sequence. Other Palaeolithic researchers shared his concern that the French scheme might not be sufficiently detailed to describe the Palaeolithic sequence of Britain.[60]

In the earliest years of the twentieth century, the French scheme was synonymous with the Palaeolithic classification of Gabriel de Mortillet (see Chapter 4). The final version of this classification had been published after his death in the third edition of *Le Préhistorique; origine et antiquité de l'homme* (1900), a book co-authored by Gabriel and his son, Adrien de Mortillet. The Palaeolithic sequence in this third edition (see Table 7.7 below) looked slightly different from the version in the first edition (1883). The Acheulian was no longer grouped with the Chellean, but occupied an epoch in its own right— although it was still described as transitional between the Chellean (with hand-axes) and the Mousterian (with flakes and flake-tools). More attention was devoted to varieties of hand-axe and their changes over time. But

---

[58] Harmer 1902, 422; Hinton 1910, 501; Bury 1916, 189; 1923, 38–39; Warren 1924b, 269
[59] Sturge 1908, 13.
[60] Sturge 1908, 13–14; 1911, 61, 65; Abbott 1911, 460.

Table 7.7. The last Palaeolithic sequence
of Gabriel de Mortillet (after de Mortillet
and de Mortillet 1900, 21).

Periods	Epochs
Paléolithique	Tourassienne
	Magdalénienne
	Solutréenne
	Moustérienne
	Acheuléenne
	Chelléenne
Eolithique	Puycournienne
	Thenaisienne

evolutionary progression still remained: the authors claimed that they could recognize the age of a hand-axe from its technical skill, development in shape, and association with other tool-types (and they still included an Eolithic before the Palaeolithic period).[61] In early twentieth-century Britain, de Mortillet's terminology was replacing the older division between tools of the Palaeolithic (or River-Drift) period and those of the Reindeer (or Cave) period.[62]

As Sturge had noticed, not all British tools had a clear equivalent in de Mortillet's sequence. Some of these problematic outliers were matched to the Belgian Palaeolithic sequence of flake-rich industries, described by Aimé Rutot in the last years of the nineteenth century. The earlier end of the Belgian sequence dipped into Eolithic times; Sollas described the flavour of their reception when he wrote that Rutot's Reutelian and Mafflian industries belonged 'to the nebulous region of "eoliths".'[63] But Rutot's Mesvinian and Strépyan industries were tolerable counterparts for the earlier Palaeolithic tools of Britain. Even Warren, wary of the earlier Belgian industries, acknowledged that some of the Mesvinian specimens might be true artefacts.[64]

The classification of stone tools in the British Museum's *Guide to the Antiquities of the Stone Age* was adjusted between the first (1902) and second (1911) editions to include Rutot's industries at the base of the Palaeolithic sequence. Hinton and Kennard connected the Mesvinian to Prestwich's Hill group: the tools that dated to a time between the earliest group from the river terraces (the Chellean) and the eoliths. The sequences of de Mortillet and Rutot were combined in various different ways to describe British industries:

---

[61] De Mortillet and de Mortillet 1900, 154–156, 163.
[62] Hinton and Kennard 1905, 99; Woodward, 1909, 78; Sollas 1910, lvii; Lubbock 1913, 419–420; Geikie 1914, 42.
[63] Sollas 1911, 109. [64] Warren 1902, 99; 1905, 341, footnote.

Table 7.8. Rutot's industrial sequence as it was perceived in Britain *c.*1911 (after Rutot 1900 and Sollas 1911).

Industry	Description
Chellean	Dominated by hand-axes; scrapers also present
Strépyan (Early/Pre-Chellean)	Many coarse scrapers; some nodules with pointed tips (primitive hand-axes)
Mesvinian (Early/Pre-Chellean)	Comprised almost solely of scrapers (*grattoirs* and *racloirs*)
Mafflian	Many scrapers; some hammer-stones
Reutelian	Dominated by hammer-stones; some scrapers made on broken hammer-stones

James Geikie explained that whilst some saw the Mesvinian and the Strépyan as two independent stages *preceding* the Chellean, others included them *within* the Chellean as 'early Chellean' sub-divisions of this industrial epoch.[65]

De Mortillet's industrial classification had penetrated British research in a piecemeal manner during the nineteenth century (see Chapter 4). The links that he had forged between the age and the character of industries had not been applied very strongly in Britain. Patterns and sequences had been identified in British tools, but they tended to be described in colloquial terms that allowed a degree of flexibility. When de Mortillet's terms were used, these might serve to indicate the date or the character of a tool—but not necessarily both. Hinton and Kennard helped to tie expectations of age *and* character to the terms 'Chellean', 'Acheulian', and 'Mousterian' when they applied these labels to the well-known industries of the Thames Valley in 1905 and associated them with specific geological stages. Tools from the 100-ft terrace, once described vaguely as the oldest class of Worthington Smith or as the pointed and sharp-rimmed implements of John Evans, became better known as Chellean or Acheulian industries; tools from the 50-ft terrace below were referred to the Mousterian industry.[66]

7.6 PUZZLING INDUSTRIES

There was one clear sign that a standard Palaeolithic sequence was starting to be accepted in Britain: exclamations at anomalies that did not match standard expectations. Such exclamations multiplied through the early twentieth

---

[65] Hinton and Kennard 1905, 91, 98–99; Read 1911, 32–33; Geikie 1914, 43.
[66] Hinton and Kennard 1905, 99; Dewey 1915, 112; Smith 1915a, 3; Kennard 1916, 257.

century as the age and the character of stone tools became bound more tightly together in the minds of British researchers. An anomaly might be seen in the layers of a gravel pit if two industries, each distinct in character, lay one above the other in the 'wrong' order for their expected sequence. Another anomaly might be seen in an industry, again of distinct, identifiable character, found with the 'wrong' geological time-marker. This time-marker would be wrong for suggesting a date either much older or much younger than was expected for an industry of this character—a date that would, most likely, already be associated with a quite different industry. In each case, either the geology or the industries would have to be re-interpreted to make the 'right' connections between the geological and Palaeolithic sequences: the connections that matched expectations.

Sturge saw an example of the first kind of anomaly in a sequence of industries from High Lodge at Mildenhall in Suffolk. This site was only a few miles from his house, which was not a coincidence: Sturge had moved to the area to be closer to rich Palaeolithic sites like High Lodge, famous for its fine flake-tools. In character, these flake-tools were often compared to the Mousterian industry.[67] The Mousterian was known to come after the hand-axe-dominated industries of the Chellean and Acheulian. But at High Lodge, sediments with hand-axes lay *above* these Mousterian deposits, which suggested that the hand-axes were *later* than the Mousterian. Sturge declared that this was an 'extraordinary phenomenon' and turned to the geology to push the Palaeolithic sequence back into the bounds of expectation. He favoured the explanation offered by Sollas, who had looked over the High Lodge section with him and suggested that these overlying hand-axe deposits were indeed older, but had been swept over the Mousterian brickearth by ice at a later date.[68]

The Palaeolithic Floors of Worthington Smith, which were thought to extend for several miles underneath north-east London, also undermined the expectations of twentieth-century researchers. In the past Smith had described the tools from these Floors as Mousterian because of the distinctive trimmed flakes.[69] But the 1911 *Guide* to the British Museum collections included the terms 'Acheulian' and 'of Chelles character' alongside the Mousterian labels: hand-axes could also be found on these Floors.[70]

The geological time-markers associated with Palaeolithic Floors also caused confusion. Researchers of the early twentieth century tended to associate the

[67] Read 1902, 28.
[68] Sturge 1911, 71. The same reasoning was followed in the revised British Museum *Guide to the Antiquities of the Stone Age* (Read 1911, 31).
[69] Smith 1894, 110–111, 169, 220; 224; Hinton and Kennard 1905, 99.
[70] Read 1911, 18–19, 21.

Chellean industry with a warm climate and southern animals like the straight-tusked elephant; the Acheulian industry saw the arrival of northern species; the Mousterian tool-makers lived in a frigid climate.[71] James Geikie pointed out that Smith's Floors, though 'of Mousterian age' because of their tools, were associated with a temperate flora: more evidence, in his view, for an interglacial episode in the history of the Thames Valley river drifts.[72]

Another '*moustérien à faune chaude*' or 'warm Mousterian' had been reported by the French researcher Victor Commont a year or two before Geikie made this point. Commont's tools from Montières in the Somme Valley were associated with bones of the warmth-loving straight-tusked elephant and hippopotamus, and his assemblage lay sandwiched between sediments containing Chellean and Early Mousterian industries.[73] Back in 1897, Reinach had also described a flake-rich industry with no hand-axes (i.e. Mousterian in character) from Taubach in Germany, and had used its association with a warm straight-tusked-elephant fauna (which dated it to the Chellean stage) to suggest that de Mortillet's industrial sequence was flawed.[74]

Commont's decision to place his warm Mousterian at the end of the Third (Riss-Würm) interglacial, in line with Boule's short chronology, was seized upon by Geikie in his attack on this model, which he thought 'crowded all the Older Palaeolithic culture-stages into the Third Interglacial'.[75] Geikie suggested that these warm Mousterian industries could be explained more neatly in terms of his long glacial chronology. If the warm Mousterian were to be placed at the *beginning* of the Riss-Würm (Third) interglacial, then the Chellean could fall back into the previous warm slot: the Mindel-Riss (Second) interglacial.[76] This explanation reconciled the existence of a warm Chellean, a warm Mousterian, and a later cold Mousterian—and also gave Geikie another chance to promote his long glacial chronology, though he failed to dent the popularity of the shorter version in Britain.

The reaction to an industry from Essex, recovered in large quantities by Warren in the early years of the twentieth century, illustrates many of the problems encountered by researchers as they tried to date ambiguous industries. Warren had been keeping a careful watch on exposures of ancient deposits near his home in Loughton ever since he arrived in the area around 1903. This activity, which had brought him to Ponder's End in 1910, also led him to an anomalous industry on the Essex coast. Warren's father had moved

[71] Duckworth 1912, 120–121; Geikie 1914, 254–255.
[72] Geikie 1914, 264. See also Smith 1894, 288–292.
[73] Commont 1912a, 246; 1912b, 299–300.
[74] Reinach 1897, 55, 57–58; Commont 1909a, 39.
[75] Commont 1912b, 300; Geikie 1914, 311–312.
[76] Geikie 1914, 312.

to Frinton-on-Sea some time before the First World War and, on his visits, Warren took the chance to extend his usual range of Quaternary observations. He was particularly interested in the exposures of an ancient river-channel at Clacton-on-Sea, not far from his father's house.[77]

The cliff and foreshore at Clacton had been celebrated for centuries for its ancient bones. In 1898, the Revd James Wright Kenworthy, Vicar of Braintree, had reported stone tools from the same area. Warren began collecting these tools around 1908.[78] In 1912, he published his first brief account of the flints in the *Essex Naturalist*: 'Palaeolithic Remains from Clacton-on-Sea.' Though Warren reported his finds to the Geological Society of London as well as the Essex Field Club, they provoked little reaction. Members might have been more interested in another artefact on Warren's table of exhibits: the tip of a wooden Palaeolithic spear he had found at Clacton in 1910: an extremely rare discovery.[79] In later decades, however, these Clacton tools were to receive more attention, and would be connected in various ways to different geological sequences. They will be used in the last half of this book as a touchstone to explore the reasoning and motivation that lay behind the diverse patterns drawn in Britain's geological and Palaeolithic past.

Many of the Clacton tools were simple, un-worked flint flakes but Warren also identified 'characteristic flint implements' such as trimmed flakes and 'rude forms of side-choppers.' He saw no hand-axes in the assemblage: 'not a single example of the usual ovate or pointed Palaeolithic types has yet been found.' He also saw no evidence of the characteristic Mousterian technique—by which he probably meant the Levallois technique of the industry commonly referred to in Britain at this time as the 'Early Mousterian'.[80] Faced with no hand-axes and no Mousterian technique either, Warren had a problem: how was he to connect this flake-dominated assemblage to the industrial sequence if it was not Chellean, Acheulian or Mousterian in character?

The age of the surrounding sediments, which might have helped to define the tools, gave no easy answers. The Clacton flints had been found in a bed of straight-tusked-elephant bones: a species associated with warm Chellean times; the mammoth that characterized cold Mousterian times was absent. But the stratigraphy and the local sequence of the river drifts—and the character of the tools—suggested to Warren that these sediments were post-Chellean and probably post-Mousterian in date. In his report to the Geological Society in January 1912, Warren used this conflict of time-markers as a warning against placing too much trust on elephant species as indicators of

[77] A. Warren to Oakley, 13 May 1958: NHM: DF140/7.
[78] Warren 1932a, 20; 1951, 108.
[79] Anon 1911, xcix; Warren 1912b.   [80] Warren 1912b, 15, 1912a, 219.

age, suggesting instead that they ought only to be used as guides for fluctuations in *climate*. This opinion was tied to his case for a number of climatic oscillations during Palaeolithic times, one of which was his Ponder's End stage. He argued that implements provided a more reliable time-scale than bones, since they seemed always to occur in the same order. But rather than returning to his problematic tools from Clacton, Warren followed this statement with a review of the Palaeolithic stages of Worthington Smith, whose work he admired.[81]

Warren's audience at the Geological Society might have been left with an impression of a post-Mousterian date for the Clacton tools. Less than a year before, though, Warren had informed the Essex Field Club that the most similar group of tools in Britain were the 'ruder surface implements of the Chalk Downs of the South of England'—apparently a reference to Prestwich's Hill group from Kent (see Table 7.6).[82] But this group of implements lay above the river valleys in Pre-Chellean times; if the character of an industry were to be tied to date, then this would cast the Clacton flints back to this early period as well. Warren was too cautious to assign an industrial label to his discovery until the early 1920s, when he recalled that he had 'obtained a large series between the years 1911 and 1916, but their affinities remained an enigma to me'.[83]

7.7 THE SOLUTION OF HUGO OBERMAIER

Hugo Obermaier (1877–1946), a Bavarian geologist and prehistorian, had also found it difficult to match the growing numbers and varieties of stone tools to the few slots available in the usual industrial sequences. His solution was to adapt de Mortillet's sequence to fit his observations. Obermaier had gained his early experience of the Palaeolithic in Central Europe where hand-axes, which formed such a prominent part of the sequences from northern France (and southern England), were lacking. This might have made him more sensitive to variety in the French tools.

Obermaier had begun his geological career in the early 1900s as field-assistant to Penck, the geologist who had developed the Alpine glacial sequence and promoted a long glacial chronology for the Palaeolithic. In 1904, Obermaier travelled from Vienna to Paris to spend a year in France. Breuil,

---

[81] Warren 1912a, 219–220.
[82] Warren 1912b, 15. See also Warren 1902, 98.
[83] Warren 1922b, 597.

who was studying at the Catholic Institute in Paris at the time, was asked by the Ambassador for France in Vienna to welcome Obermaier. They had many similarities: Obermaier was one month older than Breuil and both men were priests. More awkwardly, each was aware of the other's intention to apply for the post in Prehistory and Ethnography at the University of Fribourg in Switzerland.

The sensitive subject of the Fribourg post was, at first, avoided; Obermaier enjoyed his visit. Breuil introduced him to Boule and other French luminaries, offered advice about studying prehistory in France, and took him on excursions to the classic sites. On one of these trips Obermaier took Breuil aside to let him know that he had decided to renounce the Fribourg post in light of this kind welcome. Their long friendship survived two world wars and, although Breuil described Obermaier as his first pupil, he was eager to learn all he could from Obermaier about the Central European Palaeolithic.[84]

Obermaier also met Émile Cartailhac, a cave-enthusiast who ran a course on Prehistory at the University of Toulouse.[85] Cartailhac, a contemporary of de Mortillet, was one of the grand old men of French prehistory. He had taught Boule and now spoke glowingly of Obermaier, confiding to Sturge: 'He is marvellously familiar with Austria and Germany; he has studied in the field with Penck. In brief, I have full faith in his science and in his clear-seeing spirit. We have had very instructive discussions together. He is a good sort.' Cartailhac also shared Obermaier's solution to the problem of the warm Mousterian with Sturge: a *return* of warm fauna, following the long chronology of Penck.[86] Geikie used the same argument. But after his year in France, his observations of the classic French sites, and his work on the French Pyrenees, Obermaier abandoned the long glacial chronology of the German school and his old teacher, and switched to the short chronology of Boule and the French school.[87]

Obermaier did not lose by renouncing the post at Fribourg. He returned to Germany after his year in France and obtained a post at the University of Vienna. Soon afterwards, Breuil secured a place for his friend on the Spanish caving venture sponsored by the Prince of Monaco. Again, Obermaier acquitted himself well. In 1910, when the Prince announced his decision to set up the Institute of Human Palaeontology in Paris, a post was waiting for Obermaier alongside Breuil, under the leadership of Boule.[88]

---

[84] Breuil 1950, 105–106.   [85] Capitan 1922, 4.
[86] Cartailhac to Sturge (?1904): BM(F): Archives Box 4; Obermaier 1904, 35.
[87] Boule 1908b, 10–13; Obermaier 1909, 498–500.
[88] Breuil 1950, 106. Brodrick 1963, 152 writes that Boule and Breuil never got on very well: Boule was secretive; Breuil never turned in satisfactory accounts of his expenses.

Figure 7.2. Obermaier (in the middle) and colleagues in Spain, 1913. Burkitt sits to the right of Obermaier (postcard from Burkitt to his mother, June 1913: ULC: 7959, Box 1).

Obermaier accepted the broad divisions of the Palaeolithic established by de Mortillet.[89] But he saw more variety within each division. His version of the French Palaeolithic sequence described an alternation between industries with, and those without, hand-axes (Table 7.9). These views were outlined in 1906 and elaborated in a 1908 paper, *Die Steingeräte des französischen Altpaläolithikums*.[90] In Obermaier's scheme, hand-axes could come and go within the stretch of time that de Mortillet had filled solely with his hand-axe-dominated Chellean and Acheulian industries.

Obermaier prised apart de Mortillet's links between character and age by adapting his terminology. De Mortillet's Acheulian, for example, he split into two: an older Acheulian with no hand-axes, followed by a younger Acheulian with hand-axes. These chronological prefixes—'older' and 'younger'— allowed Obermaier to loosen the link between the old industrial terms and the *character* of an industry, whilst maintaining a strong connection to

[89] Déchelette 1908, 461, footnote 1; Obermaier 1908, 44–45; Reinach 1908, 305.
[90] Obermaier 1906b; 1908.

Table 7.9. Obermaier's classification of French Palaeolithic industries preceding the
Mousterian (after Obermaier 1908, 125).

Industry	Description / site
Younger Acheulian	Hand-axes: classic level of La Micoque and the Levallois industry with hand-axes
Older Acheulian	No hand-axes: base-level of La Micoque and Le Moustier
Developed Chellean (*Hochchelléen*)	Primitive hand-axes
Early Chellean (*Frühchelléen*)	No hand-axes

particular periods of *time*. But the terms 'Chellean' and 'Acheulian' were
so strongly associated with hand-axes in the minds of his critics in Britain
that they would outweigh Obermaier's attempts to carve out this chronological distinction. His friend Breuil would have more luck when he put
forward a similar suggestion a couple of decades later, when expectations
had changed.

7.8 INDUSTRIES AS CHRONOLOGIES AND INDUSTRIES AS CULTURES

As the presumed connections between the character of industries and their
age grew stronger in Britain, an atmosphere of expectation arose in which
anomalies could thrive. The number of puzzling tools lying outside their
allotted time slots increased. Geological and industrial sequences were
tweaked to force them back into the expected patterns—some oddities were
even seized to promote a particular geological viewpoint, like the warm
Mousterian and the long glacial chronology. But there was another solution
to these anomalies: one that muffled the insistent ticking of different time-
sequences. If industries were to be viewed as the cultural products of different
groups of peoples moving around Europe, their age and their respective
positions on the Palaeolithic sequence could be blurred. Groups making
different industries might occupy neighbouring districts; they might even
overlap in the same geographical area.

Sollas argued in *Ancient Hunters* (1911) that 'after a sufficient interval of
time' all Palaeolithic industries would exist simultaneously across the world.[91]
In his view, the earlier races that migrated into Europe would be succeeded by
a succession of later races, each bringing a more advanced industry into the

---

[91] Sollas 1911 vii, 120.

Continent. This progressive sequence of industrial waves could accommodate much overlap in time: the earlier races might persist in making their (now degenerate) industries—whether in small surviving pockets within Europe or in their original homelands—while the later races made their (more advanced) industries nearby. Sollas envisaged the Australian aborigines as the survivors of Mousterian times, writing that these 'Mousterians of the Antipodes' were contemporaneous with present-day Western civilisations.[92]

The idea that different groups, living at the same time, might make different kinds of tools was not new. Dupont had suggested in the 1870s that there had been a parallel development of two industries in Belgium during his Mammoth Age, made by two distinct populations: one by the tool-makers of the caves of Namur and Liège; the other by tool-makers from the river drifts of Hainaut (a region which included Mesvin).[93] Brown and Dawkins had also associated industries with different races. Brown had described how ancient Britain had been invaded by a succession of tool-making races, each bringing a more advanced industry with them. Dawkins's opinions about the geographical ranges occupied by Cave men and River-Drift men had meshed well with his classification of mammals (see Chapter 4).[94]

This connection between different industries and successive racial waves gave a Palaeolithic sequence the flexibility to accommodate discoveries of distinctive tools that lay outside their expected time-slots without damaging a belief in a general progression of industries up through the expected sequence. (Sollas's defence of the linear *industrial* line has some interesting parallels with his defence of a linear *hominid* line explored by Sommer.[95]) But there were drawbacks to such flexibility. The chronological value of the Palaeolithic sequence would seem to be weakened by these contemporaneous races as they blurred the positions occupied by different industries. Sollas had an answer. He explained that the most recent artefact had to be selected from the variety of industries to supply a date:

The duration of each of the several epochs may be defined on the one hand by its first appearance, and on the other by the first appearance of that next succeeding it. Thus with the advent of the Acheulean in any locality, the Chellean epoch may be regarded as closed; nevertheless the Chellean industry may have continued to exist elsewhere, a

---

[92] Sollas 1911, 162, 170. On the views of William Sollas, his shift from a linear to a branching model of human evolution, and his use of migration and racial displacement to account for cultural and biological progress, see Sommer (2005).

[93] Dupont 1873, 461, 469–470.

[94] Dawkins 1874, 367; Brown 1893, 95.

[95] On Sollas's ideas about the hominid line, see Sommer 2005, 340–344.

fact which may be expressed by the statement that the Chellean industry survived into Acheulean or even later times. Thus the industries overlap the epochs.[96]

This statement—'the industries overlap the epochs'—explains why Sollas came to disagree strongly with Obermaier's industrial sequence. Obermaier had used the term 'Chellean' to refer to a period of time that encompassed two different (though successive) cultures, and hand-axes only characterized his second culture, not his first (see Table 7.9). But Sollas placed more weight on character than date in his divisions of the Palaeolithic industries. For Sollas, it was the hand-axe that heralded the arrival of the Chellean hand-axe-making race, and this meant that he could not accept Obermaier's term 'Early Chellean' ('*frühchelléen*') for an industry that *lacked* hand-axes.[97] The distinction between hand-axe and flake industries was central to both interpretations, but one took a more cultural and the other a more chronological approach. These twin cultural and chronological aspects of Palaeolithic industries caused much confusion. They also gave the Palaeolithic sequence a dangerous ability to reflect and justify conflicting patterns in time and space.

7.9 VICTOR COMMONT AND THE SOMME VALLEY SEQUENCE

The confidence of British researchers in the French sequence of Palaeolithic industries was strengthened by Commont's work on the Palaeolithic deposits of the Somme Valley in the 1900s and early 1910s. Commont was a geologist, palaeontologist, and prehistorian who carried out this work in time spared from teaching science at the *École normale* at Amiens.[98] Commont took up the mantle of de Mortillet, subdivided his great industrial blocks and supplied a detailed description of each subdivision. But his Palaeolithic sequence retained a progressive note. This is clear in his treatment of the hand-axes. He found their distinctive shapes useful in defining different industries, and arranged them in a sequence that evolved from Pre-Chellean prototypes up to the summit of the Acheulian, each exhibiting more skill in manufacture than the one that preceded it. A similarly intimidating catalogue of hand-axes would later be described in the Thames Valley.

The crude hand-axe prototypes of Commont's '*Pré-chelléen*' industry (Fig. 7.3) were succeeded by the hand-axes of the '*Chelléen typique*' (Fig. 7.4).

---

[96] Sollas 1911, 120.   [97] Sollas 1911, 111.   [98] Reinach 1919, 197.

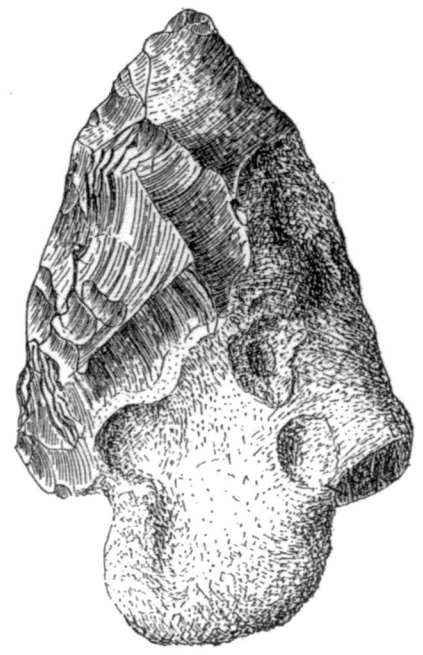

Figure 7.3. A primitive hand-axe from the Pre-Chellean levels of St Acheul (Commont 1908, 537, Fig. 8; this would become the Chellean of Breuil: see Breuil and Koslowski 1931, 465, Fig. 5).

Then came the finely worked, triangular hand-axes of Montières (Fig. 7.5), Commont's '*Chelléen évolué*.' He had initially suspected that these might have been made at the same time as the *ficrons* of the '*Chelléen typique*', perhaps by a different tribe or for a different function; Commont was not reluctant to embrace variety in his sequence. (The term '*ficron*' was applied to the long pointed hand-axe because of its resemblance to the iron point of a punt-pole.) By the time of the '*Acheuléen inférieur*' the oval '*limandes*', which had existed in the Chellean (Fig. 7.6), predominated (Fig. 7.7). (*Limandes* were named after the dab: a species of flatfish.) They were followed by the lance-like forms of the '*Acheuléen supérieur*'.[99] Around the time that Commont was describing this diversity of industries, Sturge was expressing his concern to the PSEA that the Chellean, Acheulian, and Mousterian divisions were too broad for the variety of British tools. However, when British researchers became aware of Commont's work, they had more divisions to choose from in building their own sequences.

[99] Commont 1910, 206; 1912a, 245–246; Smith 1912, 142.

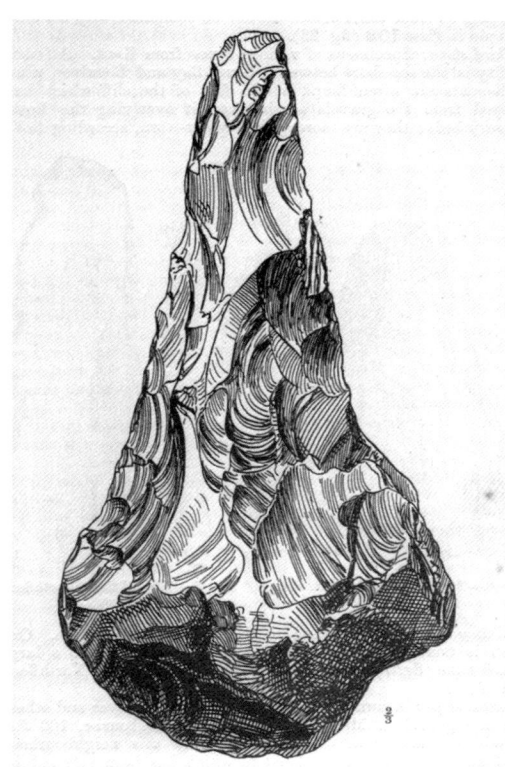

Figure 7.4. A '*ficron*', characteristic of Commont's '*chelléen typique*' (Read 1902, 17, Fig. 5; Smith 1912, 139, Fig. 12).

The changing character of tools in his Palaeolithic sequence prompted Commont to consider the relationships between their makers. He believed that the same race had lived around the Somme Valley in Chellean and Acheulian times: '*les hommes qui ont taillé les "limandes" sont les descendants de ceux qui confectionnaient les "ficrons"* ' ('the men who flaked the "*limandes*" are the descendants of those who made the "*ficrons*" ').[100] The great flakes of Levallois type, which had appeared in early Mousterian times, exhibited a completely different mode of flake production. In one paper, Commont observed that these large flakes could fulfil the same function as the finest hand-axe but were far easier to produce: he linked their arrival to the decline of hand-axes.[101] In another paper he described the difference between the specialized, varied implements of the Acheulian and those of the Mousterian as a retrogressive step, and suggested that the Mousterian had been made by

[100] Commont 1908, 571.   [101] Commont 1909b, 115, 120, 127.

Figure 7.5. A triangular hand-axe from Montières: Commont's *'chelléen evolué'* (Commont 1911, 74, Fig. 3).

a different race.[102] Whatever the interpretation, the distinction between the hand-axe industries and flake industries remained one of the most persuasive divisions in the Palaeolithic sequence.

The sediments and tools from the Somme Valley had structured the early ideas of Evans and Prestwich in the 1860s. In the early twentieth century they enabled Commont to confirm, clarify, and add further detail to the Chellean–Acheulian–Mousterian sequence popularized by de Mortillet (see Fig. 7.8 and Table 7.10). Encouraged by his papers, British researchers at last began to feel more secure about using French terms to describe the British Palaeolithic sequence. The suggestions made by Hinton and Kennard in 1905 would soon look less lonely.[103] Geikie declared that Commont's results 'lead us to believe

---

[102] Commont 1912a, 248.

[103] Sturge 1911, 99, footnote; Duckworth 1912, 98, 105; Underwood 1912, 138; Dewey 1913, 163; 1919, 49; Geikie 1914, 105–106. McNabb 1996, 35 also notes the influence of Commont on British research.

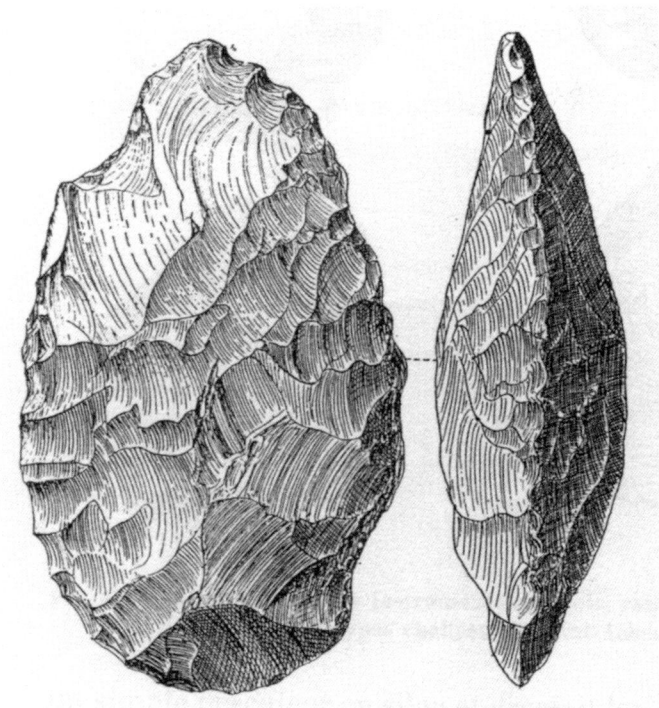

Figure 7.6. A Chellean '*limande*', from an industry otherwise dominated by elongated and pointed hand-axes (Commont 1908, 548, Fig. 32).

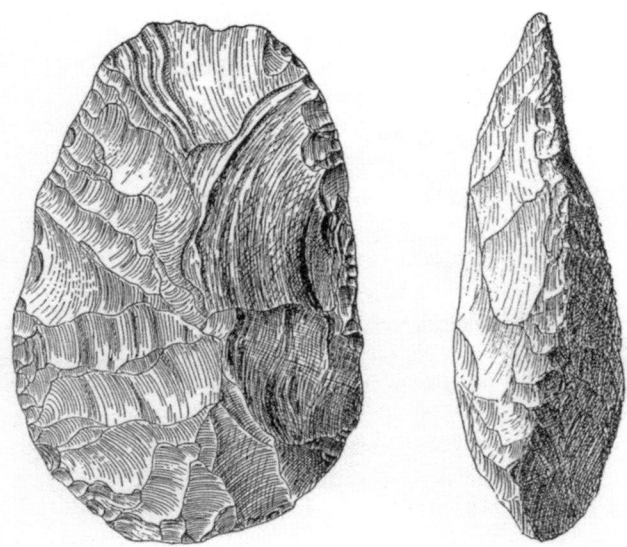

Figure 7.7. An Acheulian '*limande*', the dominant hand-axe type within Commont's '*acheuléen inférieur*' (Commont 1908, 559, Figs. 57 and 58).

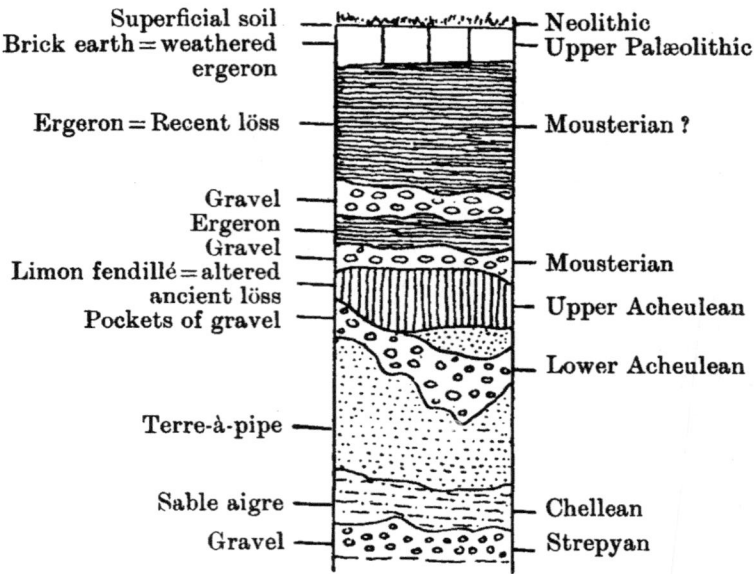

Figure 7.8. 'Palaeolithic deposits at St Acheul' (Sollas 1911, 102, Fig. 41).

Table 7.10. The industrial sequence that was often used in Britain during the early twentieth century (after Commont 1908; 1911; 1912a; Rutot 1900).

Industries	Description
Moustérien supérieur	No hand-axes
Moustérien ancién	The large Levallois flake gradually takes over from the hand-axe; smaller scrapers; Mousterian points
Acheuléen supérieur	Fine lanceolate hand-axes
Acheuléen inférieur	Various forms of finely worked hand-axes, dominated by ovates (*limandes*)
Chelléen évolué	Triangular hand-axes of Montières
Chelléen typique	Various forms of large, crude hand-axes with rough butt, amongst which the pointed *ficrons* are characteristic. Warm fauna
Pre-Chellean of Commont	Small flake-tools; coarse prototypes of the hand-axe
Strépyan (Pre-Chellean)	Nodules chipped at the point
Mesvinian (Pre-Chellean)	Flake-dominated industry with many scrapers

that when the river-drifts of such great valleys as those of the Seine and the Thames have been as assiduously investigated, they also will show a definite succession of stages'.[104] The Thames Valley river drifts had, in fact, been assiduously investigated in 1912 and 1913 by two newcomers to this Palaeolithic problem; their conclusions would justify Geikie's belief and settle the suspicions of Sturge—but would also nurture the problems associated with a standard Palaeolithic sequence.

[104] Geikie 1914, 112.

8

Swanscombe: A Standard Stone-Age Sequence for Britain?

In the early twentieth century, Palaeolithic research seemed to be flourishing on the Continent. Commont was carrying out groundbreaking work in the Somme, and rich hauls were being recovered from the reindeer-caves of France and Spain. France could also boast a research centre: the Institute of Human Palaeontology, where Boule, Breuil, and Obermaier held posts. Britain, though, was weighed down by nostalgia: unfavourable contrasts were being drawn between current research and the glorious decades of the past when Evans and Prestwich had brought such renown to British investigations.[1] This apparent loss of impetus was noted abroad. Boule considered the British to have sunk into insularity after 1875, never to regain their early brilliance; in 1912, Breuil remarked at a luncheon party in Cambridge that no one in England knew anything about prehistory.[2]

The British Museum's *Guide to the Antiquities of the Stone Age*, published in 1911 at the height of Commont's work, declared: 'the French system has now been revised in the light of recent discoveries, and is the basis of all Continental classifications'. It was regretted that the English river drifts had still not received any systematic excavations, and that the implements in these sediments still lay in confusion.[3] This *Guide* was produced by Reginald Smith of the British Museum under the direction of Charles Hercules Read (1857–1929).

In 1912, the same year that Breuil made his disparaging comment, Read arranged for Smith to excavate in one of the most productive Palaeolithic localities of the Thames Valley: Swanscombe village.[4] Smith was assisted by Henry Dewey (1876–1965) of the Geological Survey, but the negotiations that gained Dewey's help would also reveal differences of opinion between their two respective institutions about the value of Palaeolithic research. The connections drawn by Smith to the Continental sequence after working at

---

[1] Sturge 1908, 9. McNabb 1996, 32 also remarks on the early-twentieth-century perception of a decline in British Palaeolithic research.
[2] Boule 1915, 1; Burkitt typescript notes (undated), p. 14: ULC: 7959, Box 3.
[3] Read 1911, 12.
[4] McNabb 1996, gives a brief overview of the history of excavations at Swanscombe.

Swanscombe would lift the gloom about British backwardness. These connections would also help draw the Palaeolithic and geological sequences closer together.

8.1 REGINALD SMITH INITIATES INVESTIGATIONS

On a Saturday afternoon in February 1912, members of the Geologists' Association assembled under the portico of the British Museum for one of their popular field meetings. They were welcomed by Smith (Fig. 8.1), Assistant Keeper in the Department of British and Mediaeval Antiquities and Ethnography, who led them into the Prehistoric Room. Smith had worked at the British Museum since 1898. He was regarded as serious, aloof, and hardworking.[5] A junior colleague described him as 'a shortish man, bald, little moustache, pince-nez, stove-pipe collar, dark-coated' and 'an austere vegetarian', but concealing a kindly nature, 'always ready to talk about flowers and robins and suchlike, and [...] liked to be told of outstanding displays in florists windows'.[6]

Smith ran through the main types of flint implement that could be found in Britain. He explained that many of the Museum's specimens had come from collectors: their age remained mysterious because collectors were rarely interested in the geological horizons where their trophies had been found. But he told his audience of eighty-five geologists that they were equally guilty of prolonging the mystery because geologists took such little notice of the different tool-types they encountered in their delving after sediments, strata, and bones. Smith insisted that collaboration between geologists and Palaeo-lithic researchers was the only way to test current Palaeolithic classifications and to decide how well the French scheme, so popular on the Continent, matched the British evidence.[7]

The following Saturday, Smith travelled to Swanscombe in Kent to examine the Palaeolithic gravels of the largest pit in the village: the Milton Street pit. He had heard that these layers were to be removed over the next two months as the Associated Portland Cement Manufacturers (hereafter APCM), who owned the pit, worked their way down to the chalk beneath.[8] Smith was alarmed: 'One of the finest gravel-pits in the world for flint implements is in process of extinction'.[9]

---

[5] Tonnochy 1953, 85, 87–88.   [6] Kendrick 1971, 2, 5.
[7] Smith 1912, 137–138.   [8] Smith to Reid, 19 February 1912: BGS: GSM2/544.
[9] Smith and Dewey 1913a, 177.

Figure 8.1. Reginald Smith (1874–1940) of the British Museum.

Swanscombe village lay on the south bank of the River Thames in the parish of Northfleet, and had a good reputation for producing Palaeolithic tools. In 1905, Hinton and Kennard had declared: 'It is to be doubted if any locality has yielded so many implements as the neighbourhood of Swanscombe'.[10] Back in 1883, Spurrell had mentioned implements from the high-level gravels at Swanscombe.[11] Henry Lewis of Peckham, a shoemaker and collector of implements who supplied Evans with specimens, had also praised the productivity of the Swanscombe pits. In 1884, he sent Evans around a hundred palaeoliths from the Milton Street pit, and ended his letter: 'can you let me have Twenty Five Pounds for the lot. If so you will assist your obedient servt. H Lewis. PS. I would be glad to shew you all the sections'.[12]

Smith had seen enough of the Swanscombe deposits on his Saturday visit to realize that here, at last, was a chance to understand the sequence of Palaeolithic implements in Britain. As the APCM worked their way through the last of the Palaeolithic gravels in the Milton Street pit (or, as they preferred to call it, Barnfield pit), Smith decided to arrange a systematic excavation. He was

[10] Hinton and Kennard 1905, 77.
[11] Spurrell 1883, 102. Kennard 1944, 299 spotted Spurrell's early mention of implements from Swanscombe.
[12] Lewis to Evans, 5 December 1884: AMO: JE/B/2/80. Lewis was described in *Harrison of Ightham* as a shoemaker from Camberwell (which is very near Peckham) with a large collection of implements, and as an authority on stones from Finchley (Harrison 1928, 137–138).

Figure 8.2. Henry Dewey (1876–1965) of the Geological Survey. Photographed in 1912. Source: BGS Library Archive: IGS1/639, 132.

eager to enlist Dewey's geological expertise in this endeavour. If a reliable sequence of Palaeolithic culture-stages were to be established in the Thames Valley, it would have to rest on a sound geological basis; Dewey (Fig. 8.2) had been mapping the London drifts for the Geological Survey for the past two years. He had joined the Survey as a clerk in 1903 and became a Geologist in 1906. Dewey combined geology with an enthusiasm for church architecture and a range of antiquities; he was a congenial companion, a good raconteur and mimic.[13]

On the Monday after his Swanscombe excursion, Smith wrote two letters: one to Hercules Read, his superior at the British Museum; the other to Clement Reid, the Survey's District Geologist who directed Dewey's work. In these letters Smith explained his interest in a joint venture at Barnfield pit. Read was keen to support the suggestion. He had worked at the British Museum since 1880 and had helped to catalogue Christy's famous collection, replete with tools from the French reindeer caves.[14] Read also shared Smith's concern about the current state of Palaeolithic knowledge. But Smith and Dewey were not sure that their project would be welcome to both their employers. Smith assured Clement Reid:

Mr. Dewey & I are ready to take vacation if a private fund has to be raised for the purpose, but I sincerely hope that it will be recognised as proper & desirable work by both institutions, and lead to a fuller cooperation in the future. Though much time & money has been expended officially for excavating ancient sites in Egypt, Assyria, Asia

---

[13] Stubblefield 1967, 190.   [14] Tonnochy 1953, 85.

Table 8.1. Smith and his superiors in the British Museum, 1912.

Frederic Kenyon, Director of the British Museum
C. Hercules Read, Keeper of the British and Medieval Department
Reginald Smith, Assistant Keeper of the British and Medieval Department

Table 8.2. Dewey and his superiors in the Geological Survey, 1912.

Jethro J. H. Teall, Director of the Survey (retired 1914)
Aubrey Strahan, Assistant Director of the Survey of England and Wales
Clement Reid, District Geologist (retired 1913)
Henry Dewey, Geologist

Minor, Cyprus, etc., I do not think this Museum has ever undertaken similar work in England, & the time has now come to advance beyond mere collecting from workmen & others, & to ascertain the sequence of the gravels by reference to the implements & faunistic remains.[15]

This was a propitious time for such a proposal: in 1910, the Geological Survey had begun an extensive and detailed re-survey of the London district on the six-inch scale for a New Series of maps. The Old Series had been based on the one-inch Ordnance maps. In the latter decades of the nineteenth century, six-inch Ordnance maps became available for officers to take into the field. These stimulated a burst of re-surveying across the country (although the final maps were still reduced to the one-inch scale).[16]

In 1909, a field-unit of surveyors had completed their survey of the mining regions of Cornwall and Devon on the six-inch scale. The following year Dewey, his three fellow officers and their overseer, Reid, were transferred from the western counties to the London and South-eastern District. They were set to revise the areas mapped on the old one-inch scale, and entered a region renowned for its Palaeolithic tools and complicated geological history.[17] Reid, pictured in Figure 8.3, must have taken up the task of directing his field-unit with interest: he had already contributed to the debate over an interglacial Palaeolithic through his work at Hoxne at the end of the nineteenth century.

Dewey and his peers could record the complicated superficial deposits of London with great accuracy and detail on the six-inch scale. The three river terraces recognized by Whitaker (and by Pocock on his partial re-survey a few

---

[15] Smith to Reid, 19 February 1912: BGS: GSM2/544.
[16] Geikie 1898, 7, 10–11; Woodward 1922, 54.
[17] Teall 1911, 1, 24; Flett 1937, 145, 161.

Figure 8.3. Clement Reid (1853–1916), Dewey's superior on the Survey (GSL: P57/Reid).

years previously) would finally be distinguished on their New Series of drift maps. These terraces would receive distinct colours, symbols, and names based on type localities in Maidenhead, where the re-survey had begun.[18]

Although the practical importance of mapping superficial deposits tended to be emphasized in Survey memoirs—the gain to agriculture, economy, sanitation, and other fields of interest to the Government that paid for the compilation of their maps—the Palaeolithic was by no means irrelevant to Survey work. Officers mentioned stone tools in their detailed discussions of stratigraphy. They saw these implements as part of the wider geological picture and recorded their presence, just as they might record bones or shells. Horace B. Woodward did, however, warn geologists in his Survey memoir on *The Geology of the London District* (1909) that 'no chronological order such as the French scheme implies has yet been established in this country'. He took his authority from the British Museum's *Guide to the Antiquities of the Stone Age* (1902).[19]

Despite this uncertainty about the chronological value of Palaeolithic industries, Dewey's superiors might have recognised the advantage of recording the strata exposed in Barnfield pit before the sediments were quarried away. It was rare for the regimented Survey system to exploit such ephemeral opportunities—they

---

[18] Woodward 1922, 53–54; Geological Map of the London District 1927: BGS: MR 76/052 T1608—GMAP1: 10648/10.

[19] Woodward 1909, 78,79; Read 1902, 9.

usually worked over the ground systematically, map-sheet by map-sheet—but Reid and his four officers had not yet been authorized to work on the Swanscombe area (Survey Sheet 271).

The day after receiving Smith's letter, Reid raised the matter with his superior, Aubrey Strahan. Reid presented the proposal in enthusiastic terms as 'an unrivalled opportunity' to observe the deposits, obtain information for the new map, and gain a series of implements for the Geological Survey Museum (a comparative tool for officers working on Palaeolithic sediments). He asked Strahan if Dewey might be permitted to spend eight to ten days in examining the deposits at Barnfield pit as they were removed.[20] Strahan had been a keen and accurate field surveyor before he rose through the ranks of the Survey into the managerial sphere, where he became known for his tact and impartiality.[21] His reply arrived the same day:

Mr Reid
Please arrange for adequate examination of these sections to be made from time to time, especially with reference to the geological sequence and age of the deposits, and the horizons at which implements &c. occur.
Inform Mr. Reginald Smith of your proposed arrangements.[22]

This note launched an intermittent alliance between the British Museum and the Geological Survey at Barnfield pit, Swanscombe, and other Palaeolithic sites in the Thames Valley. Their alliance would last for nearly two years. Only ten days after impressing on the party of the Geologists' Association the importance of collaboration between geologists and Palaeolithic researchers, Smith had secured an opportunity to put his desire into practice. With Dewey's help he hoped to clarify the pattern of British Palaeolithic tools and relate them with greater certainty to the French Palaeolithic sequence and to British geological sequences.

8.2 THE 1912 SEASON AT SWANSCOMBE

In March 1912, the British Museum and the Geological Survey split the cost of hiring labourers to open pits and clear away debris at Swanscombe. Smith and Dewey spent ten days examining and excavating the 100-ft terrace deposits at Barnfield pit and a further two days on a small adjoining pit.[23]

---

[20] Reid to Strahan, 20 February 1912: BGS: GSM2/544.   [21] Thomas 1929, lix.
[22] Strahan to Reid, 20 February 1912: BGS: GSM2/544.
[23] Reid to Strahan, 23 February 1912 and 12 March 1912: BGS: GSM2/544; Smith and Dewey 1913a, 177.

Both institutions gained information and specimens; it seemed a productive collaboration. The *Summary of Progress of the Geological Survey* for 1912 presented the work as a valuable opportunity to observe the sequence of Palaeolithic tools in the Thames Valley, though it was regretted that bones and plant remains were too scarce to support broader connections to geological sequences. On a brighter note, the *Summary* reported that the collaboration had confirmed a correlation between the Palaeolithic sequences of the Thames Valley and the Somme Valley: a significant result for geologists. This match might calm the concerns of Woodward and justify the use of tool-types as time-markers for the correlation of geological deposits.[24]

A few months later, in April 1913, the Society of Antiquaries of London heard a longer account of the first season's work from Smith and Dewey. The Survey had granted Dewey permission to assist Smith on the geological aspects.[25] The Society of Antiquaries was an obvious venue for their report: it was officially connected to the British Museum, Read had been their President for the past few years, and he wanted the Society to play a more active part in this kind of research. The report by Smith and Dewey— 'Stratification at Swanscombe'—appeared in the Society's journal: *Archaeologia*. The Thames Valley findings took up most of their paper; the Somme Valley sequence was only mentioned in the last few pages.

Dewey described the impressive geological section that rose 40 to 50 feet high in the western side of Barnfield pit. Smith introduced the sequence of tools that coursed up through these layers of sediment. (As explained in Chapter 7, higher river terraces contained older deposits than lower terraces. But within a *single* terrace the oldest sediments were thought to lie at the deepest levels, with younger sediments being deposited in successive layers above.) In brief, and starting at the foot of the Barnfield section, their tool-bearing beds comprised: the Lower Gravel with a surprising assemblage of tools, dominated by crude flakes and a few chipped nodules, similar to the Strépyan industry of Belgium; the Middle Gravel with dark, roughly chipped, pointed hand-axes of the Chellean industry; and the Upper Loam where workmen had, in previous years, found finely made oval hand-axes of the Acheulian industry (see Table 8.3).[26]

Smith and Dewey were disappointed to have only hints and hearsay for the horizon of Acheulian hand-axes, which they had not managed to recover in

---

[24] Smith and Dewey 1913b, 83, 85; Teall 1913, 34–35.

[25] Read to Teall, 2 January 1913: BGS: GSM2/544; Strahan to Teall, 3 January 1913: BGS: GSM2/544; Smith and Dewey 1913a; 1914, 187.

[26] Smith and Dewey 1913a, 182–186, 191–192. On Dewey's interpretation of the geology, see Conway 1996, 9–16.

Table 8.3. A comparison of the industrial succession developed by Commont in the Somme Valley, by Rutot in Belgium, and the expectations and interpretations of Reginald Smith before and after the Swanscombe excavations.

The Somme Valley and Belgian sequences (Commont 1908; 1910; 1912a; Rutot 1900)	Smith's expectations (Smith 1912)	The Swanscombe sequence (Smith and Dewey 1913; 1914)	
		Industry / level	Description
Acheuléen supérieur (Commont): Lanceolate hand-axes with glossy white patina *Acheuléen inférieur* (Commont): Various hand-axe forms, dominated by oval types (*limandes*) flaked all the way round the edge, often twisted. Also varied and specialised small tools	St. Acheul II: Small, oval and slender pointed hand-axes St. Acheul I: *Limandes* dominate. Finely flaked, cutting edge all round, often twisted	St Acheul I and II: Possibly from the base of the Upper loam	Unstratified finds from Barnfield Pit (finely worked ovates), mostly with white patination, from old British Museum collections. Stratified finds from Craylands Lane pit
Chelléen évolué (Commont): Finely-worked triangular hand-axes (Montières type) *Chelléen typique* (Commont): Large hand-axes, thick butt, coarsely flaked, often retaining cortex. *Ficrons* are characteristic forms. Also a variety of small tools	Chellean: Pear-shaped or flat ovate hand-axes. Coarsely flaked, often retaining cortex. *Limandes* appear later in the Chellean	Chellean: Middle Gravel	This level yielded 'the finest implements of Chelles types, that is practically all but the ovate specimens'. Pointed, roughly chipped, more or less pear-shaped, with little secondary working. Rare ovate forms were described as heralding the later St Acheul types

(*Contd*)

Table 8.3. (*Continued*)

The Somme Valley and Belgian sequences (Commont 1908; 1910; 1912a; Rutot 1900)	Smith's expectations (Smith 1912)	The Swanscombe sequence (Smith and Dewey 1913; 1914)	
		Industry / level	Description
Pre-Chellean (Commont): Crude hand-axe prototypes and many other smaller flake instruments	Strépyan: Nodules, flaked, generally at the point	Strépyan/pre-Chellean: Lower Gravel	'The "industry" consisted almost exclusively of thick flakes' 'implements are exceptional' 'hand-axes of the ordinary type are entirely wanting'. A few nodules, possibly worked, 'correspond to the Strépy culture'
Strépyan (Rutot): Many coarse scrapers, a few nodules with pointed tips			
Mesvinian (Rutot): dominated by scrapers	Eolithic: Mesvinian, Mafflien, Reutelian. Nodules, not designedly chipped		
Mafflien (Rutot): scrapers and hammer-stones			
Reutelian (Rutot): dominated by hammer-stones			

their own investigations. They were eager to emphasize several other indications, besides the memories of the workmen, that placed the Acheulian *above* the Chellean horizon. Certain implements from the upper part of the Middle Gravel bore features that seemed transitional to the Acheulian. The collections of the British Museum proved that Acheulian hand-axes had been found at Swanscombe in the past—some had even come from the same pit—but the Museum had no record of their geological associations. Their appearance held another clue: many of the Acheulian hand-axes previously collected from this pit had a distinctive white coating (patination). Smith saw a similar colour change towards the top of the Chellean-producing Middle Gravel; white-patinated Acheulian implements had also been recovered from another pit nearby (Globe pit, Greenhithe) at a similar height in the section.[27] It was clear to Smith where the Acheulian horizon *should* be: the search for these implements would provide the main focus for their next season's work.

Mousterian flake-tools had been found above the Acheulian horizon, though by neither Smith nor Dewey, and in the Globe pit rather than Barnfield pit—and the relation between the sediments in the two pits was uncertain. Smith also observed that flake-tools of the Mousterian (Levallois) period were well known from the 50-ft terrace below.[28] Having laid out the Swanscombe sequence in detail, from the unusual flake industry at the base of the Barnfield section to the more familiar Chellean, Acheulian, and Mousterian industries, Smith explained that he had intentionally focused on Barnfield pit and the immediate neighbourhood to avoid any suspicion 'that the evidence had been twisted into agreement with supposed parallels elsewhere'.[29] Only at the end of the paper did he emphasize the similarity between Commont's results from the Somme Valley and his own findings in the Thames Valley, again protesting: 'The analogy was forced upon us at various stages of the work.' Smith was, however, jealous of Commont's sequence before he started work at Swanscombe (see Table 8.3).[30]

8.3 THE 1913 SEASON AT SWANSCOMBE

In January 1913, Read set out his plans for the forthcoming season of Palaeolithic research in the Thames Valley to the Director of the Geological Survey, Jethro Teall (1849–1924), a petrologist (Fig. 8.4). Read hoped that

---

[27] Smith and Dewey 1913a, 186–190, 193. [28] Smith and Dewey 1913a, 194–195.
[29] Smith and Dewey 1913a, 196.
[30] Smith and Dewey 1913a, 197. McNabb 1996, 33, 35 notes the strong appeal of Commont's classification, and its influence on the conclusions drawn by Smith and Dewey.

Figure 8.4. Jethro Teall (1849–1924), Director of the Geological Survey 1901–1914 (GSL: P56/Teall).

their earlier collaboration would continue. Dewey's assistance was requested at the shell-bed site of Greenhithe, rich in fossils; at Barnfield pit, Swanscombe, to clarify the horizon of Acheulian implements; and at Dartford Heath, where Chandler and Leach had been working on a tool-bearing channel that cut through the gravels. Teall replied that there would be no difficulty in extending the collaboration.[31]

The sites mentioned by Read fell within or near the area to be mapped by the Survey in the coming year. Once again, both sides could benefit from the partnership. The shell-bed at Greenhithe might illuminate the relationship between the tools and fauna of the 100-ft terrace; another visit to the upper levels of Barnfield pit might confirm the position of the Acheulian in the sediments of the Thames Valley. Read followed up his request two months later and asked Teall to lend him the services of Dewey in May to help at Swanscombe—the Museum would pay for the project. Dewey would spend four days superintending excavations at Barnfield pit with Smith. Teall agreed and offered to loan Dewey for two days at Greenhithe and another two days at Dartford Heath.[32]

In June 1913 the collaboration suffered a slight strain. With work at Greenhithe completed, Read was starting to organize the Dartford Heath excavations at Wansunt pit and asked Teall if he could spare Dewey for a week,

---

[31] Read to Teall, 2 January 1913: BGS: GSM2/544; Teall to Read, 3 January 1913: BGS: GSM2/544; Chandler and Leach 1912.

[32] Read to Teall, 5 March 1913: BGS: GSM2/544; Teall to Read, 8 March 1913: BGS: GSM2/544; Smith and Dewey 1914, 187.

perhaps longer.[33] Strahan warned Teall that this would represent a 'serious inroad upon the time available for his [Dewey's] field-work' and reminded him that the original offer had been for only two days.[34] Though Teall was not averse to Dewey's spending a day or two more, he understood Strahan's concern and suggested to Read that a whole week at the Dartford Heath section would constitute 'a rather serious interruption of Mr. Dewey's work'.[35] Read felt it necessary to re-emphasize their mutual benefit:

I quite understand your point of view on the obvious impropriety of your allowing Mr. Dewey to spend his official time in any matter that has no relation to his work. My ardent desire, however, in this matter, is to make it quite clear that you and we are working together in this determination of the sequence of the deposits where human remains exist. [A reference to a supposed skeleton, reported recently, which turned out not to exist after all.] I need not enlarge upon this to you, as you know as well as I the difficulty of persuading people that our attitude is unprejudiced whether from a geological or archaeological standpoint. If therefore, you will allow Mr. Dewey to collaborate as much as you think justifiable, that will serve my purpose, which, after all, is not ultimately confined to the acquisition of flints for the Museum, but a determination of much wider issues.[36]

This letter reveals three important points: first, Read wanted the Museum to become more involved in Palaeolithic research and to cast off its reputation as a mere hoarder of collections; second, he recognized the importance of geological support for Palaeolithic interpretations; and third, he knew that the Survey felt they would gain less from the collaboration than would the Museum. The Survey directors were more interested in completing their new maps of the London district than refining the Palaeolithic sequence. They held fixed and rather military ideas about what constituted appropriate Survey work and how this work should be carried out. Teall replied to Read's letter in conciliatory terms, but he would not be in office for much longer. The following year the collaboration between the Museum and the Survey would break down.[37]

In 1914, Read was faced with a different chain of command in the Geological Survey. When Teall retired in January, Strahan (Fig. 8.5) rose to become Director. He had taken up many of his new responsibilities the previous month and Read wrote to offer his congratulations when he learnt of the change, attempting to reinforce their earlier collaboration. Changes had also

---

[33] Read to Teall, 19 June 1913: BGS: GSM2/544. On Wansunt pit, see also Bridgland 1994, 190.
[34] Strahan to Read, 20 June 1913: BGS: GSM2/544.
[35] Teall to Read, 23 June 1913: BGS: GSM2/544.
[36] Read to Teall, 24 June 1913: BGS: GSM2/544.
[37] Teall to Read, 27 June 1913: BGS: GSM2/544.

Figure 8.5. Aubrey Strahan (1852–1928), Director of the Geological Survey 1914–1920 (GSL: P53/46).

occurred further down the hierarchy: Reid had retired as District Geologist in 1913 and George Barrow (1853–1932) now oversaw the work of Dewey and his three fellow officers in the London and South-eastern District.[38] Barrow would be less sympathetic to Palaeolithic research than his predecessor.

When Read asked Strahan if Dewey could spend a morning examining a series of tools from Dartford Heath (Wansunt pit), in preparation for a second report to the Society of Antiquaries, Strahan passed the request to Barrow.[39] Reluctantly, Barrow agreed, but only because Dewey had personally examined the site where the tools had been found. He protested to Strahan that Dewey ought not be allowed to exceed his limited responsibilities: 'it is as well to guard against the idea that members of the staff may be officially called in as experts on flint-implements, which have not been collected by members of our staff, more especially when the ground has not been officially surveyed by them'.[40]

The aims of the Museum did not mesh well with the traditional practices of the Survey. Read and Smith were trying to draw together the Palaeolithic and geological sequences of the Thames Valley; the Survey had divided the same area into distinct territories, and assigned each to different officers. There was still some truth in the complaint made by Wood to Fisher in 1869: that the

[38] Read to Strahan, 16 December 1913: BGS: GSM2/544; Strahan 1914, 27.
[39] Read to Strahan, 16 December 1913: BGS: GSM2/544; Strahan to Barrow, 18 December 1913: BGS: GSM2/544.
[40] Barrow to Strahan, 19 December 1913: BGS: GSM2/544.

Survey men were 'like Tailors working at a coat; one takes the skirt, another the sleeve, another the collar [...] & they do not [...] even pay much attention, often none, to what their colleagues close by are doing'.[41]

Barrow was a fierce guardian of these boundaries. The Museum researchers (including the co-opted Dewey) wanted to understand the relationship between tools from Dartford Heath (Wansunt pit) and those from Worthington Smith's Palaeolithic Floor at Caddington. Barrow pointed out to Strahan that Caddington fell within the area allotted to Pocock, and the district of Swanscombe and Wansunt would in all probability be assigned to Cyril Bromehead (1885–1952) for the 1914 season. In response to Read's request, the Survey sent all three officers to the British Museum to examine the tools: Dewey, Pocock, and Bromehead.[42]

This was not the first time that Barrow, habitually forthright, had provoked controversy. He owed his position on the Geological Survey of England and Wales to irresolvable differences with his former superiors on the Scottish branch. Barrow's interpretations of the structure and sequence of rocks in the Highlands had caused such uproar in the late nineteenth and early twentieth centuries that he had arranged to be transferred south to less acrimonious counties.[43]

8.4 1914: THE END OF COLLABORATION

In January 1914, Smith was preparing a paper for the Society of Antiquaries on the second season's work in the Thames Valley and was trying to draw together observations from Swanscombe, the Greenhithe shell-bed, and Dartford Heath (Wansunt pit). Since Dewey had worked with Smith on some of these sites, Read asked Strahan if Dewey could spend two or three days on his contribution to their joint report.[44] Again, Strahan passed the letter to Barrow, who could not understand why Dewey wanted a week to write up his work—unless he was trying to stray beyond the remit of Survey requirements.

Barrow cautioned Strahan that this second report was 'the result of a compact between Dewey and Smith'. Driven by territorial concerns, he complained that Dewey's contribution to the previous paper on the first season's work was

[41] Wood to Fisher, 11 May 1869: ULC: 7652/V/P/56.
[42] Strahan to Read, 19 December 1913: BGS: GSM2/544. Bromehead later became known for his sword dances amongst the cutlery at the annual Geologists' Dinner (Wilson 1985, 121).
[43] Green and Wooldridge 1933, 111; Bailey 1952, 157–158; Oldroyd 2004c. These disputes and their origins are explored by Oldroyd in *The Highlands Controversy* (1990).
[44] Read to Strahan, 22 January 1914: BGS: GSM2/544.

'largely a series of incursions into areas not allotted to him & with which he has at present nothing to do'. Barrow insisted that Dewey should have no more than two or three days to write his report; this should be a purely descriptive account of the geology and organic remains; any further generalizations should be made by Dewey in his spare time; and 'the references to the area shortly to be surveyed by his colleagues & outside the pits examined should be as brief as possible'.[45] The Survey's restrictive policy on publications by its officers was an old custom and a longstanding grievance: the information they gathered was government property, which might be guarded possessively by their superiors until it had appeared in an official Survey Memoir.[46]

In his letter to Strahan about Dewey's role in the publication, Read had also made suggestions for future work. Strahan, having absorbed Barrow's comments, agreed that Dewey could write up his geological notes for the second article in *Archaeologia*, but was wary about committing the Survey to further investigations:

With regard to future work I should hardly feel justified in continuing to assist in making excavations for the purpose of finding implements. To do so would endanger the carrying out of our programme of field-surveying. At the same time I should be prepared to send one of our staff to see any excavations which you may make with a view to assisting you in any geological questions which may arise.[47]

This spelt the end of two years' official collaboration between Smith and Dewey. Read took this retraction on the part of the Survey with resignation, but was determined to continue the work. He must have been stung by the suggestion that the Museum was only interested in finding implements. In his reply to Strahan, Read emphasized the importance of Palaeolithic research for geological purposes:

In view of your decision for this year, we must be content with a minimum of geological assistance, but shall continue the field-work, as very little is being done privately and for the credit of the country an organised effort must be made to keep pace with continental progress. We have as good a field as anyone abroad, and if our own countrymen neglect the work, foreigners will take the opportunity, and reveal to us our own resources. There has been a good deal of stagnation in recent years, and we cannot afford to give up the undertaking, even if we did not think it as important as any other branch of departmental work. If implements can be regarded as fossils, this seems the only way of classifying the gravels and throwing light on recent geological changes.[48]

[45] Barrow to Strahan, 27 January 1914: BGS: GSM2/544.
[46] Oldroyd and McKenna 2005, 203, 205–206, 216, 221.
[47] Strahan to Read, 28 January 1914: BGS: GSM2/544.
[48] Read to Strahan, 11 February 1914: BGS: GSM2/544.

Smith and Dewey read their second report to the Society of Antiquaries on 'The High Terrace of the Thames' in April 1914. The main object of the second season had been to find the horizon of the Acheulian hand-axes. These implements had emerged from the expected stratigraphical level in Craylands Lane pit, just across the road from Barnfield pit, which seemed to solve the Acheulian problem of the previous year. In the deposits above, Smith and Dewey had found a large number of flake-tools, similar to the Mousterian (Levallois) flake-tools from Spurrell's old site at Northfleet (Baker's Hole).[49] In conclusion, after two seasons' work: 'The earlier palaeolithic sequence seems therefore to be completely represented in the gravels of the 100-ft. terrace of the Thames, the two pits bordering Craylands Lane being complementary to each other in this respect, and the deposits ranging from pre-Chelles to Le Moustier times'.[50]

Dewey and Smith announced the same conclusions to a party of the Geologists' Association who had visited Swanscombe during their second season of work. A complete sequence of Palaeolithic industries from the River-Drift period had been recovered from the 100-ft terrace of the Thames, from the Pre-Chellean (or Strépyan), to the Chellean, to the Acheulian—and finally to the Mousterian, which might be better placed in the Cave period. The sequence from the Swanscombe pits proved that different tool-types were confined to different layers in the Thames Valley, just as Commont and Rutot had reported from the river valleys of France and Belgium. Earlier suspicions had apparently been justified: British tools of distinct character could indeed be matched to distinct periods of time.[51]

With the publication of their Swanscombe reports, Smith and Dewey established a standard Palaeolithic sequence in Britain. They seemed finally to have proved what their peers had only suspected. Although Hinton and Kennard had suggested a simpler sequence in 1905, Smith and Dewey had an advantage that helped them to develop a more detailed and more readily accepted scheme: Commont's sequence. They acknowledged his inspiration and example, while claiming that the evidence from the Swanscombe pits had not 'been twisted into agreement with supposed parallels elsewhere.' But Smith could not stifle his expectations of the Thames Valley sequence. He had explained these to the Geologists' Association at the British Museum a week before he instigated the Swanscombe collaboration. Smith's expectations were derived directly from the French and Belgian sequences, and they were confirmed by his findings at Swanscombe (see Table 8.3).[52]

---

[49] Smith and Dewey 1913a, 195; 1914, 187, 190.   [50] Smith and Dewey 1914, 190.
[51] Dewey and Smith 1914, 90, 92–94, 96.   [52] Smith and Dewey 1913a, 196.

Despite the strong appeal of the Swanscombe sequence in Britain, it is interesting to find that Commont did not agree entirely with Smith's interpretation. In a letter written to Sollas around the time when Smith was preparing his first Swanscombe publication, Commont remarked: '*à Swanscombe, à Barnfield, peut-être que M. Smith a des tendances à établir trop de subdivisions dans les alluvions de la terrasse au 100 ft*' ('at Swanscombe, at Barnfield, it is possible that Mr. Smith has a tendency to establish too many sub-divisions in the deposits of the 100-ft terrace').[53]

The Swanscombe results illustrate the influence of Continental sequences on British Palaeolithic researchers, but the collaboration also highlights the advantages and disadvantages of employment within institutions that paid for work on Quaternary topics. Few of those interested in the Palaeolithic managed to earn a living in a related sphere during the early twentieth century. Moir, desperate to find a post in prehistoric archaeology, was told by Lankester that he had no chance at all; Hinton tried and failed to find a geological position early in his (unpaid) geological career, though he did succeed later.[54] Dewey might have been the envy of non-professional colleagues, who snatched Quaternary research in their spare time, but he too laboured under restrictions and might, in turn, have envied Reginald Smith.

The attention devoted to stone tools was closely related to the level of interest in Palaeolithic questions amongst influential men on the Survey. A comparison can be drawn to the Geological Survey of Portugal, where stone tools and palaeoanthropology had a prominent place in Survey work between the late 1870s to early 1880s, thanks to the interest of Ribeiro, discoverer of the Portuguese eoliths. After Ribeiro's death in 1882, the focus of the Portuguese Geological Survey reverted to stratigraphy and palaeontology.[55]

The British Museum was certainly the more enthusiastic and flexible of the two partners in the Swanscombe collaboration. Smith was fortunate in the enthusiasm of his superior, Read, to promote British Palaeolithic research and catch up with the Continent. Although the Museum would benefit in practical terms from collections made during this venture, Read also wanted to contribute to research. He was keen for Smith to search for a Palaeolithic sequence in the field and recognised the importance of solid geological backing for observations made by Museum staff.[56]

The Geological Survey, the largest geological employer in Britain, had less to gain from the partnership. Territoriality over geographical areas and intellectual

[53] Commont to Sollas, 1 January 1913: BGS: GSM1/445.
[54] Hinton to A. Geikie, 3 June 1900: ULE: Gen 521/3; Lankester to Moir, 12 August 1911: BLL: 44968/53.
[55] Carneiro 2005, 159–161.
[56] Read to Teall, 24 June 1913: BGS: GSM2/544.

property, the fluctuating opinions of colleagues about the relevance of the Palaeolithic to Survey work, and personal differences all restricted Dewey's activities. Looking beyond Swanscombe, Dewey had plenty more to complain about. Pay and promotion were high on the list of Survey grievances. When Dewey's colleague Pocock joined in the 1890s, a few years before Dewey entered the Survey, the pay for an Assistant Geologist—the position first occupied by new field surveyors—was said to have been on a level with that of a dock labourer.[57] A few years after the end of the Swanscombe collaboration, with an increase in staff and a rise in the cost of living, the situation had not improved. Dewey sent a memorandum to the Director asking for better conditions.[58] When compiling this formal complaint, he explained the problem to a friend on the Survey:

> It is not only the fact that we are underpaid but the want of proper respect for our work that causes indignation. It is within the knowledge of many of us that men of about our own age who hold positions as lecturers or professors at the Universities are regarded by our Superior Officers with more respect than we are.
>
> In some cases they are men who were formerly on the Survey & who left us for their own advancement so that the irritation is increased by the thought that our loyalty is regarded as either timidity or stupidity.[59]

Dawkins, Geikie, and Sollas had all occupied Survey positions before taking up their university chairs. But the Survey was not short of new recruits and continued to carry out fieldwork in the traditional manner, undismayed by the expansion of geological posts and courses in universities.[60] At Swanscombe, Dewey may have been removed from official involvement with Smith's Palaeolithic work but he continued to publish on Palaeolithic topics. He eventually rose to become District Geologist for the London and South-eastern District in 1920 and, the following year, worked again with Smith on the Sturry gravels in the Stour Valley, Kent. Their findings there seemed to complement and justify their earlier conclusions in the Swanscombe pits, and a joint report by Smith and Dewey appeared once more in *Archaeologia*.[61] In 1932, when Dewey was awarded the Lyell Medal by the Geological Society of London, particular mention was made of his unofficial work on the Palaeolithic.[62]

---

[57] Kitchin 1927, 91.
[58] Memorandum draft, sent back to Dewey by Dinham, 18 March 1918: BGS: GSM1/291.
[59] Dewey to Dinham, 17 November 1917: BGS: GSM1/291.
[60] On the professionalization of geology in the late nineteenth and early twentieth centuries, see O'Connor and Meadows 1976 and Porter 1978, 1982. Oldroyd and McKenna 2005 discuss the conditions of employment on the Geological Survey.
[61] Dewey and Smith 1924, 135–136.   [62] Garwood 1932, lvi.

The Palaeolithic would not feature prominently in the Survey publications of later years. Although Bromehead had been sympathetic to the idea of using implements to distinguish between the 50-ft and 100-ft terraces, he made little attempt to base geological classifications on implements in his 1922 revision of the memoir on *The Geology of the London District*. Stone tools were consigned to the end of his account of river drifts 'because of the large number of unsolved problems involved'.[63] In memoirs on the London district published between the 1930s and 1960, the correlations drawn by Smith and Dewey between implement types and layers of terrace material were mentioned only briefly. Implements were not used to assist geological interpretations and little effort was made to follow developments in Palaeolithic research.[64]

8.5 PROBLEMS WITH THE PRE-CHELLEAN

Smith and Dewey presented the Swanscombe sequence as a justification of Continental classifications. Commont had supplied a template of distinctive tool-types, dominated by a progression of hand-axes until the arrival of the flake-rich Mousterian industry. It was not difficult for Smith to see a similar pattern in the British types. But he did have one problem. Fired by these expectations, Smith admitted that the assemblage of tools from the Lower Gravel 'was rather a surprise both as to its quantity and quality'.[65] His reaction to these tools—the earliest in the Swanscombe section, but rich in flakes and with few distinguishing tool-types—recalls the treatment of similarly puzzling industries by his peers described in Chapter 7. Like many others, Smith would turn to the Belgian sequence to mop up this problematic early, flake-dominated industry that had no precise equivalent in the French classification.

The Lower Gravel assemblage presented Smith with two difficulties. First, these tools could not be consigned to the best-known flake-rich industry—the Mousterian—because they lay *beneath* the Chellean in the Barnfield pit section: they had to date to Early Chellean or Pre-Chellean times. Smith called them 'Pre-Chellean', a popular term for awkward early industries. Commont had a Pre-Chellean at the base of his sequence; in Britain, this label embraced a cluster of varied assemblages. As he searched for a more specific label, Smith encountered his second difficulty: there seemed to be no distinctive retouched tool-types in the Lower Gravel on which to pin an

---

[63] Bromehead in Teall 1912, 74; Woodward 1922, 66.
[64] Sherlock 1935, 57–61; 1960, 49, 52–53.   [65] Smith and Dewey 1913a, 182.

analogy: 'implements are exceptional at this horizon, while hand-axes of the ordinary type are entirely wanting'.[66]

Hand-axes had borne a heavy typological burden in Commont's classifications; they made an equally impressive show in the Swanscombe sequence (see Table 8.3). But for Smith and Dewey, the tools from the Lower Gravel did not even qualify as 'implements': a term that usually referred to trimmed or 'retouched' tools, and was sometimes restricted to retouched tools made from a stone nodule or 'core', and not from a flake.[67] 'The "industry" consisted almost exclusively of thick flakes, with prominent bulbs of percussion and a minimum of flaking, due to use or shaping, on the edges'.[68] The workmen, used to finding hand-axes and other trimmed tools favoured by collectors, believed this horizon to be barren.

Smith, eager to find a diagnostic tool-type among these nondescript flakes, seized on: 'A few nodules trimmed at the point and squared at the butt [...] which, if accepted as human work, would correspond to the Strépy culture' (see Figure 8.6).[69] Once again the Belgian sequence offered reassurance to British researchers, wanting to find a label for early flake-rich assemblages, in its few extra industries that hooked onto the base of the French classification. The Strépyan was rich in flakes and scrapers, and Rutot regarded his chipped 'Strépy nodules' as the prototypes of Chellean hand-axes.[70] This industry provided a good analogy for Smith, and the Strépyan took its place at the base of the British sequence.[71]

British researchers would not be able to use the Belgian sequence to label Pre-Chellean industries for much longer. Rutot's industries were re-dated to later times not long after Smith and Dewey had settled on a Strépyan analogy for the earliest of their Swanscombe industries. Warren had noticed back in 1905 that Rutot was using a different industrial timescale to most researchers: his 'Chellean', for example, was the 'Acheulian' of most authors; his 'Mesvinian' was not Pre-Chellean in age and was no earlier than Palaeolithic industries in Britain.[72] But Warren tucked away his warning in a footnote.

Commont confided similar suspicions to Sollas in 1913, around the time when Smith was working on the first Swanscombe report. He wrote that Rutot would do well to leave his imagination aside: he had long considered Rutot's system with its supposedly ancient Strépyan and Mesvinian to be flawed, but had not come out with this earlier because he liked Rutot and

---

[66] Smith and Dewey 1913a, 182–183.   [67] Hinton and Kennard 1905, 91.
[68] Smith and Dewey 1913a, 182.   [69] Smith and Dewey 1913a, 183.
[70] Rutot 1903, 434; MacCurdy 1905, 452.
[71] Dewey and Smith 1914, 93; Kennard 1916, 253; Dewey and Bromehead 1921, 55; Warren 1924c, 68–69.
[72] Warren 1905, 341, footnote.

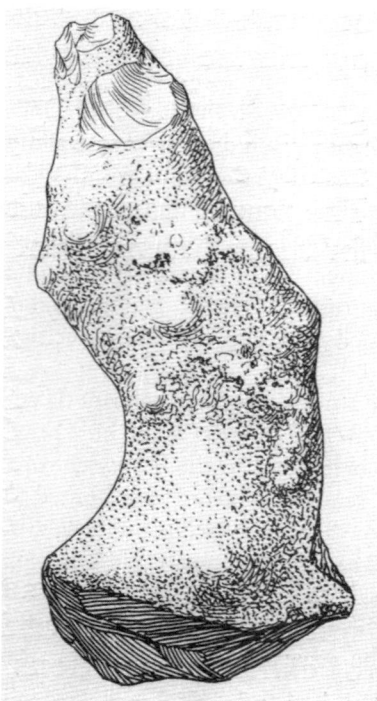

Figure 8.6. 'Nodule chipped at point, perhaps of Strépy type' from the Lower Gravel at Barnfield pit, Swanscombe (Smith and Dewey 1913a, 183, Fig. 10).

hoped to find some points of agreement with him. Commont disagreed with Rutot's geological sequence. He believed that the sediments identified by Rutot as older loess were in fact younger loess: a re-identification that would draw Rutot's industries forward into much later times.[73] Commont explained to Sollas:

Quant au Strépyan il n'existe pas à mon avis. Sans doute il y a une industrie plus vieille que le chelleen typique francais à Chelles, mais cette industrie préchelléenne ne peut correspondre au niveau qualifié de Strépyan par M. Rutot. Son Strépyan est imaginaire, il serait d'ailleurs post Würmian, des alluvions des basses terrasses belges était Würmiennes. (As for the Strépyan, it does not exist in my opinion. There certainly is an industry older than the typical French Chellean at Chelles, but that Pre-Chellean industry cannot correspond to the level designated as Strépyan by M. Rutot. His Strépyan is imaginary and in any case it would be post-Würmian, the deposits of the Belgian low terraces being Würmian.)[74]

---

[73] Commont to Sollas, 1 January 1913; 23 January 1913: BGS: GSM1/445; Commont 1912c, 168–170; Sollas 1915, 134.

[74] Commont to Sollas, 1 January 1913: BGS: GSM1/445.

When Sollas received this letter he was working on the second (1915) edition of *Ancient Hunters*. Commont's opinions had a prominent place in his revisions, and Sollas changed his assessment of the Belgian sequence from 'the remarkable section at Helin, [...] so well described by M. Rutot' to a 'discordant note' that 'requires reinvestigation'.[75] In Britain, the term 'Strépyan' survived as a label for the earliest Swanscombe industry even after the Belgian industries had been pushed forward in time. The Somme Valley classification came to be seen as a more reliable series of analogies for early Palaeolithic industries.[76] But the flake-dominated Belgian industries, unhooked from their Pre-Chellean slot, floated awkwardly alongside the hand-axe industries of the standard Palaeolithic sequence. Not all researchers accepted them back into a later slot with the other flake-rich Mousterian industries.

8.6 A NAME FOR THE INDUSTRY FROM CLACTON-ON-SEA

As Commont wrote his letter to Sollas, and Smith decided to label his Swanscombe industry the 'Strépyan', Warren regarded his flake-dominated assemblage from Clacton-on-Sea with perplexity, and abandoned the attempt to give it a name (see Chapter 7). Nearly a decade later, Breuil looked at Warren's tools and assigned them to the 'Mesvinian' industry.[77] The Mesvinian was another of Rutot's Belgian industries that had, like the Strépyan, been ejected from its old Pre-Chellean position and moved forward in time. But when Warren suggested in the early 1920s that his flake-rich industry was *contemporary* with the Acheulian, this opinion was not accepted or even understood by most of his peers, schooled in the standard linear sequences of Swanscombe and the Somme Valley.

In character, Warren's assemblage from Clacton matched the Mesvinian well. Rutot's industry was dominated by flakes and flake-tools: end-scrapers and side-scrapers (Figs. 8.7 and 8.8), flakes with concave notches, and natural or waste flakes.[78] Warren's industry from Clacton was also dominated by crude flakes (Fig. 8.9) with some scrapers, as well as core-tools that Warren called side-choppers (Fig. 8.10). There were, of course, no hand-axes. In 1922, Warren announced to the PSEA that Breuil's opinion was 'obviously correct,

---

[75] Sollas 1911, 107; 1915, 132.
[76] Sollas 1915, 137–140; Kennard 1916, 253; Woodward 1922, 67; Marston 1937, 340.
[77] Breuil 1930, 221.   [78] Rutot 1900, 714.

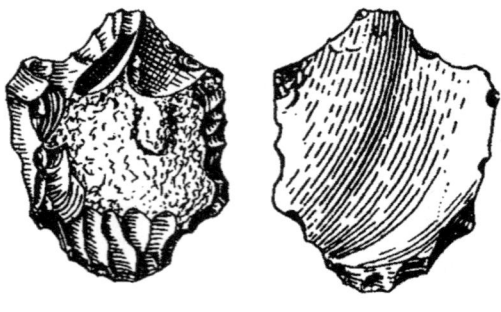

Figure 8.7. Mesvinian '*grattoir*' (end-scraper) from Belgium (Rutot 1900, 727, Fig. 12).

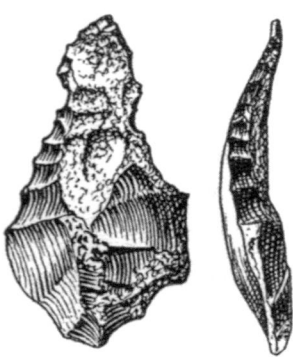

Figure 8.8. Mesvinian '*pointe-racloir*' (pointed side-scraper) from Belgium (Rutot 1900, 727, Fig. 14).

and it has since been confirmed by several continental authorities who are intimately acquainted with the Mesvinian of Belgium'.[79]

Warren had refined his ideas about the age of this Mesvinian industry from Clacton. He now thought that the bones and stratigraphy corresponded with 'some part of the Acheulian stage, or possibly a little earlier. This is in agreement with the present continental dating of the Mesvinian industry'.[80] His tool-bearing sediments at Clacton had gathered in a former river-channel (referred to hereafter as the 'Clacton Channel'), which also held the remains of warmth-loving species. It was known as the '*Elephas antiquus*' bed because straight-tusked-elephant bones were so plentiful. Reid's analysis of the seeds also supported a warm climate, pre-dating the arrival of cold Mousterian times. Warren believed the Clacton Channel to be a tributary of the Thames that had cut through the 100-ft terrace deposits; Kennard, who worked on the shells, connected the bed to an early stage of the 50-ft terrace.[81] All these geological indications placed the Clacton industry in Chellean or Acheulian

[79] Warren 1922b, 597, 598–602.
[80] Warren 1922b, 602. See also Rutot 1921, 55.
[81] Warren 1922b, 597.

Figure 8.9. Flake collected by Warren from Clacton (PRM: 1940.3.26).

times, but Warren believed that the character of his industry was so different that it might have been made by a separate group of tool-makers:

On taking a general survey of the Mesvinian industry, as so admirably displayed at Clacton, one cannot help feeling that it might well be the precursor of the Mousterian industry, but that it has no cultural connection with the Chellian and Acheulian stages.

As knowledge of the Palaeolithic period increases, we are realizing more fully the divergence of races and cultures which were living contemporaneously together. I believe it is in this light that the Mesvinian industry is to be understood. But we have yet so much to learn that no conclusions can be regarded as more than tentative.[82]

The following year Warren gave stronger voice to these cautious views in a paper to the Geological Society: 'The *Elephas antiquus* Bed of Clacton-on-Sea (Essex) and its Flora and Fauna.' He placed the Mesvinian at the base of a sequence of three flake industries (Mesvinian, Levalloisian (Early Mouster-ian), Mousterian). This sequence of flake industries was separated from the sequence of hand-axe industries (Chellean, Acheulian). Warren set them next to each other—to run in parallel through time. He added that the early date of the Clacton industry confirmed the suggestion that the Mousterian was not related to the hand-axe cultures but had been developed independently by a different race.[83]

[82] Warren 1922b, 602.   [83] Warren 1923d, 614.

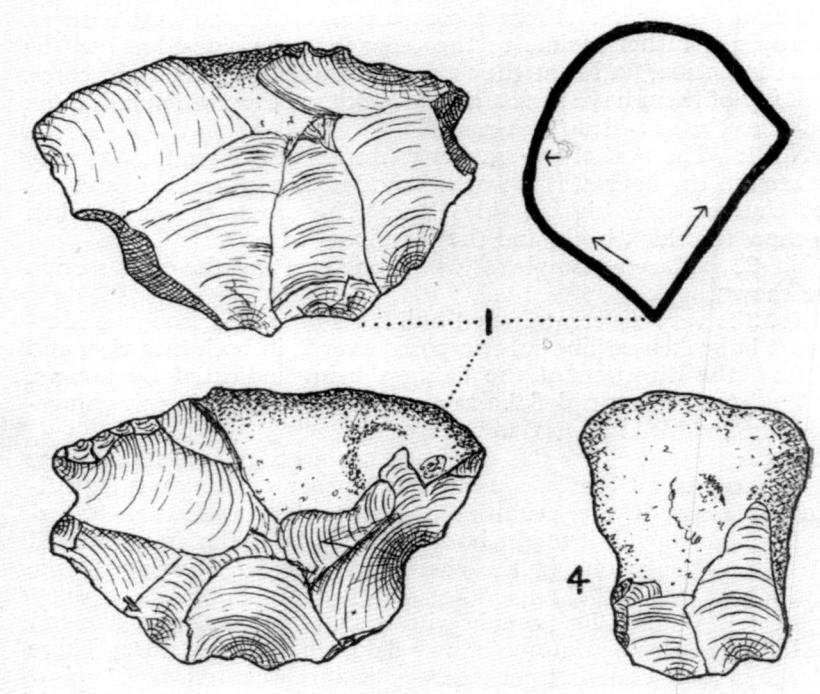

Figure 8.10. A Mesvinian side-chopper from Clacton (Fig. 1) and a small flint nodule with an axe edge at one end (Fig. 4) (Warren 1922b, 600).

As Warren observed, he was not the first to propose that different industries had been made by different races. Commont had suggested in 1912 that an intrusive race had brought the Mousterian industry into the Somme Valley.[84] The Mousterian had been connected to the Neanderthals since the nineteenth century. A few years before Commont gave this opinion, Boule had separated the Neanderthals from the human line (see Chapter 4). A branching hominid tree had then become popular. The fine Acheulian hand-axes would often be linked to our own branch.[85] But this did not mean that Warren's peers would be happy with his picture of contemporary Acheulian and pre-Mousterian industries. The sequence of industries was still expected to follow one line, not two. As explained in Chapter 7, the connections between industries and mobile groups of tool-makers offered a way to mask overlap on the standard Palaeolithic sequence by stretching the time-slots occupied by each industry.

[84] Commont 1912a, 248.   [85] Boule 1908a, 524–525.

Even if the Mousterian was described as the degraded product of a race who, for a brief period, pushed our ancestors out of the area, the Mousterian industry still came *after* the Acheulian industry in Britain.[86]

Warren was not alone in his suspicions that there had been a more permanent overlap between different groups of tool-makers. Obermaier, who had described an alternation of French hand-axe and non-hand-axe industries earlier in the twentieth century, presented a grander view of Palaeolithic geography in 1919. He plotted the different areas occupied by contemporary hand-axe-making and non-hand-axe-making groups on a map of Europe (Fig. 8.11). Obermaier's Pre-Mousterian tool-makers spread across central and eastern Europe; their hand-axe-making contemporaries occupied western Europe and the Mediterranean region.[87] Obermaier could trace these lines back to the Pre-Chellean, but his zones retained their distinctive patterning in later times. In Britain, this German paper passed largely un-noticed.

Obermaier worked on this map in Madrid: he had been in Spain when the First World War broke out and was unable, as a German, to return to Paris. In any case, he soon lost his job in France: his friend Breuil never forgave their colleague and Director, Boule, for writing the letter that forced Obermaier out of the Institute. And after all these problems, Obermaier was denounced as a Francophile in Spain.[88] Commont's death was another tragic outcome of the war. Forced to abandon his house, books, and collections during a bombing raid, Commont's demise in 1918 was hastened by the trauma of the Abbeville evacuation.[89] It had not been an easy war for his old correspondent Sollas either. If the account of an acquaintance is to be believed, he narrowly avoided a messy death. Sollas, by then in his mid-sixties, had joined the Volunteer Labour Force. Apparently, he 'was sent down with a party to Didcot where they were loading high explosives into trucks. Professor Sollas was found smoking a cigarette, surrounded by notices or posters saying, "DANGER, NO SMOKING," and he was sacked on the spot'.[90]

Sollas was present at the Geological Society a few years later to hear Warren read his paper on the Mesvinian from Clacton. In the discussion, Sollas suggested that this assemblage might represent the encroachment of a Neanderthal race of flake-tool makers onto the hunting grounds of hand-axe makers, whom

---

[86] Read 1911, 8; Sollas 1911, 162, 170; Sturge 1912, 213–215; Geikie 1914, 46.

[87] Obermaier 1908, 125; 1919, 146–147.

[88] Breuil 1950, 107–108.

[89] Boule 1918–19, 162.

[90] Douglas, transcript of an interview recorded in 1976: UMO: WJS, Box 10/17. This reminiscence must be treated with caution—the relationship between Douglas and Sollas had been strained (Marianne Sommer, pers. comm. 2006).

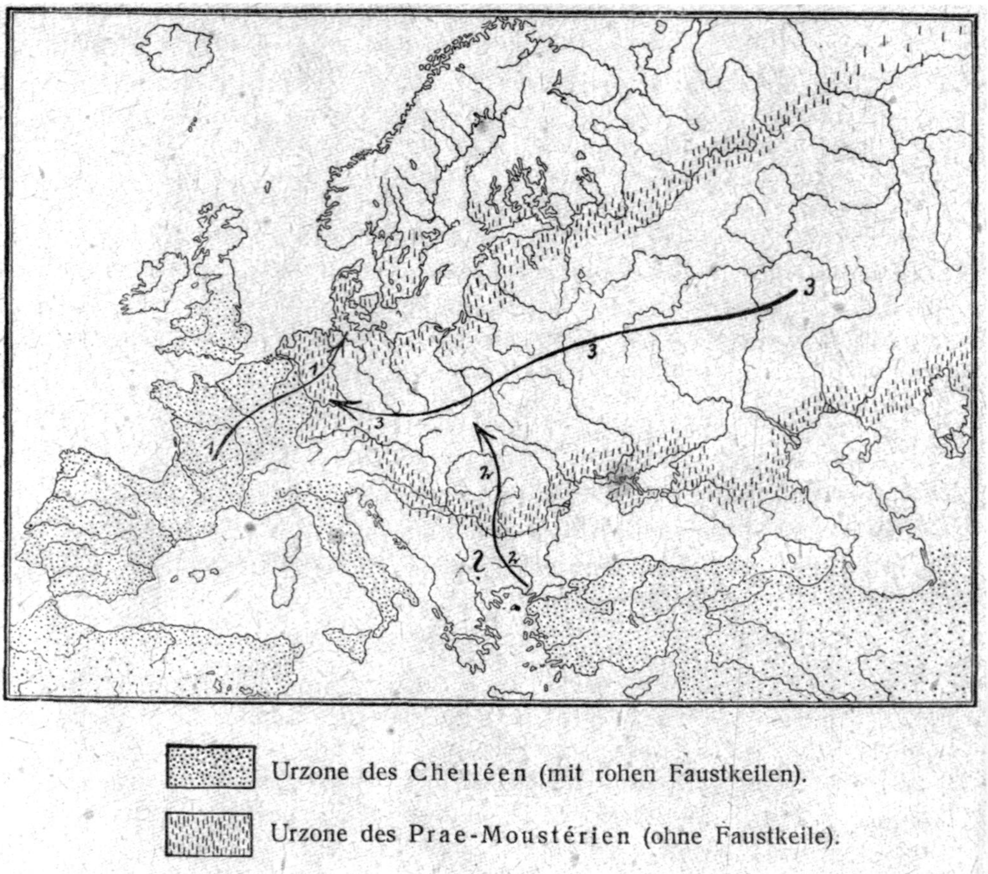

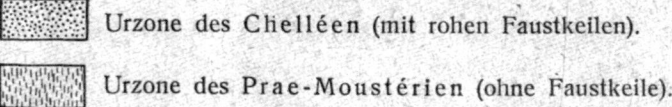

Urzone des Chelléen (mit rohen Faustkeilen).

Urzone des Prae-Moustérien (ohne Faustkeile).

Figure 8.11. Obermaier's map of Palaeolithic Europe, showing the Chellean zones (with crude hand-axes), the Pre-Mousterian zones (with no hand-axes), and the routes taken by the Acheulian: 1. West Acheulian, 2. South Acheulian (substantially identical), 3. East Acheulian (Obermaier 1919, 147, Fig. 1).

he associated with the Piltdown hominid. But he insisted on describing the tools as 'good examples of the Mousterian industry' in an anomalous position. Once again, the standard Palaeolithic sequence had been maintained by a temporary encroachment of tool-makers.[91]

The difficulty of reconciling the age and the character of Warren's Mesvinian industry led another member of Warren's audience to observe: 'With respect to the Mesvinian implements, they seemed to represent types which

[91] Warren 1923d, 634.

are found in several periods, and unless associated as a group, they would require further investigation before the date assigned could be accepted'.[92] This remark captured the incomprehension that lay between Warren and his audience. The Institute of Archaeology in London holds a copy of the published paper with Warren's annotations. He had underlined the words 'unless associated as a group' and written in the margin: 'certainly they are so associated at Clacton—that is the point'.[93]

The point, for Warren, was that his Mesvinian was *not* a fortuitous mixture of types from several periods; his assemblage *was* a discrete industry even if it did not fit the pigeon-holes of the standard sequence. Flakes and flake-tools might be present, but these were not Mousterian flakes. Implements worked on both faces (biface-tools) were also present, but they had no resemblance to hand-axes and did not deserve the glib label 'Acheulian.' For his peers, however, Warren's industry did not match expectations forged by the Swanscombe sequence or by Commont's work in the Somme Valley. The Mousterian industry followed the Acheulian industry; if it did not, there was an anomaly to be explained. They might well assume that this assemblage comprised a mix of elements from various different ages, or represented a temporary incursion by another tool-making race. Warren and Obermaier had little influence on the comprehension of the British Palaeolithic in the early 1920s.

8.7 HOXNE RE-VISITED: THE INTERGLACIAL PALAEOLITHIC OF THE 1920S

The Palaeolithic sequence established by Smith and Dewey at Swanscombe seemed to have dispersed the confusion that some had seen hovering over British research in the first decade of the twentieth century. This growth of confidence reinforced the rejection of anomalous industries. The close connection between Britain and France in their patterns of tools also encouraged a closer scrutiny of the relationship between these industries and the glacial epoch. In the conclusion to their first Swanscombe report, Smith and Dewey had agreed with prevailing opinion that the 100-ft terrace and its Palaeolithic contents were post-glacial (i.e. post-Chalky Boulder Clay) in date.[94] But by the early 1920s, British researchers were looking querulously across the

---

[92] Anon 1923a, 56.
[93] Warren, marginal note (undated): IAL: DAA410.E/7WAR. My thanks to John McNabb for telling me about this source.
[94] Smith and Dewey 1913a, 200, 202; Smith 1915b, 223–224.

Channel, where Palaeolithic industries were interspersed through several glacial and interglacial periods, and wondering whether events had been so different in Britain. Perhaps Geikie and Skertchly had been right, after all.[95] A few old British sites were re-opened in the 1920s to clarify the position of the Palaeolithic in relation to the glacial sequence. The results from three sites—High Lodge, Foxhall Road, and Hoxne—would be cited frequently in later years.

In 1920, Marr declared in his Presidential Address to the Prehistoric Society of East Anglia: 'whereas a few years ago it was universally maintained that man only appeared in post-glacial times, there are now few, if any, who would subscribe to this belief'.[96] Marr, who had made a small geological contribution to the case of the East Anglian pre-palaeoliths, was now Professor of Geology at the University of Cambridge and President of the PSEA. That year, 1920, Marr dug at High Lodge, Suffolk, to discover the connection between implements and glacial episodes. In the previous century, Skertchly had included High Lodge alongside a number of other East Anglian sites that he thought proved the interglacial date of the Palaeolithic: he had described how the boulder clay rested *on* the tool-bearing beds as well as underlying them. Back then, Skertchly had not gained much support for this opinion; now Marr confirmed his interpretation. The Mousterian implements lay *between* two boulder clays at High Lodge.[97]

Nina Layard, another prominent member of the PSEA, had excavated the site of Foxhall Road, Ipswich, in the early 1900s and had found Acheulian implements lying above the boulder clay.[98] (Foxhall Road is a different site to the Crag pit at Foxhall Hall, mentioned in connection with the Pre-Palaeolithic in Chapter 6.) In 1922, Boswell (Fig. 8.12) and Moir excavated at the same site. In his report, Boswell moved away from his old monoglacial stance and linked the boulder clay from Foxhall Road to the earlier of *two* boulder clays. He divided the Chalky Boulder Clay between two different glaciations: one had left the Chalky-Kimmeridgic; the other, the Upper Chalky Boulder Clay (see Table 8.4).[99] Boswell now spoke of interglacial episodes rather than climatic ameliorations.

Hoxne, the most celebrated of the three sites, had provided the classic example of a post-glacial Palaeolithic ever since Prestwich had read his report to the Royal Society in 1859. A few decades later, when Geikie and Skertchly

[95] Moir 1920b, 221; Marr 1921, 353.   [96] Marr 1920, 190.
[97] Whitaker *et al.* 1891, 56–57; Marr 1921, 362.
[98] Layard 1903, 43. On Layard and the Foxhall Road excavations, see Plunkett (1999) and White and Plunkett (2004).
[99] Boswell and Moir 1923, 243.

Figure 8.12. Percy Boswell (1886–1960), glacial geologist (GSL: P53/60).

had disquieted their peers with interglacial arguments, the site had been re-visited and the old view upheld.[100] The gaze turned once again to Hoxne in the uncertain atmosphere of the 1920s. Moir superintended excavations at the site between 1924 and 1926, funded by the British Association. His observations supported the picture from High Lodge and Foxhall Road: there were two boulder clays at Hoxne, not one; the Acheulian and Early Mousterian industries lay between them.[101] The old vision of a post-glacial Palaeolithic lying above a single Chalky Boulder Clay faded.

Table 8.4. Boswell's opinion on interglacials changed between 1914 and the 1920s: he altered his terminology and his correlations to the East Anglia glacial sequence (after Boswell 1914; Boswell and Moir 1923, 240–244; Moir 1927a, 162).

East Anglian glacial sequence: Boswell, 1914	East Anglian glacial sequence: Boswell, 1920s
Chalky Boulder Clay	Upper Chalky Boulder Clay (Mousterian)
Middle Glacial series. Brief climatic amelioration, not an interglacial	Interglacial (Acheulian, Early Mousterian)
North Sea Drift (Cromer Till, Contorted Drift, Norwich Brickearth)	Lower (Chalky-Kimmeridgic) Boulder Clay. Probably contemporaneous with the North Sea Drift
Cromer Forest Bed	Cromer Forest Bed

[100] Prestwich 1860a, 308–309; Reid 1896, 411.

[101] Moir 1927a, 141–142. Boswell later claimed that he organised the re-opening of the Hoxne sections, and recalled: 'Altho' I turned the geology over to Reid Moir, it was really *my*

8.8 A LONG OR A SHORT GLACIAL CHRONOLOGY?

Though an interglacial age for the British Palaeolithic was soon widely accepted, there was another question to be answered: how were these two glaciations to be related to the Alpine glacial sequence? The sequence of four glaciations from the Alps supplied four fixed points on which to slot other British sequences: a way to connect the localized and varied columns of boulder clays, river terraces, bones, and tools. The industries were usually correlated to glaciations by counting back through these four glacial episodes, beginning with the Mousterian industry. Those who followed the short glacial chronology of Boule linked the Mousterian to the last glaciation of the four: the Würm. Adherents of the long glacial chronology of Penck and Geikie connected the Mousterian to the previous glaciation: the Riss, and started counting back from this earlier point.

In 1924, Warren reported that the 'orthodox' theory was to link the Mousterian to the Würm, and to follow the version proposed by Boule.[102] The connections between different geological and Palaeolithic sequences had always been contentious, but two sequences had now gained credibility: the standard Palaeolithic sequence established at Swanscombe and the short glacial chronology. They offered a tempting guide that could connect geologists' confusing sediments: a distinctive stone tool might decide the age (in glacial terms) of a boulder clay or bed of river sediment, and establish its relation to other tool-bearing strata. In the 1920s it was not only the industries but also the glacial deposits of East Anglia and the river terraces of the Thames Valley that became tied to the short glacial chronology.

The habit of counting back through industries and glaciations became a characteristic feature of Quaternary research. But there was more than one way to count. Disagreement swirled around the orthodox 'short chronology' of the 1920s. A few followed Penck and Geikie, and associated the Mousterian (and the Upper Chalky Boulder Clay) with the Riss rather than the Würm (see Table 8.5). Sollas suggested a lengthier chronology for the Somme Valley industries in 1923.[103] Boswell was too cautious to make any firm connections. Warren did not think the Mousterian was associated with a glaciation at all: he described a much later period of maximum cold, associated with the Magdalenian industry (of the old Reindeer period) and his Ponder's End stage.[104]

---

geology he reproduced, being no geologist himself' (Boswell to Baden-Powell, 14 December 1951: UMO: DB-P: K14).

[102] Warren 1924b, 269. On the 'orthodox' version of industrial–glacial correlation, see Peake 1922, 126 and Burkitt 1925, 47, 52.

[103] Sollas 1923, 333–334.   [104] Moir 1920b, 222; 1927a, 162; Warren 1924b, 270, 280.

Table 8.5. A simplified summary of the long and the short glacial chronologies that were used in Britain during the 1910s and 1920s. The short chronology was popular in the 1920s.

Industries	Short glacial chronology	Long glacial chronology
Magdalenian		Würm
Upper Mousterian		Riss-Würm
Lower Mousterian	Würm	Riss
Acheulian	Riss-Würm	Mindel-Riss
Chellean		
Pre-Chellean	Riss	Mindel
		Günz-Mindel

Many palaeontologists would have agreed with him—although Hinton and Kennard would not have had any earlier glaciations either. In 1920, Kennard wrote to Breuil's old pupil, Burkitt: 'the fauna is absolutely dead against more than one cold period. *All* palaeontologists are agreed upon that [. . .] You give up the multiple Ice Age theories & stick to our views & you will be right'.[105]

Reports of early Palaeolithic industries in East Anglia during the late 1910s and the 1920s encouraged some geologists to consider a longer glacial chronology than the version accepted by most of their peers. James Edward Sainty (1882–1967), a Norfolk schoolmaster, geologist, and archaeologist, found a Chellean hand-axe on one of his frequent visits to the coastal exposures of the Cromer Forest Bed near his house. He believed that this hand-axe, from Sidestrand, had been derived from these very early deposits.[106] A great many more early tools were found beneath the boulder clays by Moir—indeed, his discoveries had led to the re-excavations at High Lodge and Hoxne by reinforcing suspicions of an interglacial Palaeolithic.[107]

In 1913, Moir had placed the arrival of Palaeolithic tool-makers after the time of the Chalky Boulder Clay. He reconsidered this opinion in 1920 and arranged the Palaeolithic industries of East Anglia along a lengthier glacial chronology (see Table 8.6). Loathe to compress the Acheulian and Chellean into a single interglacial, Moir used Sainty's hand-axe from Sidestrand and his own discoveries of an Early Chellean industry from the Cromer Forest Bed to date these deposits to the interglacial *before* the Acheulian industries from Hoxne and Foxhall Road. His Pre-Palaeolithic industry lay even further back in time.[108] Moir's 'long chronology' for British Palaeolithic industries aroused interest in the 1920s, but attracted few converts. Most continued to place the

---

[105] Kennard to Burkitt, 13 April 1920: ULC: 7959, Box 3.
[106] Anon 1923b, 126–127; Baden-Powell 1968, 267.
[107] Marr 1920, 181; Moir 1927a, 141; Boswell 1922, 303; 1940, 269.
[108] Moir 1913a, 319; 1920b, 222–223; 1923, 136–137; 1924b, 236–237; Boswell and Moir 1923, 260; Sainty 1927, 189–190.

Table 8.6. Moir's long industrial–glacial chronology, with connections to industries and East Anglian deposits (after Moir 1920b, 222–223; and Boswell and Moir 1923, 260).

Industries	Alpine glacial sequence	East Anglian deposits
Solutrean	Würm	
Upper Mousterian	Riss-Würm	
Mousterian	Riss	Upper Chalky Boulder Clay
Early Mousterian; Acheulian; Late Chellean	Mindel-Riss	Industries of Hoxne; Foxhall Road; High Lodge
	Mindel	North Sea Drift; Lower Chalky Boulder Clay
Early Chellean	Günz-Mindel	Cromer Forest Bed
	Günz	
Pre-Palaeolithic (Pre-Chellean)		Sub-Crag

Chellean and Acheulian in the Riss–Würm interglacial episode, even if they did add another bunch of Early Chellean or Pre-Chellean industries to the older end of the glacial sequence.

Numerous texts on prehistory written in the 1920s attached pre-palaeoliths to the base of the Palaeolithic sequence, and assigned to them a variety of labels that ranged from 'Pre-Chellean' to 'Foxhall', 'Ipswich' or 'Crag'. The credibility of a Pre-Palaeolithic stage had been strengthened by Breuil's conversion in 1920, but was enhanced still further by reports of rostro-carinates from Africa and the Near East.[109] Although pre-palaeoliths were now rarely associated with the old term 'eolith', the debate about their origin had not ended. Warren and other sceptics continued to argue that they had been flaked by natural causes; Moir maintained a robust defence.[110]

One geologist who worked closely with Moir and observed his long industrial chronology with curiosity was Boswell. Harmer had died in 1923 at the age of eighty-eight, leaving Boswell to lead the field of glacial and industrial correlations in East Anglia, assisted by Moir and John Solomon.[111] Their publications would form the central core of reference for research on geological sequences in the 1920s and 1930s. After he became Professor of Geology at Liverpool University in 1917, Boswell retained an interest in the Palaeolithic discoveries made by Moir in Suffolk, and by Sainty and others in Norfolk. It was through Sainty that Boswell met Solomon in the 1920s

---

[109] Sollas 1924, 102–106; Burkitt 1925, 82–88; Peake and Fleure 1927, 92–95, 99; Reinach 1929, 327.

[110] Warren 1925, 1928, 1929, Barnes and Moir 1926; Moir 1927b, 7–51.

[111] Wright 1937, 82.

(Solomon's family lived at Runton, near Cromer), and the three men would gather on field trips in the area. In 1930, when Boswell left Liverpool to take up the chair of geology at Imperial College, London, Solomon joined him as a research student, his subject being the glacial sequence of East Anglia with special reference to Palaeolithic industries. A year or two before, when he was at Trinity College, Cambridge, Solomon had worked with Louis Leakey (1903–1972) on the geological context of East African industries.[112]

During the late 1920s, Boswell became troubled by discrepancies between the British glacial sequence and the Alpine chronology. His concern was stimulated in part by the patterning of industries through the East Anglian boulder clays. On the Continent, the Chellean and Acheulian occupied the same interglacial episode; in Moir's version the Chellean reached back into the preceding interglacial, straddling a glacial stage. Boswell also wondered if the Lower Chalky Boulder Clay and the North Sea Drift were really contemporaneous, or if these two deposits might instead reflect *two* glacial episodes—which would add another early glaciation to the British glacial sequence.[113] In his paper on 'Early Man and the Correlation of Glacial Deposits,' presented to the 1930 Meeting of the British Association for the Advancement of Science in Bristol, Boswell declared:

As compared with early Alpine glaciations there is apparently one cold period too many in the British area. The case at present is worse than that of the mythical Irishman's waistcoat, which had one button too many at the top and one buttonhole too many at the bottom, for here we have one button too many in the middle.[114]

Some geologists disregarded the Alpine sequence because of problems like these; Boswell was still dubious about the connections between the Alps and British deposits.[115] Plagued by too many boulder clays, Boswell had also been tied willingly into his confusion by the Palaeolithic sequence, which had helped him to correlate British and Alpine glaciations. Solomon remarked in 1932: 'The finding of flint implements *in situ* in various deposits in the Cromer district stimulated both geologists and, more especially, archaeologists, to a perfect orgy of correlation, much of it ill-informed.' This comment was directed mainly at Moir, Boswell's helpmate.[116]

[112] Solomon to Baden-Powell, 1 May 1930: UMO: DB-P: K126; Myres to Fleure, 26 December 1930: BLO: Myres papers 14/67; Boswell to Myres, 15 January 1931: BLO: Myres papers 5/52; Boswell, notes, December 1960, pp. 1–2: ULL: D4/10.
[113] Sainty 1927, 213; Boswell 1930, 380.
[114] Boswell 1930, 380.
[115] Boswell 1930, 380; 1932a, 83–85.
[116] Solomon 1932, 243; Solomon to Baden-Powell, 23 January 1955: UMO: DB-P: K126.

In 1929, Boswell discussed Moir's industries and his own uncertainties about the East Anglian glacial sequence with Breuil as they sailed together to the British Association conference in South Africa. Unlike most of Boswell's British colleagues, Breuil supported a long glacial chronology, but even he was puzzled by Moir's Early Chellean industries from the Cromer Forest Bed.[117] Boswell wondered if some of his problems might be solved by a longer chronology, despite the difficulties this would cause to British sequences, which so many geologists had aligned to the shorter chronology during the 1920s. He remained convinced that the Palaeolithic sequence offered a guide through his mess of East Anglian sediments.

In 1930, Boswell sent a copy of his paper on 'Early Man and the Correlation of Glacial Deposits' to Burkitt, with the remark: 'From this you will see that there are snags in the East Anglian succession'.[118] Burkitt still taught prehistoric archaeology at Cambridge, now in the official capacity of a University Lecturer. Boswell referred to the problem of correlating British industries and boulder clays to the industrial–glacial sequence used on the Continent, and thought that Burkitt might be interested in the solution he had proposed in this paper: 'it would seem that we may expect most from the human industries themselves. If we use these industries for detailed correlation, we must regard them as contemporaneous, notwithstanding the time occupied in the migration of the peoples responsible for them, or in the diffusion of technique'.[119]

Boswell argued that Palaeolithic industries supplied more specific time-markers than bones, river-terraces, glacial deposits, or the Alpine sequence. Animals had changed slowly through the Quaternary. The height of river terraces and remnants of sea level change were unreliable guides to correlation because of the patchy nature of these records and disparities between different districts in the rise of land after weighty glaciers had melted away. The times of maximum glaciation varied with the geographical location of the nearest local ice-sheet, and had differed across Europe, and even across different parts of Britain. Geologists were starting to complain about the difficulties of matching the Alpine glacial sequence to regions far from the Alps.[120] Stone tools, however, were seen by Boswell as a more reliable guide to age. They could help geologists to order and correlate the complicated glacial deposits

---

[117] Boswell 1930, 380.

[118] Boswell to Burkitt, 6 October 1930: ULC: 7959, Box 2. On Burkitt's teaching at Cambridge see his typescript notes (undated), pp. 16–17: ULC: 7959, Box 3; and Smith (2004, 53–68).

[119] Boswell 1930, 379.

[120] Gregory 1930, 376, 378; Sandford 1930, 379; Boswell 1930, 379; Solomon 1930, 381; Peake 1930, 383; Sandford 1932, 2.

of East Anglia, link the geological history of the Thames Valley into the glacial scheme, and establish connections between Britain and the Continent.

In making this claim, Boswell trusted to the standard Palaeolithic sequence established by Smith and Dewey and later elaborated by Moir and others. Industries like the Mesvinian of Clacton had no part in his vision. Boswell wanted none of the complications that might arise from industrial overlap. But the linear Palaeolithic line would soon be fractured, a new industrial–glacial framework would arise, and a trail of geological alterations would follow.

9

The Advent of the Abbé Breuil

In 1930, Boswell made a compelling statement of his faith in the British Palaeolithic sequence as a reliable guide to geological time. The archaeologist Harold Peake (1867–1946), honorary curator of Newbury Museum whose interests ranged from earliest prehistory to the Bronze Age, had attended the same session at the annual meeting of the British Association for the Advancement of Science.[1] He was provoked by Boswell's conviction to offer a cautious warning:

As a geologist he [Boswell] is sceptical of the possibility of solving the problem [of placing the East Anglian glacial deposits in sequence] by geological means, and turns to archaeological evidence as supplying more reliable data for the purpose. As an archaeologist I have similar doubts as to the efficacy of my own subject, though I am inclined to believe that the possibilities of the geological approach have been underrated.

I would submit that the true succession of types of the Lower and Middle Palaeolithic phases, with which alone we are concerned, appears today to be by no means as certain as it did ten years ago. Broadly speaking we have evidence of successive stages of two industries, a core industry and a flake industry.[2]

Peake explained that some stages of the flake industry, which included 'the types known as Levallois and Le Moustier and perhaps others', seemed to have existed in Britain before the core industry went out of use. ('Core' industries were those like the Chellean and Acheulian: with hand-axes that were often made on nodules or 'cores' of flint.) This meant that 'the simple succession, Early Chelles, Chelles, Evolved Chelles, St Acheul, and Le Moustier no longer holds good'.[3]

Early flake industries, like Warren's Mesvinian from Clacton, had attracted more interest of late. By appearing *alongside* the hand-axe industries of the simple, standard sequence, they added greater variety to the character of stone tools that had existed at any one period of time, but they also reduced the chronological value of the old Palaeolithic sequence. Boswell, though he was

---

[1] Fleure 2004, 268–269.   [2] Peake 1930, 382.   [3] Peake 1930, 383.

Figure 9.1. The Abbé Henri Breuil (1877–1961), influential prehistorian (ULC: 7959, Box 1).

absent from this meeting of 1930 (his paper had been read for him), learnt of Peake's concern. He complained the following year: 'If, as Mr. H. Peake has recently said, "...the simple succession Early Chelles, Chelles, Evolved Chelles, St Acheul, and Le Moustier no longer holds good," I personally almost despair of a solution'.[4]

One author of Boswell's troubles was Breuil (Fig. 9.1). He had encouraged the loss of confidence among British researchers in the simple linearity of their Swanscombe sequence. A decade earlier, Warren and Obermaier had provoked little reaction with their accounts of contemporary flake- and hand-axe-making peoples. When Breuil later made similar suggestions, interest and concern had risen dramatically.[5] But Breuil also offered a solution: a more intricate Palaeolithic framework that incorporated problematic flake industries, like Warren's Mesvinian, *alongside* core (or hand-axe) industries. The details supplied by Breuil about the character and date of these different tool-making traditions, and the relationships between them, would restore confidence in Palaeolithic industries as a guide to geological time. But Breuil's time ticked at a different rate to the old British clocks.

[4] Boswell 1931, 107.
[5] Childe 1935, 4–5; 1951, 234; Garrod 1938a, 2; Dennell 1990, 553; McNabb 1996, 36–38; Cohen 1999, 306.

9.1 BREUIL, PARALLEL CULTURES, AND THE GLACIAL CHRONOLOGY

Burkitt described Breuil as a man of 'an electric, not to say impatient, temperament': an exciting but exhausting house-guest, who 'smoked without ceasing and would only discuss prehistory'.[6] For the first two decades of the twentieth century, Breuil's Stone-Age interests centred upon the industries of the reindeer-caves.[7] Sackett has remarked on the contrast between Breuil's flexible approach to the tools of the Reindeer period in France, and the rigid classification he would later develop for the Palaeolithic period—a classification that would become much more popular in Britain than in France.[8]

Breuil disagreed with de Mortillet's progressive sequence of industries from the Reindeer period (or Cave period; now also becoming known as the Upper Palaeolithic period), and saw finer sub-divisions within each of de Mortillet's discrete industrial stages. In 1912, he went to Geneva to give a detailed description of his ideas to the International Congress of Prehistoric Anthropology and Archaeology. He explained that the character of industries from different regions had varied according to the influences, infiltrations, and invasions of different tribes. Interestingly, as Breuil was promoting greater variety in the tools of the Reindeer period, Boule was dividing the single evolutionary line of tool-makers into multiple branches; his divisions, of course, occurred much earlier in Palaeolithic time.[9]

Some joked that being '*jeune et mince*' (young and slender), Breuil was well-suited to cave-work. But he had spent many of his youthful autumns amid the Palaeolithic sites of the Somme Valley under the tutelage of Commont, after being introduced to geology by d'Ault du Mesnil.[10] Breuil inherited Commont's old interest in the river-drift industries of the Somme Valley; by the early 1920s, he was scrutinizing these tools and their sediments closely.[11] Breuil's interpretation of the Somme Valley industrial–glacial sequence would have a great influence on British research, and the inspiration for the glacial aspects of his scheme can be traced, in part, to visits he made to Britain.

---

[6] Burkitt, transcript of radio broadcast, transmitted on 5 March 1962: ULC: 7959, Box 3. See also Boyle 1963, 13.

[7] Breuil 1937a, 54.

[8] Sackett 1991, 139, note 2.

[9] Boule 1908a, 524–525; Breuil 1913b, 169; Hammond 1982, 20; Sackett 1991, 117–121; Dubois and Bon in Coye (ed) 2006, 135–147. My thanks to Marianne Sommer for drawing my attention to these palaeoanthropological connections.

[10] Breuil 1921a, 162; 1937a, 61; 1948, 65; 1949, 12; Brodrick 1963, 26–30, 131.

[11] Kenyon 1937, 255; Breuil 1939, 33; Garrod 1961, 206.

The French regarded British glacial deposits with almost as much enthusiasm as the British felt for the industrial sequence of the Somme Valley. The famous French Palaeolithic sites, though rich in stone tools, lay far from those districts that had been covered by ice-sheets. French researchers looked elsewhere when they wanted to link their tools to a glacial sequence.[12] Boswell once observed that the British Palaeolithic record might not be able to compete with the breathtaking painted caves of France and Spain, but it held valuable evidence for dating stone tools because of its position on the margin of successive glaciations.[13]

Breuil had made several trips to Britain after the First World War as his interest deepened in the Palaeolithic industries of the Somme Valley. In 1919, he was still using the short glacial chronology of Boule and the French school to date the tool-bearing deposits. The following year, when lecturing on southern England at the Institute of Human Palaeontology, Paris, Breuil found it necessary to distribute the industries across several glacial and interglacial episodes. But it was the year after, on a visit to England in 1921, that he found the key to correlating his Somme Valley sediments with glacial cycles.[14]

Various Palaeolithic sites in southern England were on Breuil's itinerary for 1921. His hosts included Warren, Chandler, Dewey, Marr, and Moir. In retrospect, Breuil characterized them as an insular group, concerned only with local matters and lacking any broader conception. When he was presented with the Gold Medal of the Society of Antiquaries in 1937, Breuil declared: '*je m'efforçai de coudre ensemble les recherches un peu dispersés de vos excellents chercheurs locaux*' ('I forced myself to knit together the somewhat scattered researches of your excellent local researchers').[15] But it was in their company that Breuil became excited by solifluxion deposits in the periglacial Thames Valley. Like the Somme Valley, this district lay to the south of the ice-sheets, but periods of severe climate were reflected nonetheless in the contorted and tumbled deposits that his British hosts called by names like 'Trail' and 'Coombe Rock'.

Breuil returned to the Somme Valley with the ability to recognize the mark of glaciations hidden in solifluxion deposits. In the past, connections between the Somme industrial sequence and the Penck–Brückner Alpine glacial sequence had relied upon cycles of valley erosion or changes in fauna. Now, Breuil had the means to make more confident correlations and, after a few more years of research, he produced his great industrial–glacial synthesis (see Tables 9.1 and 9.2). This clear link to the Alpine sequence—a popular

---

[12] Boule 1888, 133, 281; Breuil 1929, 104.   [13] Boswell 1932a, 64.
[14] Breuil 1948, 66.   [15] Breuil 1937b, 259. See also Breuil 1948, 66.

chronology across much of Europe—would allow researchers working in other regions to compare their own industrial–glacial sequences with Breuil's synthesis.

Breuil connected his glacial sequence to an equally influential interpretation of the stone tools. In 1926 he published a short note on contemporary flake-dominated and hand-axe-dominated industries in *Man* (the anthropological equivalent of *Nature*). Breuil suggested that certain discoveries of early flakes and flake-tools, once grouped with the Mousterian industry, had been contemporary with the late Acheulian industry. He mentioned several assemblages that had puzzled earlier researchers: the flake-tools from High Lodge, for example, which lay beneath a horizon of hand-axes; and Commont's 'warm Mousterian' from Montières, which Breuil termed the Levalloisian. Breuil concluded that the industries of late Acheulian peoples had been interrupted occasionally by a cycle of industries of Levallois type, made by a different cultural stream of tool-makers.[16]

Breuil was not the first to publish such thoughts in a British journal.[17] But when British researchers looked back, from the vantage point of the 1930s and 1940s, at earlier opinions on the Palaeolithic, they associated the shift from a linear sequence to parallel tool-making cultures with Breuil's note of 1926. One other publication was sometimes included in this assessment: a Presidential Address to the PSEA, delivered two years later by Dorothy Garrod (1892–1968).[18] Like Burkitt before her, Garrod had studied under Breuil in Paris at the Institute of Human Palaeontology.

Garrod had been the only pupil to work regularly at the Institute during her two years' stay. Upon her arrival in 1922, she had been presented with a box of Commont's papers on the Somme Valley. Breuil had told her to come back if there was anything she didn't understand. Garrod had returned with many queries about the complicated geology. She learned later that this had been a test: Breuil would have been uninterested in any pupil who claimed to have no difficulties, and would have classed them as unpromising.[19] When she gave her paper on parallel cultures in 1928, Garrod took up Breuil's ideas and described three converging industrial lines between Acheulian and Mousterian times: Chellean-Acheulian, Levalloisian, and Pre-Mousterian.[20] She drew a genealogical tree to illustrate their relationship (Fig. 9.2): a device common

---

[16] Breuil 1926, 178.

[17] Kendall 1915, 138; Kennard 1916, 256; Bury 1916, 176–177; Moir 1918, 508. Such opinions explain Warren's remark that the Levallois industry 'is sometimes considered to be of Upper Acheulian date' (Warren 1923d, 614).

[18] Childe 1935, 4; 1944, 18; Clark 1941, 147.

[19] Garrod 1961, 206; Boyle 1963, 14.

[20] Garrod 1928, 267.

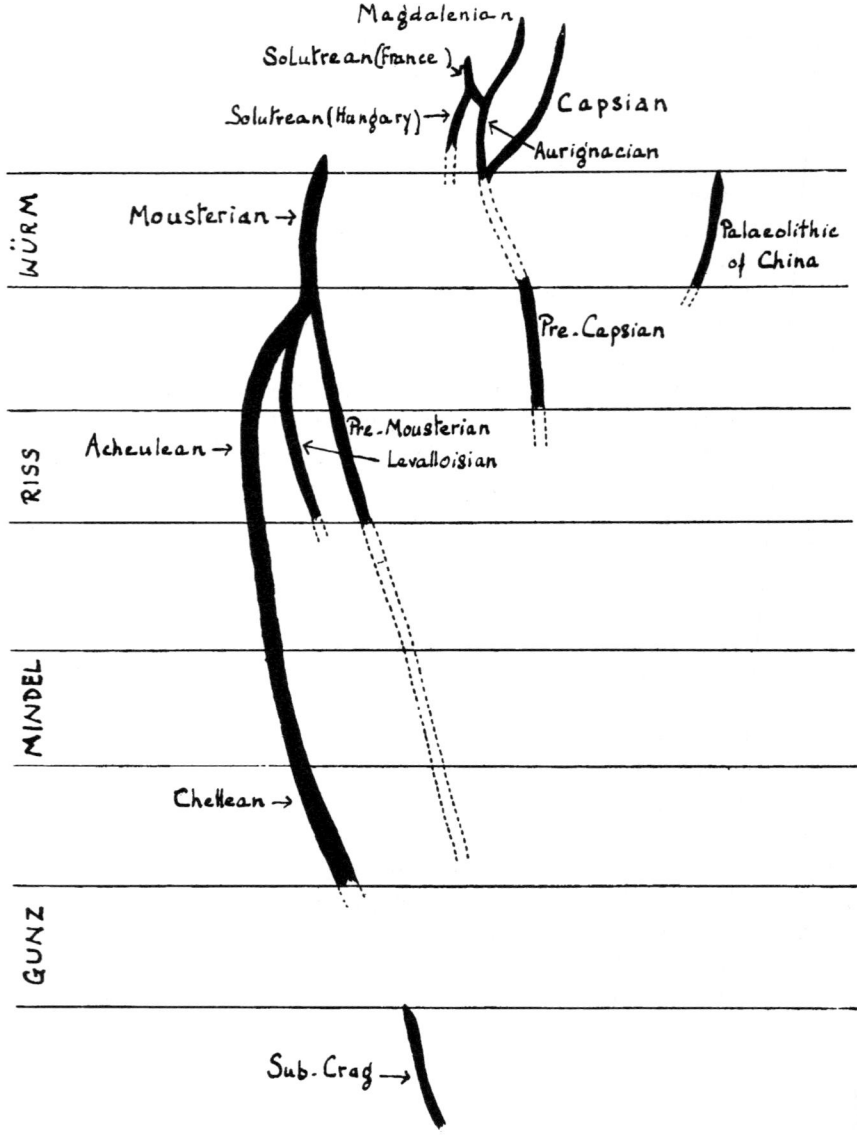

Figure 9.2. Garrod's chart showing the relations between Palaeolithic industries and their connections to the glacial sequence (Garrod 1928, 262).

among palaeoanthropologists. Palaeolithic researchers rarely used these trees to depict the relationship between industries unless accompanied by consideration of the races that had made them—and Garrod did discuss the tool-makers in her article.[21]

Breuil, meanwhile, continued to work out his own Palaeolithic patterns. He stretched his cultural streams further back in time, clarified their geographical extent, and published his opinions in two noteworthy articles in 1929 and 1932.[22] By this time, Breuil had been elected Professor at the Collège de France. The connections he drew in the second, and more detailed, of these two papers, 'Le Paléolithique Ancien en Europe Occidentale et sa Chronologie', are summarized in Table 9.1. Breuil placed the line of hand-axe industries alongside the line of flake industries. Each line was linked to a different Palaeolithic 'civilisation' and a different geographical territory. Whilst his hand-axe-making group (Chellean, Acheulian, Micoquian) had occupied the south and west of Europe, his flake-making group (Crag, Clactonian, Levalloisian, Mousterian) had lived in the north and east of Europe. Breuil's sequence of flake industries began with the Pre-Palaeolithic flakes discovered by Moir at Foxhall Hall, Ipswich, in the Red Crag (see Chapter 6). Warren's industry from Clacton came next, and the flake stem then branched into two: one branch progressed to the Levalloisian via the Mesvinian of Belgium; the other developed into the Mousterian via the Tayacian.[23]

Unlike many British researchers of the 1920s, who had lumped most of the Palaeolithic industries into the Riss–Würm interglacial, Breuil used a long glacial chronology in his paper of 1932. This allowed him to draw detailed time-connections between industries on his different lines. The climatic changes of the glacial–interglacial sequence also offered Breuil an explanation for an alternating pattern he had observed in the flake and hand-axe industries of northern France and southern England: a zone that lay between his two Palaeolithic territories. Flake industries seemed to appear at the approach of glacial episodes, and to persist into the beginning of interglacial episodes; hand-axe industries were truly interglacial in date. Breuil saw in this pattern the reflection of distinct hand-axe-making and flake-making culture-groups, periodically pushed into these borderlands from their heartlands by cyclical changes in climate.[24]

---

[21] Genealogical trees have a long history and were easily adapted for palaeoanthropological purposes. Smith 1894, 13 wrote of tool-makers at the top of a main stem and apes on the lower branches. Schwalbe 1906, 14 used simple genealogical trees to illustrate two alternative versions of hominid evolution: a linear line from *Pithecanthropus* to Neanderthals to modern humans; and a branching tree that placed the two former species on separate side-branches.

[22] Breuil 1929; 1932a.

[23] Breuil 1932a, 571–572; 1932b, 131.   [24] Breuil 1932a, 570–573.

Table 9.1. These were the connections drawn by Breuil between his parallel industrial lines and the glacial sequence (after Breuil 1932a, 573).

Glacial sequence	Flake-making populations	Hand-axe-making populations
Würm	Mousterian (pre-Würm to post-Würm maximum)	
Riss-Würm	Levalloisian	Micoquian
Riss	Levalloisian (pre-Riss to mid-Würm)	
Mindel-Riss		Acheulian
Mindel	Clactonian (end of Günz-Mindel to beginning of Mindel-Riss)	
Gunz-Mindel		Chellean (Abbevillian)
Günz		
	Ipswich (Crag)	

In this paper Breuil introduced the patterns, sequences, and explanations that would dominate perceptions of the British Palaeolithic in the 1930s. He also mentioned some unfamiliar industrial names: terminology in Britain, as well as France, had changed by the 1930s. Warren's 'Mesvinian' from Clacton and Smith and Dewey's 'Strépyan' (or 'Pre-Chellean') from Swanscombe were both becoming better known as examples of the 'Clactonian' industry. Breuil did not call the earliest true hand-axe industry the 'Chellean,' but preferred the term 'Abbevillian': he believed that the type-site of Chelles contained a relatively late industry.[25] The 'Micoquian' referred to a late Acheulian industry, remarkable for the character of its hand-axes, which were extremely pointed.

Although these new terms entered British literature, the alteration provoked some grumbling. Dewey suggested in 1931 that the names he and Smith had applied to the Swanscombe industries might as well be abandoned: such changes had rendered them meaningless.[26] Terminological change was a favourite grievance of Warren. He captured the confusion in 1932, when he wrote: 'I hope I make this clear: the Chelles industry (= the Grays Inn group) is no longer "Chellian", but a re-constituted "Acheulian"; the present "Chellian" is the former "Pre-Chellian" or "Strépyan"; while the dates at which different authors made this big change-over are not uniform'.[27]

---

[25] Breuil 1932a, 571. Warren named the industry from Clacton-on-Sea 'Clactonian' in 1926 after Breuil re-interpreted the industry from the type-site at Mesvin (Warren 1926, 47, footnote; Chandler 1930, 81, footnote 2; Breuil 1926, 178, footnote). Breuil seemed unaware of Warren's decision; nonetheless, he also named the industry 'Clactonienne' in 1929 (Chandler 1930, 87; Breuil 1932b, 125, footnote, 132).

[26] Dewey 1931, 147. See also McNabb 1996, 38.

[27] Warren 1932a, 7. See also Warren 1924b, 279; 1924c, 68–69; 1926, 39–41.

9.2 INFLUENCES ON BREUIL'S SCHEME

Breuil was adamant that he had been the first to conceive of distinct, contemporary industrial cultures in the Palaeolithic: those with hand-axes and
flakes, and pure flake cultures with no hand-axes. When Louis Leakey,
renowned for his work on the tools and tool-makers of East Africa, used a
similar classification, Breuil suggested that he ought to have been cited. But
such ideas cannot be pinned on a single person. Before Breuil published his
papers on parallel Palaeolithic cultures, Warren, Obermaier, and others had
made similar suggestions.[28]

Ever since 1904, when they first met, Obermaier and Breuil had each
exerted a strong influence on the views of the other. When Obermaier
suggested in 1908 that hand-axe industries had alternated with non-hand-
axe industries in France (see Table 7.9), Breuil defended him in the *Revue
archéologique* and admitted that his own opinions were very similar:

*j'ai souvent causé avec Obermaier des diverses coupures du y Paléolithique ancien, et
autant que j'ai pu suivre son texte, il ne s'éloigne pas notablement de mes propres
opinions* (I often talked with Obermaier on the various divisions of the Lower
Palaeolithic, and insofar as I could follow his [1908] text, it does not move away
notably from my own opinions).[29]

Breuil agreed with Obermaier that early industries lacking hand-axes had
existed in the period that was generally assigned to hand-axe-makers of the
Acheulian epoch: '*Il y a certainement des gisements antérieurs à l'Acheuléen
supérieur et qui n'ont pas de coup de poing (Micoque inférieur)*' ('There are
certainly layers preceding the upper Acheulean and which do not have the
hand-axe (lower Micoque)').[30]

In the early 1910s, Obermaier and Breuil (Fig. 9.3) worked together as
fellow Professors at the Paris Institute. They maintained their friendship
through the First World War and were still in contact after Obermaier
moved to Madrid. It was there, in 1919, that Obermaier produced his paper
on the geographical zones of Europe that he thought had been occupied by
hand-axe makers and non-hand-axe makers (Fig. 8.11). Obermaier's description of alternating industries and regional variation, and his use of a glacial
sequence to order his industries had close parallels in the ideas outlined by
Breuil some years later.

---

[28] Breuil 1936, 208. Narr (in Collins 1969, 309) observes that Obermaier set out the idea of
two different tool-making traditions before Breuil.
[29] Breuil 1908, 416.   [30] Breuil 1908, 417.

Figure 9.3. Obermaier (middle) and Breuil (right), *c.* 1913 (ULC: 7959, Box 1).

By the early 1920s, when Obermaier adopted Spanish nationality and began to focus on the Spanish Palaeolithic, Breuil had started work on Commont's sequence in the Somme Valley.[31] This was a significant diversion of their paths—at least from the perspective of British researchers. Perceptions of the British Palaeolithic had already been saturated with Commont's conclusions through the work of Smith and Dewey at Swanscombe. In the Somme, Breuil had inherited a persuasive foundation for his own Palaeolithic elaborations.

Breuil met Warren several times in the early 1920s, so they too very likely exchanged views on parallel cultures. Warren's opinion on the independent origin of hand-axe and flake industries has been set out in Chapter 8 and needs no further elaboration.[32] But the thoughts of Denis Peyrony (1869–1954) on this subject must be mentioned. Peyrony had been born in the Dordogne in south-west France; he excavated several caves in this classic Palaeolithic region and was the curator of the National Museum in Les Eyzies. Like Warren, Peyrony worked on a smaller geographical scale than either Obermaier (the whole of Europe) or Breuil (western Europe). His ideas about parallel Palaeolithic cultures were based on observations at Le Moustier and

---

[31] Breuil 1939, 33; 1950, 108; Garrod 1961, 206.
[32] Warren 1922b, 602; 1923d, 614.

La Micoque, and Breuil would integrate Peyrony's work within his broader Palaeolithic picture.[33]

Warren met Peyrony in June 1923, when a party of the Geologists' Association travelled to France.[34] This was some time before Peyrony decided that there were two Mousterian industries at the site of Le Moustier: the 'Mousterian of Acheulian Tradition' and the 'Typical Mousterian'. He argued in 1930 that these two industries were the products of two contemporaneous Mousterian tribes, and traced them both back to the end of Acheulian times.[35] The following year, when he gave his report on the 1929 season of excavations at La Micoque, Peyrony concluded that the Typical Mousterian was even earlier than he had suspected: it must have been made at the height of the Acheulian industry. The Typical Mousterian of La Micoque (level H) lay beneath the Micoquian (final Acheulian) levels:

Il a dû se différencier de bonne heure sous la forme qu'on lui connaît et se développer parallèlement au Clactonian de Breuil, au Levalloisian, à Acheuléen et au Moustérien de tradition acheuléene. Ce sont diverses techniques industrielles qui ont été contemporaines. (It must have differentiated early on into the form in which it is now known, and developed parallel to the Clactonian of Breuil, to the Levalloisian, to the Acheulean and to the Mousterian of Acheulian Tradition. These are various industrial techniques which were contemporaneous).[36]

Although Breuil gave a clear account of parallel cultures in 1932, similar thoughts had been expressed before. Nonetheless, Breuil was far more successful in winning recognition for his views than these earlier researchers, who often remained nameless. There are several reasons for his greater success. One lies in the time when Breuil presented his ideas. In the early 1920s, anomalously early flake industries were still regarded as such—as anomalies—and were jostled back into the single standard line of Palaeolithic culture-stages. Warren's peers listened to his case, but were not yet doubtful enough about the Swanscombe sequence to adopt his alternative industrial pattern, or even to fully comprehend his suggestions. Doubts had grown by the time Breuil gave his account— indeed, he had nurtured some of those doubts himself—and British researchers were more prepared to hear and understand a different Palaeolithic framework that could solve their difficulties.

Breuil's esteem in British learned society encouraged a high regard for his views. In 1920, his opinion had been powerful enough to provoke the dramatic change of mind over Moir's pre-palaeoliths, described in Chapter 6, which had given even Warren pause to think twice. Breuil's ideas infiltrated the

[33] Peyrony 1930; 1931; Breuil 1932a, 572.   [34] Warren 1924d, 137.
[35] Peyrony 1930, 172–173.   [36] Peyrony 1933, 441.

'Cambridge school' of Palaeolithic research, which had started to expand in the 1920s into an important centre for the study of prehistory in Britain. Burkitt, Breuil's first English pupil, publicized his master's views in his Cambridge lectures, which he began just after the First World War and continued for several decades afterwards.[37] By 1934, Breuil was President of the Prehistoric Society (which had lost its old regional tag: 'of East Anglia'). In 1939, Dorothy Garrod, who had also studied under Breuil at the Paris Institute, became Disney Professor of Archaeology at Cambridge. She was, however, a less ardent popularizer of Breuil's opinions than Burkitt. After publishing her much-cited 1928 article on parallel cultures, Garrod moved on to later Palaeolithic industries, into the Near East, and out of our gaze.

Another key to the sympathetic reception of Breuil's Palaeolithic scheme lay in the scale of his vision and the links he made between industries and the glacial sequence. Support for an interglacial Palaeolithic had grown in Britain over the 1920s, assisted by new interpretations of High Lodge, Foxhall Road, and Hoxne (discussed in Chapter 8). Long glacial chronologies like those propounded by Geikie and Penck had also gained more adherents: Boswell had been intrigued by Moir's long industrial–glacial sequences from East Anglia.[38] There was, however, still uncertainty about the connection between the sequence of boulder clays and industries, argument about their relation to the Thames Valley sequence, and no standard, accepted links between these varied patterns and the Alpine sequence. Some geologists were attracted by the promise of Breuil's scheme to guide their correlations.

9.3 CULTURES, CHRONOLOGIES, AND THE CLACTONIAN

As the old Palaeolithic sequence was carved up to incorporate a parallel line in the 1930s, the flake industries began to be seen differently in Britain. These changes are illustrated by the treatment of the flake-rich Clactonian industry. Back in the early 1920s, Breuil had named Warren's assemblage from Clacton the 'Mesvinian.' These tools soon gained wider connections and a new name: the 'Clactonian'. Such changes cannot be attributed solely to the influence of Breuil. British researchers had continued to make detailed observations and interpretations of Palaeolithic sites throughout the 1920s, and some of their ideas meshed well with Breuil's scheme of the 1930s.

---

[37] Boyle 1963, 13. On the role of Cambridge in Palaeolithic research, see Clark's *Prehistory at Cambridge and Beyond* (1989). See also Smith (2004).

[38] Marr 1921, 362; Boswell and Moir 1923, 243; Moir 1927a, 141–142.

It was Chandler, Leach's co-worker in the Thames Valley, who noticed the similarity between the Strépyan (or Pre-Chellean) tools from the Lower Gravel of Swanscombe, and the tools found by Warren at Clacton. Chandler had been keeping an eye on the Swanscombe exposures since 1921, visiting the site between twenty or thirty times each year to observe the sediments and tools. Other flake-rich assemblages, scattered across England and further afield, would also be re-classified and incorporated into the Clactonian over the 1930s, but this connection to Swanscombe was particularly important: the Swanscombe sequence still held a pivotal place in British Palaeolithic classifications. Breuil had made the same connection: Chandler remembered showing him around the site in 1925 and again in 1928, and they may have spoken on the subject then.[39]

It might seem odd that Warren and Reginald Smith had not drawn any analogies between their discoveries. Smith, for his part, avoided personal communication with Warren, with whom he had clashed over the eoliths (see Chapter 6). His assistant Thomas Kendrick, who joined the British Museum in 1922, remembered how Smith kept firmly to a small circle of 'approved friends' in the Museum and disapproved of any staff who were friendly towards those 'with contrary views to ours. I think immediately of Hazzledine Warren and my visits to his anti-eolith collection at Loughton'.[40]

Chandler argued that the Swanscombe Clactonian was contemporary with the Acheulian: a date that matched Warren's earlier estimation. Initially, Chandler had described these tools as 'pre-Acheulian' in age because they lay in the Lower Gravel, beneath the industry of the Middle Gravel (the 'Chellean' of Smith and Dewey, now known as the 'Acheulian').[41] Chandler then altered this date to 'Early St.-Acheul.' He wrote to Ernest Dixon (1876–1963), a Senior Geologist on the Survey: 'to call the Clactonian "Pre St Acheul" was too comprehensive. It is more likely to be early St Acheul perhaps a different race as there is damned little connection between the Clactonian & the St Acheul'.[42]

Chandler's consideration of whether the Clactonian ought to be assigned to 'Pre' or 'early' Acheulian times offers a reminder that Palaeolithic labels could refer to a period of time as well as to a distinct tool-making culture. The switch between the chronological and cultural aspects of industries, though confusing, illustrates the chronological value invested in tools of distinct character: whether their affinities were deduced from tool-types, like

[39] Breuil 1929, 105; Chandler 1930, 81; 1931a, 175; Bull 1942, 21. See also Warren 1926, 47, footnote and Dewey 1932, 39.
[40] Kendrick 1971, 4–5. On the favour shown by Smith to the eoliths, see Smith 1922, 25–26.
[41] Chandler 1930, 79, 92, footnote 4; Dewey 1932, 43.
[42] Chandler to Dixon, 21 November 1930: BGS: GSM1/295; Chandler 1931b, 250.

hand-axes; or techniques, like the distinctive preparation of Levallois flakes. Breuil placed the Clactonian industry on his line of flake cultures: between the Pre-Palaeolithic industry of East Anglia, and the two branches that later sprouted into the Levalloisian and Mousterian industries. This interpretation was rapidly accepted in Britain during the early 1930s. Warren had failed to convince his audience at the Geological Society in 1923 that the Clactonian was an industry in its own right, quite separate from the Chellean or Acheulian. But now that his industry had gained such respectable relations in the Thames Valley, and the blessing of Breuil, Warren's problems seemed to have ended.[43]

9.4 TOOLS AND TOOL-MAKERS

As stone tools flowed from their old single stream into the multiple channels of parallel culture-stages, they became matched to different branches on the family tree of their makers. The single line of tool-makers had been split by Boule at the beginning of the century, when he cast the Neanderthals off the human line and onto a separate branch of the hominid tree (see Chapter 4). The divergence of Palaeolithic sequences in the 1930s reinforced the attacks on those who still supported a simpler line of human evolution. In 1937, Kenneth Oakley (1911–1981), who worked at the Department of Geology in the British Museum (Natural History), criticized Robert Rudolph Schmidt's *The Dawn of the Human Mind* for assuming 'that the known types of fossil man represent a true line of descent: *Pithecanthropus—Sinanthropus—Homo heidelbergensis—H. primigenius [sic]—H. sapiens.* No allowance seems to be made for the possibility of parallel or divergent lines of descent'.[44] Oakley had studied geology and anthropology at University College London before moving to the Museum in 1935, and he had also held a post on the Geological Survey.[45]

This is not the place to scrutinize past theories about tool-makers in detail. Other authors have examined ideas about the relationships between different hominids, and this book is concerned with the classification of their products rather than their bones. Brian Regal's *Human Evolution: A Guide to the Debates* provides a comprehensive account, and Marianne Sommer's articles on Sollas and Boule give a comprehensive explanation of the shift from a linear to a branching hominid tree.[46] But the lines of descent assigned to

---

[43] Chandler 1932, 70; Breuil 1932a, 571–572.   [44] Oakley 1937b, 187.
[45] Molleson 1983, 95; Roe 2004a.   [46] Regal 2004; Sommer 2005; 2006.

tool-makers were closely linked to the paths traced in their industries. This relationship is illustrated by the speculations about fragments of hominid skull found at Swanscombe in the 1930s. Hominids had occupied a branching evolutionary tree for the past few decades, but the idea of branching industrial cultures, which had caught the imagination of British Palaeolithic researchers more recently, permeated the discussion of Swanscombe Man.

On a Saturday in the summer of 1935, Alvan Theophilus Marston (1889– 1971) sent the following night letter telegram to Dewey at the Geological Survey:

> 11.44 LONDON T 37
> TEL LTR H DEWEY GEOLOGICAL SURVEY MUSEUMS
> SOUTH KENSINGTON
> I REPORT FINDING HUMAN PALEOLITHIC OCCIPUT UNROLLED
> MIDDLE GRAVEL BARNFIELD SWANSCOMBE STOP LETTER
> IN POST STOP ADVISE COMMUNICATE WITH SWANSCOMBE TO
> INSTITUTE PROPER LOOK OUT FOR REMAINDER
> MARSTON CLAPHAM.[47]

Marston, a dentist from Clapham, had been searching the Swanscombe pits since 1933. He had a deep interest in the British Palaeolithic: patients seated in his dental chair might have been soothed by the splash of water falling in his palaeolith-lined fountain, and they could have examined a case of Swanscombe implements in his waiting room.[48] This large fragment of hominid skull from Barnfield pit was Marston's prize for long hours of hunting. Dewey travelled to Swanscombe on the Monday after receiving the telegram, accompanied by the Director of the Geological Survey, John Flett. They confirmed the authenticity of the find. The following year, Marston recovered a further large fragment of skull from Barnfield pit. This belonged to the same tool-maker: the parietal bone now joined the occiput of what became known as Swanscombe Man.[49]

There was no single, simple hominid analogy for these Swanscombe skull fragments. Neanderthals, the now notorious Piltdown Man, and modern humans offered the most obvious alternatives in the 1930s. But the features of the skull were ambiguous: Swanscombe Man seemed to be modern in aspect, but had some Neanderthaloid features as well. The tools at Swanscombe did not offer much assistance either: Acheulian and Clactonian tools had both been found near the skull fragments.

Different industries had been linked to different races of tool-makers since the earliest years of Palaeolithic research. A division between races of 'Cave

[47] Marston to Dewey, 30 June 1935: BGS: GSM2/556.   [48] Carreck 1973, 119.
[49] Marston 1937, 340, 342.

men' (or 'reindeer-folk') and River-Drift men had been popular in the nineteenth century. The flake-rich Mousterian was thought to have been made by Neanderthals.[50] Crude hand-axes of the Chellean or Abbevillian had been linked to more primitive hominids by the discovery in 1907 of a massive, ape-like jaw with human-like teeth in the sands of Mauer, near Heidelberg in Germany: large, crude Chellean hand-axes had been found nearby. This *Homo heidelbergensis* was often considered to be a precursor to the Neanderthals.[51]

When flake industries crept back in time from the Mousterian epoch, they became associated with the pre-Mousterian ancestors of Neanderthals. The beautifully crafted Acheulian hand-axes, which had been linked to our own ancestors from the early years of the twentieth century, were set alongside the flake industries of these pre-Mousterian peoples.[52] Breuil, like Obermaier, had located these two great groups of tool-makers in different parts of Europe: the pre-Mousterians in the north and east; the hand-axe-makers in the south and west.[53] The big-brained *Eoanthropus* from Piltdown was sometimes thought of as an ancestor of the hand-axe makers. The forged Piltdown Man, with its human-like cranium and ape-like jaw, had been associated with a range of implements: some were described as Eolithic; others, as Chellean or Pre-Chellean.[54]

After the rise of parallel industries in the 1930s, the makers of crude hand-axes tended to be recalled to our own line of hominid ancestors. Some suspected that these hand-axe makers would have looked far more modern than *Homo heidelbergensis*. Several discoveries had been hailed as the remains of hand-axe peoples, but were later proved to be modern bones that had intruded into ancient deposits. In the early 1930s, Louis Leakey claimed that he had discovered modern-looking humans in East Africa, lying in early Quaternary deposits. They were associated with tools that ranged from Pre-Chellean to Acheulian types. At that time Leakey was a Research Fellow at St John's College, Cambridge. The ancient date of these bones from Olduvai, Kanam, and Kanjera failed to win widespread support, and Leakey's geological evidence was criticized severely by Boswell.[55]

---

[50] Boule 1908a, 519; Sturge 1912, 213–214; Bowler 1986, 91; Sommer 2005, 330.
[51] Schoetensack 1908, 1; Garrod 1928, 266; Burkitt 1933, 126.
[52] Read 1911, 8; Sturge 1912, 213–214; Geikie 1914, 46.
[53] Obermaier 1908, 125; 1919, 146–147; Breuil 1932a, 571.
[54] Dawson and Woodward 1913, 122.
[55] Boswell 1932b; 1935; Leakey 1932; 1933; 1935, 14; Anon 1933; Spencer 1990, 121–123. The most notorious of the supposedly ancient *H. sapiens* finds was the Ipswich Man from beneath the East Anglian boulder clay (Keith 1912, 209; Moir 1912d), and the Galley Hill skull, found near Northfleet, Kent in the 100-ft terrace (Read 1911, 9; Sturge 1912, 214).

Most of these hominid species were connected in some way to the Swanscombe skull. Arthur Smith Woodward (1864–1944) of the British Museum (Natural History) initially associated the skull with the Neanderthals. Marston argued for a link to *Eoanthropus*: he saw his find as a precursor to the Piltdown skull (which had been re-dated by Oakley to a relatively recent period).[56] Leakey disagreed with Marston and considered the find to be similar to his own discoveries in Africa: a primitive *Homo sapiens*. Garrod also thought that Swanscombe Man might support Leakey's African evidence, despite the problems associated with his finds.[57] This connection to *Homo sapiens* was popular.[58]

Then there were the tools. Marston was eager to emphasize that the skull fragments came from the Acheulian horizons of the Middle Gravel, not the earlier Lower Gravel that contained 'only implements of the Clacton and Strépy types'.[59] Christopher Hawkes (1905–1992) of the British Museum also drew attention to the abundant Acheulian implements near the skull. He admitted that a late variant of the Clactonian industry was sometimes found in the Middle Gravel, but declared that no examples had been encountered near the skull.[60] Oakley and Louis Leakey disagreed, and reported: 'unabraded Clactonian flakes and tools are almost as common as Acheulian ones' in the vicinity of the 1935 find.[61]

Amidst this variety of opinion, a preference for an Acheulian/*Homo sapiens* connection was stated most frequently. At the time of these Swanscombe arguments, some researchers were also starting to link early European flake industries with the ape-like *Pithecanthropus* group from East Asia, because of a similarity between the Clactonian and the Asian chopping-tool industries.[62] As the branching pattern of industries was aligned to the complicated evolutionary tree of hominids, the cultural distinction between industries on

---

[56] Woodward to Marston (undated, but must be 1935): NHM: DF140/5; Marston 1936, 200; 1937, 394; Spencer 1990, 123–126.

[57] Leakey to Woodward, 26 January 1937: NHM: DF140/5; Garrod 1938b, 469.

[58] Clark and Morant 1938, 469; Keith 1938.

[59] Marston's 1937, 340 reference to 'Strépy types' was regarded as outmoded by this time. Hawkes, Oakley, and Warren stated (in Hinton, Oakley *et al.* 1938, 31): 'the conception of a distinct Strépy industry has long been obsolete', and cited its absence in the authoritative article by Breuil and Koslowski 1934.

[60] Hawkes 1938, 468. The 'Clactonian' from the Middle Gravels was thought to be a late phase of the Clactonian industry (Clactonian III, High Lodge), distinct from the Clactonian found by Warren and by Smith and Dewey in the 1910s and 1920s. These numbered industrial phases are introduced later in this chapter.

[61] Oakley and Leakey 1935, 916; McNabb 1996, 43–44. By 1938, Oakley was emphasising the connection between the Acheulian and the *Homo sapiens* Swanscombe skull (Oakley to Baden-Powell, 29 March 1938: UMO: DB-P: K99).

[62] Hopwood 1935, 57; Oakley and Leakey 1937, 217; Heekeren 1948, 29.

different lines (such as the Acheulian and the Clactonian) was reinforced by the biological differences assigned to their makers.

9.5 PALAEOLITHIC CLASSIFICATIONS IN THE 1930S

Perceptions of the character of Palaeolithic industries began to change as the division between hand-axe and non-hand-axe cultures grew more popular over the 1930s, and the flake line received more attention. These changes are seen clearly in the reaction to two definitions of the Clactonian: one offered by Breuil, and the other by Warren. If a casual observer were to be confronted with a Clactonian assemblage, they might gain the impression that this industry was dominated by crude, simple, flint flakes, which had been struck off the original nodule or 'core' of flint with a weighty blow and not thereafter trimmed (or 'retouched') into implements. (Smith and Dewey had looked hard for a diagnostic implement in the Swanscombe Lower Gravel before they eventually settled on the dubious 'Strépy' nodules.) But if Warren and Breuil had looked at the same tools, the image held by each would have been quite dissimilar.

Warren agreed that the Clactonian was 'characterized above all by the strength and freedom of the flaking,' but he also founded his definition firmly on distinctive implements. He drew attention to trimmed flake-tools (scrapers and other retouched forms) and claimed that one particular re-touched core-tool—the side-chopper—was definitively Clactonian in char-acter.[63] In 1932, Warren published a list of eleven Clactonian tool-types. He ranked them in order of relative importance: these side-choppers took first place. They were followed by the waste flakes and then by the trimmed flakes. Further down the list, at number seven, came a group of crude, bifacial, pointed implements: their shape suggested to Warren that the Clactonian peoples might have been influenced by the hand-axe makers.[64]

Breuil gave a different account of the Clactonian industry. He was eager to stress the identity of the Clactonian as a flake culture and to emphasize its distinction from hand-axe cultures. Breuil presented his definition in two papers: 'Le Clactonien et sa place dans la chronologie' (1930) and 'Les Industries a éclats du paléolithique ancien, 1.—Le Clactonien' (1932).[65] Both accounts focused on flaking technique; Warren's tool-types had a neg-ligible role. Tool-making techniques allowed Breuil to relate this industry to others on the flake-culture line. He explained that the makers of Clactonian

---

[63] Warren 1924e, 38.   [64] Warren 1932b, 69.   [65] Breuil 1930; 1932b.

Figure 9.4. Warren at work in his field clothes. Photographed just before the Second World War (© The Natural History Museum, London: DF140/7).

flakes and cores, for example, had not prepared the surfaces of their flints in the characteristic manner of Levalloisian or Mousterian flake industries.[66]

The differences between these definitions reflect the different aims of Warren and Breuil, the different questions that they asked, and the different scales on which they worked. Warren had been studying the tools from Clacton for over twenty years. He visited the Palaeolithic exposures regularly and built up a large personal collection of these flints; he observed the sediments, bones, and shells that came from these deposits; he wondered about their date and their makers. Warren read widely: he was aware of discoveries made by his peers in Britain and on the Continent, and presented his own findings in local and national societies. But despite considering these broader industrial patterns and links to the Alpine glacial sequence, he was eager to understand the character of his particular assemblage from Clacton and its connection to the Thames Valley sequence.

Breuil has been described as more of a synthesiser than a fieldworker: a remark inspired by the grand Palaeolithic schemes that he drew together from a mass of scattered discoveries.[67] (Those who studied the later Palaeolithic of

[66] Breuil 1930, 221–222.   [67] Smith 1962, 205; McNabb 1996, 38.

the caves might have disagreed, and remembered Breuil better for his extensive fieldwork on cave art.) Warren's tools from Clacton and Chandler's Clactonian from Swanscombe were just two of many assemblages that Breuil saw on his travels between different regions to view Palaeolithic sites and collections. He was looking for points of general similarity that would help him understand Palaeolithic patterning across the whole of western Europe.

Tool-making techniques, so useful in connecting different flake industries, took a prominent place in Breuil's definition of the Clactonian. His belief that Palaeolithic Europe could be divided between two great tool-making groups—a hand-axe culture, and a flake culture with no hand-axes—led him to emphasize industrial characteristics that reinforced, and even caricatured, his division between two cultural lines. Warren's Clactonian core-tools blurred this distinction; so they were one casualty. Breuil did not believe that Warren's side-choppers and bifacial forms reminiscent of the hand-axe were true implements. Instead, he described them as retouched '*nucléi*' or cores: '*Les nucléi sont assez fréquemment réutilisés, mais sans idée systématique d'aboutir à un biface régulier comme ceux du Chelléen et de l'Acheuléen*' ('the cores are quite frequently re-utilized [re-worked], but without the systematic idea that leads to a regular biface like those of the Chellean and the Acheulian').[68]

Like Breuil, Warren promoted a distinction between hand-axe and flake lines, but his detailed observations of the Clacton assemblage convinced him that the Clactonian industry was more varied than Breuil would admit. In 1932, Warren asserted that his side-choppers were *not* simply cores—although he also made it clear that his bifacial implements were distinct from Chellean or Acheulian hand-axes. He was not overly disturbed by Breuil's argument: 'Even if their use as choppers be open to some doubt, the form is the most important feature of the Clactonian industry, and it is convenient to give it a name'.[69] But Warren would soon be provoked to strengthen his defence of core-tools within the Clactonian: when his peers adopted Breuil's scale of vision, as well as the division between hand-axe and flake lines, they were to join the attack on his side-choppers.

Warren's friend Oakley was one of the new generation of Palaeolithic researchers sympathetic to Breuil's views. In August 1932, Oakley recalled seeing 'this small, dynamic figure in clerical garb' bobbing up and down as he examined a pit at Swanscombe. This was his first meeting with Breuil.[70] During the 1930s, Oakley added an interest in the British Palaeolithic to his geological pursuits. He excavated at Jaywick Sands, Clacton, in 1934 with Mary Nicol (1913–1996), soon to be Mary Leakey; he published a classic study of the geological and Palaeolithic history of the Thames Valley in 1936

---

[68] Breuil 1932b, 132.   [69] Warren 1932a, 21; Warren 1932b.   [70] Boyle 1963, 13.

with William King (1889–1963), Professor of Geology at University College London; and he became Secretary of the Swanscombe Research Committee when it was set up in 1937.

As the idea of parallel cultures gained in popularity and early flake-industries multiplied, Oakley and Mary Leakey became particularly interested in the geology and tools of Warren's Clacton Channel. They wanted to clarify the relationship between this Channel and the Thames Valley, and to understand the place of the Clactonian in the British Palaeolithic sequence. By the 1930s, a drift of beach material had covered the exposures where Warren had made his early collections. Although the same Channel could be found lying near the surface two miles to the south-west—at Jaywick Sands, near Lion Point— Warren warned Oakley and Leakey that this second site would shortly be covered by a new housing estate, so they ought to start excavating soon. The three of them discussed their plans over several picnic lunches, Leakey recalled Mrs Warren's coffee with a shudder, and work began in the autumn of 1934.[71]

This was the first excavation to be directed by Mary Leakey, then aged twenty-one. The great-great-great-grandaughter of John Frere (discoverer of hand-axes at Hoxne in the late eighteenth century), Leakey had attended a few lectures on geology and archaeology in London in 1930, after her mother failed to persuade Sollas to let her into Oxford, and she had already assisted on several excavations. In the summer of 1934, Mary Leakey helped her husband-to-be, Louis Leakey, with his exploration of the Clactonian industries at Swanscombe.[72] The year after her own excavations at Clacton, she joined him in East Africa where she built a powerful reputation in Palaeolithic research.

In their joint report on the excavations at Clacton, Oakley and Leakey interpreted Warren's core-*implements*, including the side-choppers, as mere *cores*: 'we maintain that the majority were in the first place cores, even if they were eventually utilized'.[73] They shared Breuil's view. The two authors were happy that 'the industry belongs to what has been termed the flake-culture group': these cores were the by-products of flake production and were not converted into implements. They observed that hand-axes were absent, and emphasized that Warren's bifacial implements formed only 'a very subsidiary element in the early Clactonian industries'.[74] Oakley and Hawkes expressed similar opinions in the Report of the Swanscombe Research Committee. Warren remained convinced that his core-tools formed a definitive part of

---

[71] Oakley and Leakey 1937, 217–218; Leakey 1984, 48.
[72] Leakey 1974, 5; 1984, 35–40, 46–48.
[73] Oakley and Leakey 1937, 227.
[74] Oakley and Leakey 1937, 235.

the Clactonian industry, and was now so concerned about the core *versus* core-tool controversy that he tried to alter the Committee's final printed 'Report on the Swanscombe Skull' (1938).

The Swanscombe Research Committee had been appointed in 1937, in the aftermath of Marston's discovery of Swanscombe Man, to explore the geo-logical and industrial context of this great find. Hawkes wrote the section of the Report that dealt with 'The Industries of the Barnfield Pit', but he had the co-operation of Oakley and Warren.[75] All three were members of the Com-mittee. Hawkes had been Assistant Keeper in the Department of British and Medieval Antiquities at the British Museum since 1928. He worked under Reginald Smith, Keeper of the Department until his retirement in 1937. This was a difficult relationship: Smith had grown irritable through the 1920s and his sniping contributed to Hawkes's nervous breakdown in 1935.[76]

Although no Acheulian hand-axes had been found in the Clacton Channel assemblage or in the Lower Gravel at Swanscombe (which lay far below the level of the skull fragments), Warren insisted that the Report must include the observation that 'crude core implements of pointed and other forms are of frequent occurrence'.[77] When the Swanscombe Report was being drafted, Warren wrote to Hawkes in concern about the character of the Clactonian industry:

after 'cores' I would suggest in brackets '(or perhaps core-implements)'. In my own mind, I am satisfied that the Clacton industry includes many primitive core-imple-ments, that I call 'barbarous imitations of the Acheulian hand-axe'—I agree that there should be a 'perhaps' or a '?' in the case of a joint report, although if I were writing only for myself I should call them bi-face core-implements without qualification.[78]

His co-authors included Warren's sentence in the published version. They accepted core-implements as 'a normal constituent part of the Clactonian', but added the telling proviso: 'though a subsidiary one in the sense that they do not upset the basic conception of it as a "flake-culture"'.[79]

These arguments over Clactonian core-tools might seem esoteric, but they illustrate perceptions of British Palaeolithic industries that pervaded publica-tions of the 1930s. Oakley, Leakey, Hawkes, and others wanted to incorporate the varied flake industries of Britain into a new general framework, now that the old standard sequence from Swanscombe could no longer accommodate

[75] Hinton, Oakley *et al.* 1938, 30–47.
[76] Webster 1991, 165, 167–168, 190, 203, 205–212.
[77] Warren, draft report on the Lower Gravels of Barnfield Pit, for the Swanscombe Committee (undated, *c.* 1938): BM(F): Swanscombe I.
[78] Warren to Hawkes, 2 January 1938: BM(F): Swanscombe II.
[79] Hinton, Oakley *et al.* 1938, 31, 32.

them. Impressed with the grand scope of Breuil's classification, they began to re-classify British industries. In the process, they widened the division between industries with flakes and flake-tools, and industries described variously as hand-axe, biface, or core cultures.[80] For them, one of the most important features of the Clactonian was its identity as a flake culture. But the distinction between flake and hand-axe cultures was growing so strong that some researchers felt compelled to remind their peers that hand-axe cultures *did have flakes*, even if the earlier flake cultures did not have hand-axes.[81]

9.6 GEOLOGICAL CHRONOLOGIES AND THE BREUIL-KOSLOWSKI PAPERS

The dissatisfaction of Oakley, Leakey, and Hawkes with the old Palaeolithic sequence was shared by geologists who had been using industries to date and correlate tool-bearing sediments. Disturbed by the suspicion that hand-axe and flake industries overlapped in time, Boswell had complained in 1931: 'If [...], "...the simple succession Early Chelles, Chelles, Evolved Chelles, St Acheul and Le Moustier no longer holds good," I personally almost despair of a solution'.[82] Boswell was more hopeful the following year. He asserted in his Address to the Geological Section of the British Association: 'If the time-succession of human industries recognised by our archaeological colleagues holds good (and in general it is becoming more firmly established every year), we should expect the sequence pre-Chellian, Chellian, Acheulian (Clactonian-Levalloisian), Mousterian [...]'.[83]

Boswell's industrial terminology had changed since the previous year. The most notable alteration was his abandonment of 'Le Moustier,' and his addition of 'Clactonian' and 'Levalloisian' alongside 'Mousterian'. As more flake industries were recognized over the 1930s, the old umbrella references to 'Mousterian' or 'Early Mousterian' industries disappeared.[84] By inserting 'Clactonian-Levalloisian' in brackets after 'Acheulian,' Boswell acknowledged

---

[80] The three terms—'hand-axe', 'biface', and 'core'—were closely associated. Breuil 1932a, 571 spoke of *biface* and flake cultures, but noted that the term 'bifaces' was often applied to hand-axes: '*souvent appelés haches ou coups-de-poing*'; Burkitt 1933, 56 observed, 'essentially coups-de-poing are core-tools'.

[81] Garrod 1928, 266; Leakey 1931, 35; Kelley 1937, 15; Paterson 1945, 1. Breuil 1932a, 571; 1936, 208 had distinguished between pure flake cultures with no hand-axes, and hand-axe-*and-flake* cultures.

[82] Boswell 1931, 107.

[83] Boswell 1932a, 65.

[84] Burchell 1932, 258, 262; Moir and Burchell 1935, 129.

their contemporaneity and managed to retain the industries as a guide to the date of tool-bearing sediments. But this would remain only a crude guide until he understood the flake sequence in more detail.

In 1936, stimulated by recent work on flake cultures, Boswell urged the Prehistoric Society to help: 'all the older records of "Mousterian" implements should now be re-examined and re-defined before they can be used for correlation or for dating'.[85] A similar comment was made the following year by Frederick Everard Zeuner (1905–1963), the eminent German geologist and palaeontologist:

The middle Pleistocene of East Anglia has up to the present furnished very few implements, and these, moreover, do not help us in correlating. Finds of Clactonian, Levalloisian or Acheulian implements in this period are of little value unless the cultural phase can be reliably determined. All three techniques continue on the Continent until after the Saale (Riss 2) glaciation.[86]

Zeuner's request for 'the cultural phase' and Boswell's desire for a re-definition of the Mousterian were inspired by a precise classification of industries published in *L'Anthropologie* between 1931 and 1934: 'Études de stratigraphie paléolithique dans le nord de la France, la Belgique et l'Angleterre.' Breuil had written these four papers with Professor Koslowski of the University of Lvow, Poland.[87] Their classification was attractive to geologists, like Boswell and Zeuner, who were trying to correlate the sediments of southern England.

Zeuner had started searching for chronological patterns in the sediments and bones of Britain after arriving from Germany in 1934: Hitler's policies had forced Zeuner, and his Jewish wife Etta, out of the country. He became a Research Associate in Palaeontology at the British Museum (Natural History), and also worked for a short while at Imperial College, London, under Boswell. Zeuner finally settled at the newly opened London University Institute of Archaeology: first as Honorary Lecturer in Geochronology and later as Professor of Environmental Archaeology.[88] Boswell hoped that the Institute of Archaeology would at last give Britain an opportunity for teaching and research comparable to the Institute of Human Palaeontology in Paris.[89]

The Palaeolithic classification presented by Breuil and Koslowski in the early 1930s offered a valuable alternative to the vague Palaeolithic terminology and

---

[85] Boswell 1936, 157.   [86] Zeuner 1937, 152.
[87] Breuil and Koslowski 1931; 1932; 1934. The reference to 'Angleterre' suggests that a fifth paper was planned. This never appeared, perhaps because King and Oakley 1936 took up the ideas of the first four papers in their own detailed paper on the Thames Valley succession, which made such an addition redundant (a suggestion made by Roger Jacobi, pers. comm. 2003).
[88] Clark 1937, 166; Cornwall 1964, 117–118; Sutcliffe 1998, 131.
[89] Boswell 1936, 150.

Table **9.2.** A comparison between Commont's industrial sequence and the correlations between industries, glaciations and fauna published by Breuil and Koslowski (1931; 1932).

Industries		Glacial phases	Selected Fauna
Commont	Breuil and Koslowski		
	Aurignacian; Solutrean; Magdalenian	Würm, 2nd phase	
Mousterian	Levalloisian V–VI; Upper Levalloisian	Würm, 1st phase	*Elephas primigenius; Rhinoceros tichorhinus; Rangifer tarandus*
Acheulian; Warm Mousterian	Final Acheulian VI, and VII (Micoquian); Levalloisian III–IV	Riss-Würm	*E. antiquus; Rh. mercki; Hippopotamus* (associated with Levalloisian III–IV)
Acheulian	Upper Acheulian V; Levalloisian I–II (derived)	Riss	*E. primigenius; Rh. tichorhinus*
Evolved Chellean	Middle Acheulian IV	Mindel-Riss, 3rd phase	
	Middle Acheulian II–III	Mindel-Riss, 2nd phase	*Rh. mercki; E. antiquus; Hippopotamus*
Chellean	Early Acheulian; Evolved Clactonian	Mindel-Riss, 1st phase	*E. antiquus; E. primigenius*
Pre-Chellean	Chellean; Early Clactonian of England	Günz-Mindel	Fauna with Pliocene affinities: *E. antiquus; E. meridionalis*

crude timescale used hitherto in Britain. Each industry—Acheulian, Clactonian, Levalloisian, and so on—was divided into a lengthy sequence of numbered phases (see Table 9.2). Commont and Smith had distinguished between an earlier 'Acheulian I' and a later 'Acheulian II' (see Table 8.3), but Breuil and Koslowski described many more phases for this and for other industries. They also gave each of their phases a specific cultural and chronological position, placing them in relation to the phases of other contemporary industries that ran in parallel through time, and in relation to various geological sequences. Their industrial framework was anchored by the 'long chronology' of Alpine glaciations, but was also linked to sequences of river sediments and bones.

So, while the names of industries (such as 'Acheulian') suggested cultural affinities (a link to the hand-axe-making peoples), the phase (such as 'Acheulian I') provided a precise chronological position. The old assumption of progress infiltrated this varied picture of the Palaeolithic through the gradual shift from (say) Acheulian I to II to III (and so on). Once-puzzling industries acquired specific identities: Commont's old 'warm Mousterian,' for example, became Breuil and Koslowski's 'Levalloisian III-IV'.[90] Though masked by these new industrial phases, the debt to Commont's work in the Somme Valley is evident in the content, references, and even the illustrations of Breuil and Koslowski's papers.

To clarify the industrial terminology: Acheulian II was an 'industrial phase' of the Acheulian 'industry,' which was one of the hand-axe 'cultures.' This use of 'culture' to describe the two great groups of hand-axe industries and flake industries (also known as 'culture-groups', 'traditions', 'families', or 'complexes') would later be criticized for its breadth of meaning. The connection between industries and these parallel 'culture' streams, which were thought to have spread across Europe and beyond, would be held responsible for devaluing an older, more specific connection between industries and the 'culture' of a group of interacting tool-makers located more discretely in time and space.

In 1933, Breuil's old pupil, Burkitt, gave effusive praise to 'the recent brilliant researches of Professor Breuil on the industries of the Somme valley', which were 'enabling prehistorians to subdivide minutely the lower palaeolithic industries in the area'.[91] This detailed industrial classification was also welcomed heartily in Britain. Geologists valued the fine-grained timescale offered by the various industrial phases. By the late 1930s, the British Palaeolithic was described in terms of the Breuil–Koslowski classification; the old Palaeolithic sequence had been re-structured, and old assemblages re-classified. Breuil himself was sometimes at hand to help with the new labels. He found no problem in matching industries to his framework, and the regard for his expertise in industrial identification strengthened his grip on Palaeolithic patterns in Britain.[92]

---

[90] Breuil and Koslowski 1932, 304. On Breuil's views of progressive evolution, see Dennell (1990, 553).

[91] Burkitt 1933, vi.

[92] References to Breuil's industrial phases were made, for example, by Dewey 1932, 45, Bury 1935, 63, Smith 1935, 417, Wade and Smith 1935, 353, King and Oakley 1936, 52–53, Lacaille 1936, 430, Kelley 1937, 15, Wright 1937, xv, Zeuner 1937, 151. Breuil's role in the identification of British industries was mentioned frequently in British articles of the late 1920s and 1930s: see, for example, Sainty 1927, 187, Dines 1929, 24, and Chandler 1930, 92.

9.7 FINE-TUNING THE BRITISH PALAEOLITHIC CHRONOLOGY

The Clactonian was one of many industries to be re-classified and sub-divided over the 1930s by British researchers, eager to follow Breuil's example and elaborate their old Palaeolithic sequence. The framework set out by Breuil and Koslowski included only an early Clactonian and an evolved Clactonian; Britain soon boasted four Clactonian phases: I, IIA, IIB, and III (see Table 9.3).[93] And as the hitherto neglected sequence of flake industries gained in detail, geologists established finer correlations between tool-bearing sediments.

The sub-division of the British Clactonian began in the Thames Valley in 1929, when Chandler divided the assemblage from the Swanscombe Lower Gravel into two phases of different ages. Some of the tools seemed to have been derived from older deposits: these, he called Clactonian I. The rest were contemporary with the Lower Gravel and had affinities to the industry from the type-site at Clacton: Chandler called them Clactonian II. Breuil had visited Chandler at Swanscombe in 1928; they may have discussed these Clactonian phases then.[94] In any case, by 1932 Breuil was also distinguishing between '*le Clacton I*' and '*Le Clacton évolué, que je note II provisoirement*' ('Clactonian I' and 'The evolved Clactonian, which I provisionally term II').[95]

The Clactonian industry gained a later phase from High Lodge. Finely worked flake-tools had been gathered from this site for decades. They had often been compared to the tools from Le Moustier, and were termed 'Early Mousterian' in the 1920s.[96] In 1932, Breuil saw a connection to the Clactonian II from Swanscombe in the flaking technique, which led him to describe the High Lodge assemblage as the latest Clactonian in England. The High Lodge industry became Clactonian III.[97] This relationship seemed to be reinforced by Oakley's observation of High Lodge-type (Clactonian III) flake-tools in the Middle Gravel of Barnfield pit—above, and therefore later than, the Clactonian II assemblage in the Lower Gravel.[98]

Oakley made a final refinement to this sequence of Clactonian phases after excavating at Jaywick Sands, Clacton, in 1934. Until then, the 'Clactonian II' phase had embraced the tools from Clacton and those that Chandler thought contemporary with the Lower Gravel at Swanscombe. Oakley and Warren both believed the Clacton tools to be later than the Swanscombe examples,

---

[93] Breuil and Koslowski 1934, 256–257, 313.   [94] Chandler 1930, 81, 84; 1931a, 175.
[95] Breuil 1932b, 129–130.   [96] Sturge 1911, 69; Moir 1921d, 367; 1927a, 143.
[97] Breuil 1932b, 160–162; Leakey 1934, 119–122; King and Oakley 1936, 60. See also Garrod 1928, 266–267.
[98] Oakley and Leakey 1937, 240–242.

Table 9.3. The Clactonian phases used by Oakley, with associated sites (after King and Oakley 1936 and Oakley and Leakey 1937).

Phases of the Clactonian industry	Sites
Clactonian III	High Lodge; Stoke Newington; Swanscombe Middle Gravel
Clactonian IIB	Clacton Channel, Clacton-on-Sea
Clactonian IIA	Swanscombe Lower Gravel
Clactonian I (associated with Abbevillian and Acheulian I–II)	Found in derived condition in the Swanscombe Lower Gravel

and Oakley decided to sub-divide the Clactonian II phase.[99] He explained to Hawkes: 'It is proposed to refer to the "Lower Gravel" industry (Swansc.) as *Clacton IIa*, and to the Type industry of Jaywick and Clacton as *Clacton IIb*'.[100]

Four phases of the Clactonian had been established in just a few years; other British industries were sub-divided in a similar manner over the 1930s. But behind the proliferation of new labels lay older assumptions of industrial progression. Oakley, for example, expected that tool-working techniques had become more refined and skilful through time, and that tool-types had become more standardized and specialized.[101] Technique and typology helped Oakley and others decide on the age and connections of tools from different sites. They were assisted by the blueprint of industrial age and character supplied by Breuil and Koslowski.

Not everyone welcomed the new multiplicity of industrial phases that invaded publications on the British Palaeolithic. Warren was uneasy about the new terminology. He reflected on past confusions caused by the shifting meaning of de Mortillet's 'Chellean' and 'Acheulian' labels, and shared his concerns about the latest Clactonian phases with the Swanscombe Committee.[102] Its Report revealed: 'one of us (S.H.W.), viewing such serial sub-division with some misgiving, would prefer to speak of "the Clacton industry" for that of the type-site, and to call the main Lower Gravel industry "the Swanscombe Clactonian"'.[103] The original draft of this Report, in Warren's handwriting, gives more detail of his reasoning. (The words that are absent in the published version are given in italics below.) Here, Warren viewed such sub-division by numbers:

[99] Warren 1932a, 25; 1932b; Oakley and Leakey 1937, 240.
[100] Oakley to Hawkes, 22 October 1935: BM(F): misc. correspondence.
[101] Oakley and Leakey 1937, 240.   [102] Warren 1924b, 279; 1926, 41–43; 1932a, 5–7.
[103] Hinton, Oakley *et al.* 1938, 31.

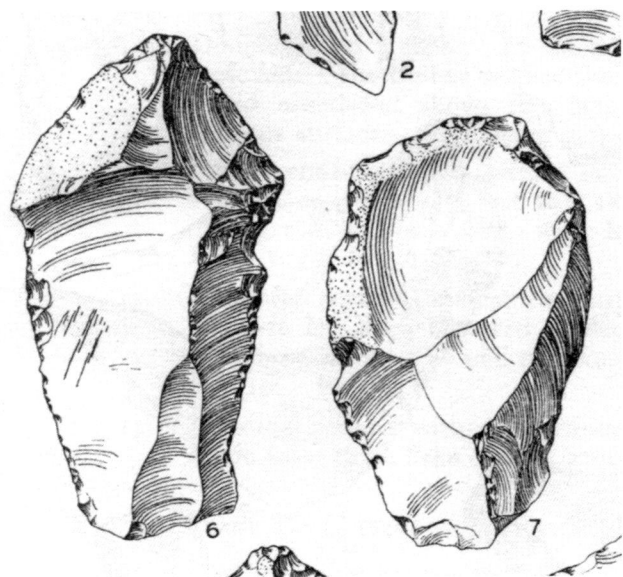

Figure 9.5. 'Clactonian IIB flake-tools from Jaywick Sands' (Oakley and Leakey 1937, 229, Fig. 3).

with some misgiving, *as these in course of time are too readily transferred from one industry to another (which confuses the literature), and* would prefer to speak of 'the Clacton industry' for that of the type-site, and to call the main Lower Gravel industry 'the Swanscombe Clactonian'. *This does not connote any undue implication of relative date.*[104]

But the implication of relative date was an attractive prospect for many of Warren's peers who were working on geological correlations, such as Oakley, Boswell, and Zeuner.

Several problems of the British Palaeolithic seemed to have been solved by the advent of Breuil in the 1930s. His successful promotion of parallel cultures had enabled researchers to explain once-puzzling flake industries that fell outside the single line of the old Swanscombe sequence. Worthington Smith's Palaeolithic Floor from Stoke Newington, for example, which had caused problems in the past because of its mixture of trimmed flakes and fine ovate hand-axes, could now be described as Clactonian III, with advanced Middle Acheulian hand-axes.[105]

---

[104] Warren, draft report on the Lower Gravel of Barnfield Pit, for the Swanscombe Committee (undated, *c.* 1938): BM(F): Swanscombe I. My italics: applied to words not included in the published report. See also Hinton, Oakley *et al.* 1938, 31.

[105] Smith 1894, 220; Read 1902, 13; 1911, 18–19, 21; King and Oakley 1936, 60.

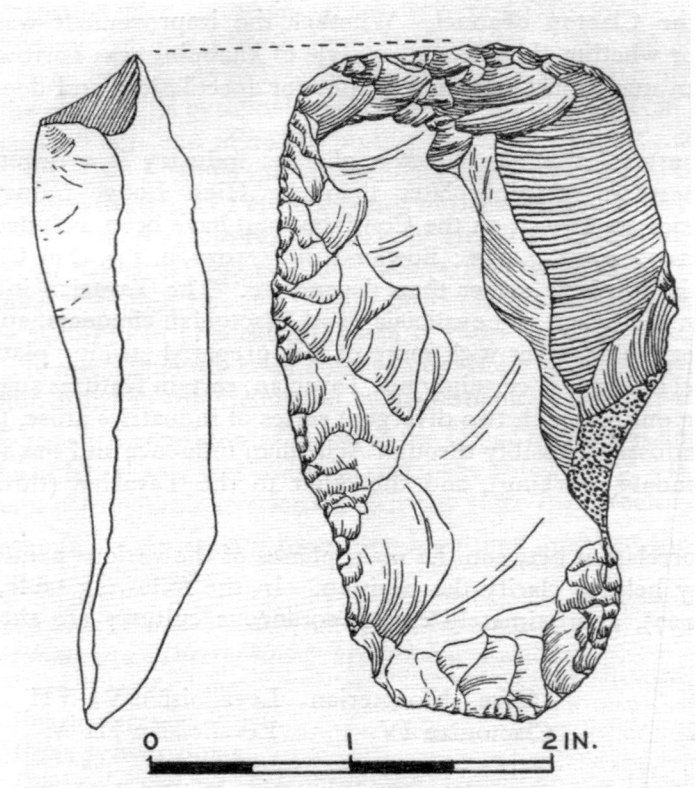

Figure 9.6. 'Flake-tool of High Lodge (Clactonian III) type from the Middle Gravel, Barnfield Pit, Swanscombe' (Oakley and Leakey 1937, 241, Fig. 7).

Geologists like Boswell might, as Peake had warned, be faced with 'evidence of successive stages of two industries, a core industry and a flake industry'.[106] But Boswell's fear that the Palaeolithic sequence would have to be dismissed as an unreliable timescale had been quietened by the detailed industrial classification of Breuil and Koslowski, which clarified the connections between these two cultures. The flake industries might have burst out of the narrow time limits to which they had once been assigned; but they were soon subdivided into discrete industrial phases, after the example of Breuil. Confidence in the chronological value of the British Palaeolithic seemed to have been restored. One problem remained: Breuil had set his industrial phases to run on a long glacial chronology. If British tools and sediments were to keep time with Breuil's industrial–glacial sequence, a drastic re-alignment would be required.

---

[106] Peake 1930, 382.

10

Geological Re-shuffling and the
Growth of Suspicion

I still feel that Pleistocene chronology is the clock by which archaeological, or rather Palaeolithic, time is to be measured, and it makes
nonsense of time-counting to use as clocks the very objects it is desired
to time.

(Zeuner, 1959)

A song written for the last day of the 1973 conference on Quaternary geology
held at Burg Wartenstein, Austria, offers an unintentionally accurate description of the atmosphere that surrounded British geological research in the
1930s. The words were intended to be sung to the tune of *Clementine*:

> In a valley in die Alpen in neunteen-hundred neun
> Oh, there werkte Penck and Brückner, carving up the Pleistoceun.
> They had Günzes, they had Würmses in the Stages on their list,
> But the biggest and the Grösstest—Interglacial Mindel–Riss
> CHORUS (repeat after each verse)
> *Correlation, constipation. Oh, now what are we to do?*
> *All these bedses without dateses. Oh we're really in the stew!*[1]

In the archives of British geologists who were working about four decades
before this song was written, lie scraps of paper covered with scribbled
suggestions about geological correlations. They testify to an intense interest
in the connections between boulder clays, river terraces, and stone tools; and
the great question of how they could all be correlated to the master-sequence
of Alpine glaciations established by Penck and Brückner. This was a complicated problem. In 1930 Chandler, who had been examining the Clactonian
industries at Swanscombe, confessed to Dixon of the Geological Survey:

I once (before the War) was impertinent enough to try & apply Penck's classification
to Raised Beaches & Terraces inland but I read & read & took notes till I could not see
the forest for trees & gave it up.

---

[1] Leakey 1984, 195.

Now, I don't see why we should not work out our own glaciations without reference to what happened on the Continent. [...]

Stratification is the key, I believe & palaeontology may be helpful or, it may be a bloody nuisance, as we have seen.

E. Anglia is the place, but very difficult, & I confess I have never been able to read S.V. Wood & understand him.[2]

Chandler's peers would return with vigour to the task of Alpine correlation as the 1930s drew on. Their attempts to weave a coherent pattern of time through the Quaternary period drew on threads of evidence familiar from previous chapters: the reflections of glacial cycles in East Anglia, the Thames Valley, and the Alps; the history of rivers and the remains of animals which once lived alongside them; and the tool-makers who discarded their tools with more abandon than their bones. But they now had a new pattern to follow: Breuil's industrial–glacial sequence, founded on his observations in the Somme Valley.

Geologists were stimulated by recent developments in Palaeolithic research to re-examine their old assumptions and correlations. Their efforts to match British sequences to these new expectations peaked in 1936 and 1937, with the publication of three important papers in the *Proceedings of the Prehistoric Society*. Boswell dealt with 'Problems on the Borderland of Archaeology and Geology in Britain' in his Presidential Address for 1936, and re-examined the correlations of boulder clays made by Wood and Harmer. In the same volume, King and Oakley published a new synthesis of the river sediments, bones, and tools in the Thames Valley. The following year Zeuner elaborated upon Boswell's work in 'A Comparison of the Pleistocene of East Anglia with that of Germany'.[3] These three articles set the tone for geological discussion in the late 1930s.

10.1 PROBLEMS WITH INDUSTRIAL–GLACIAL CORRELATIONS IN THE 1930S

In Breuil's opinion, developments in Britain during the 1930s were an immense improvement on British research of the early 1920s. He praised '*un renouveau de ces recherches de ce côté-ci du Channel, auquel les noms de King, Oakley, Palmer, Paterson, Burchell se sont brillamment attachés*' ('the renewal of researches from this side of the Channel, with which the names of King,

---

[2] Chandler to Dixon, 21 November 1930: BGS: GSM1/295.
[3] Boswell 1936; King and Oakley 1936; Zeuner 1937.

Oakley, Palmer, Paterson, Burchell were splendidly associated').[4] Despite his poor opinion of earlier research, it had been in the company of British geologists during the 1920s that Breuil had gathered inspiration for his industrial–glacial sequence. In 1921, Breuil had returned from England to his tools in the Somme Valley with the chronological key of the solifluxion deposits (see Chapter 9); in the 1930s, British researchers saw in Breuil's industrial framework the key to connect their scattered sediments.

Breuil's industrial–glacial sequence inspired complicated re-interpretations of British geology during the late 1930s. Convoluted arguments over correlations, and daunting quantities of geological detail were characteristic of this time. Some of the details are summarized in Table 10.2, but the atmosphere and the reasoning are more interesting than the geological specifics. The central problem was the Alpine glacial chronology. British sequences had been set to run on a short chronology; Breuil, however, had arranged his industries down a long chronology.

When the 1930s dawned, the Upper Chalky Boulder Clay of East Anglia was usually linked to the Mousterian industry and the Würm glacial episode, whilst the Lower Chalky Boulder Clay (which underlay Acheulian deposits at Foxhall Road and Hoxne) was generally associated with the Riss. These correlations had gained authority in the 1920s after the excavations at High Lodge, Foxhall Road, and Hoxne (see Chapter 8).[5] In the Thames Valley, equivalent deposits were identified in the Coombe Rock: a tumble of chalk-rich solifluxion deposits, generally associated with the Upper Boulder Clay and the Würm; and in the boulder clay that lay beneath the 100-ft terrace of the Thames, which was identified with the Lower Boulder Clay and the Riss. The presence of Coombe Rock above Mousterian industries at Baker's Hole, Northfleet, reinforced its connection to the Würm.[6] The short glacial chronology seemed secure; most Palaeolithic industries clustered amongst the last two glaciations: the Riss and the Würm.

Local legend proclaimed that Baker's Hole was named after a drunken Mr Baker, who had fallen down an old working many years before.[7] This site played an important part in the industrial–glacial correlations of the 1930s and 1940s. In the early 1930s, it was known as the particular province of Major James P. T. Burchell. He used the Coombe Rock to suggest that river terraces were more complicated than the Geological Survey had claimed.

---

[4] Breuil 1937b, 259.
[5] Boswell and Moir 1923, 233; Moir 1927a, 142–143; Solomon 1930, 382; Burkitt 1933, 138; Boswell 1936, 160; Zeuner 1937, 139, 153.
[6] Boswell 1932a, 80–81; Dewey 1932, 52.
[7] Abbott 1911, 467.

Table 10.1. A comparison between the industrial–glacial correlations often used in Britain during the 1920s, and the correlations suggested by Breuil in 1932 (after Burkitt 1925, 47; Peake and Fleure 1927, 99; Breuil and Koslowski 1931; 1932; Breuil 1932a, 573; 1932b, 127). Note the bunching of industries in the British scheme of the 1920s: most Palaeolithic industries cluster in the Riss-Würm; Moir's controversial Early Chellean and Pre-Chellean industries cluster much earlier in the glacial sequence.

Alpine glacial sequence	British industrial correlations, 1920s	Breuil's correlations, 1932	
		Flake	*Hand-axe*
Würm	Mousterian	Mousterian; Levalloisian V–VI	
Riss-Würm	Early Mousterian; Acheulian; Chellean	Levalloisian III–IV	Acheulian VI–VII (Micoquian)
Riss		Levalloisian I–II (derived)	Acheulian V
Mindel-Riss Mindel		Clactonian (II)	Acheulian I–IV
Günz-Mindel	Moir's Early Chellean/ Pre-Chellean	Clactonian (I)	Chellean (Abbevillian)
Günz		Pre-Palaeolithic	
	Moir's Pre-Palaeolithic (Foxhall Hall)		

Chandler crowed in delight to his Survey friend, Dixon: 'The Survey might as well shut up now he [Burchell] is scampering round the country correlating everything!'[8]

Despite all the connections that had been drawn to the short glacial chronology in East Anglia and the Thames Valley, geologists of the 1930s again found themselves counting back through sequences of glaciations and industries. This time, they were hunting for a match to the long industrial–glacial chronology favoured by Breuil (and by a few others in the 1920s, such as Moir). But they became tied up in a way that recalled Boswell's comment of 1930 about the 'Irishman's waistcoat, which had one button too many at the top and one buttonhole too many at the bottom, for here we have one button too many in the middle'.[9]

The correlating geologists were faced with two main problems. The first was that the industries of the 100-ft terrace (Acheulian I–IV and Clactonian II–III) had to be moved back by one interglacial slot, from the Riss–Würm to

[8] Chandler to Dixon, 22 December 1932: BGS: GSM1/295. See also Burchell to Baden-Powell, 23 June 1934: UMO: DB-P: K18; Burchell 1934.
[9] Boswell 1930, 380.

the preceding Mindel–Riss of the long glacial chronology.[10] The second problem concerned the Mousterian. Enthusiasts of the short chronology had associated the old ubiquitous Mousterian with the Würm. But 'Mousterian' assemblages had now been dispersed amongst several flake industries and several glaciations: the Clactonian and some early phases of the Levalloisian *pre-dated* the Würm (see Table 10.1).[11] If the Coombe Rock and the Upper Boulder Clay continued in their old connection, the Mousterian and the Levalloisian would be compressed within the same glacial episode, and this did not match new expectations.

In 1931, Breuil wrote to Sollas to suggest a correlation that harmonized with his recent views. He told Sollas that his ideas had changed since 1926, 'when I was thinking the upper Chalky boulder clay and the coombe rock of Northfleet were Würm; now I am convinced they are *Riss*'.[12] In 1926 Breuil had described the Baker's Hole industry as Levalloisian (rather than Mousterian), but he now decided that it dated to the very earliest phases of the Levalloisian (I-II), which he associated with the Riss glaciation. Counting back from a Coombe Rock that dated to the Riss rather than the Würm, the deposits from the 100-ft terrace above fell neatly into the Mindel-Riss slot predicted by Breuil and Koslowski (see Tables 10.1 and 10.4). Breuil's two letters to Sollas on this subject were published in the *Geological Magazine* for 1932: a clear indication of the value attached to his opinion and the interest inspired by such correlations in Britain at this time.[13]

Despite Breuil's solution, geologists were left with several problems as they tried to connect the sequences in the Thames Valley to those in East Anglia. It seemed to Boswell, writing in 1936 (and referring to the arguments of Oakley and Zeuner), that the Levalloisian of the Thames Valley *had* to share the same cold slot as the Mousterian, and be matched to the Upper Boulder Clay. He could see no way out of the straitjacket that bound together the sequences of industries and boulder clays.[14] If the Upper Boulder Clay were associated with the early Levalloisian instead of the Mousterian, this would cause problems further up the East Anglian sequence. Moir had found Aurignacian artefacts in the Hunstanton (the fifth and last) Boulder Clay of East Anglia, which seemed to have been swept up in the Magdalenian glaciation. If such an alteration were made, the Mousterian would then be left without a place,

---

[10] Moir 1920b, 222–223; Boswell and Moir 1923, 260; Breuil 1930, 223; Breuil and Koslowski 1932, 304.

[11] Boswell 1930, 380; 1931, 109.

[12] Breuil to Sollas, 24 August 1931: BGS: GSM1/445. See also Breuil 1926, 177.

[13] Breuil to Sollas, 24 August 1931 and 8 September 1931, quoted in Sandford 1932, 17–18. See also Breuil 1931.

[14] Boswell 1936, 160.

because Moir's connection between the Aurignacian and the Hunstanton Boulder Clay put a cap on the glacial sequence of East Anglia (see Table 10.2).[15] In short, there seemed to be a glacial phase missing.

Boswell drew a gloomy conclusion: although the Lower Boulder Clay = Riss; Upper Boulder Clay = Würm I correlation fitted some aspects of the case, it did not match the French sequence and left no space for the Levalloisian industry.[16] Nonetheless, by 1937 there was a consensus that the 100-ft Thames gravels dated to the Mindel–Riss interglacial, a connection that slid many of the industries into their expected positions on the long glacial chronology. That year, Zeuner suggested various solutions to Boswell's mismatches: industries were re-identified and boulder clays re-labelled. These changes enabled him to link the boulder clay underlying the 100-ft terrace with the Mindel (and with an East Anglian glacial deposit that *preceded* the Lower Boulder Clay: the North Sea Drift). Zeuner emphasized the value of bones and tools in building a reliable glacial correlation, and believed that all three lines of evidence pointed independently to the same result. He saw additional support in the conclusions of French researchers: they too had associated the Clactonian II and Early Acheulian industries with the Mindel–Riss. Breuil's demonstration that the Chellean (Abbevillian) was of Günz–Mindel age had likewise led Zeuner to shift this industry back a stage from the Mindel–Riss, with which he had correlated it two years previously.[17]

After Zeuner's paper was published in 1937, Oakley cautiously followed a Mindel–Riss date for the 100-ft gravels and tentatively correlated the pre-100-ft Lower Boulder Clay with a phase of the same (Mindel) glaciation that had left the North Sea Drift.[18] Everything seemed to be falling into place. Zeuner even noted that Moir's primitive Pre-Palaeolithic (now often called Pre-Chellean) industries of the Crags 'lose much of their peculiarity' if the Crags were partly contemporaneous with the Günz glaciation.[19]

The classic site of Hoxne, a familiar source of consolation in troubled times since 1859, was also drawn into the re-alignment of these old sequences. In 1930, Breuil had assigned the boulder clays of Hoxne to the Mindel and Riss glacial episodes, and described the flake industries of the latter as early Levalloisian.[20] Zeuner noticed from Boswell's paper of 1936 that his previous

---

[15] Moir in Moir and Burchell 1930, 361–371; Boswell 1931, 98; 1936, 158–159. See also Moir to Baden-Powell, 18 September 1933: UMO: DB-P: K93.

[16] Boswell 1936, 160.

[17] Zeuner 1937, 151, 154–155; Zeuner to Baden-Powell, 10 December 1937: UMO: DB-P: K158.

[18] Hinton, Oakley *et al.* 1938, 56–58; Oakley to Moir, 29 October 1938: UMO: DB-P: K94; Oakley 1939, 357.

[19] Zeuner 1937, 151.   [20] Breuil 1930, 222.

Table 10.2. A comparison of several different industrial–glacial correlations suggested in the 1930s. As researchers tried to connect the glacial deposits of East Anglia to the cold stages of the Thames Valley, they found it difficult to match the sequences of industries from each area (after Peake 1922, 126; Burkitt 1925, 47; Breuil and Koslowski 1931; 1932; Breuil 1932a, 573; Solomon 1932; King and Oakley 1936; Boswell 1936; Zeuner 1937).

Alpine glacial sequence	Breuil, 1932		East Anglian sequence: Solomon, 1932	Thames Valley sequence: King and Oakley, 1936*	East Anglian sequence: Boswell, 1936; Zeuner, 1937
	Hand-axe	Flake			
Würm II				Slades Green Trail	Hunstanton (Brown) Boulder Clay (Upper Palaeolithic)
Würm		Mousterian; Upper Levalloisian	Upper Chalky Boulder Clay; Little Eastern (Mousterian)	Baker's Hole Coombe Rock (Levalloisian)	Upper Chalky Drift (Boswell's Mousterian and Levalloisian)
Riss-Würm	Final Acheulian (Micoquian)	Middle Levalloisian		(Acheulian; Clactonian)	
Riss	Upper Acheulian	Early Levalloisian	Lower Chalky Boulder Clay; Great Eastern (?)	Chalky Jurassic Boulder Clay	Lower Chalky Boulder Clay
Mindel-Riss	Early and Middle Acheulian	Evolved Clactonian		(Clactonian I; Early Acheulian, Abbevillian)	(Acheulian; Clactonian)
Mindel		Clactonian	North Sea Drift; Norwich Brickearth	Plateau Drift	North Sea Drift; Cromer Till (? Chellean)
Günz-Mindel	Abbevillian	Early Clactonian	Cromer Forest Bed (Abbevillian)		Cromer Forest Bed (Abbevillian)
Günz		Pre-Chellean of Foxhall Hall	(Pre-Chellean)		Later Crag deposits (Pre-Chellean)

*Not all the British schemes related to the Alpine sequence given in the first column. King and Oakley (1936) referred only to Solomon (1932).

identification of the Hoxne industries 'has been modified so as to make the implements older, in order to fit Hoxne into the Mindel–Riss interglacial'.[21] Both Boswell and Zeuner stressed the importance of knowing the exact identity of the 'Mousterian' from the boulder clay above the Hoxne lake deposits. But whilst Boswell, like Breuil, suggested an early Levalloisian, Zeuner did not see any conclusive evidence for the age of the site in the Hoxne industries. He thought that the geology still supported a Riss–Würm age.[22]

With the renewal of confidence in the relationship between industries and glaciations, cases that failed to fit expectations stood out prominently as problems to be resolved. Industries, boulder clays, and solifluxion deposits were re-labelled and re-shuffled to smooth over these many irregularities, and the lines between them were re-drawn. But an alteration in one area could raise problems in another. The complexity of these arguments, though daunting, explains why the archives of geologists who worked on this giant puzzle in the 1930s contain so many jottings about possible correlations.

10.2 PROBLEMS WITH INDUSTRIAL–FLUVIAL CORRELATIONS IN THE 1930S

As Boswell and Zeuner argued about the identity of the glaciations that bracketed the 100-ft and 50-ft river terraces of the Thames Valley, others scrutinized the sequence of fluvial sediments that had been left by the ancient river *between* those two glaciations. In 1935, Oakley was considering the patterns revealed in his recent excavations with Mary Leakey at Jaywick Sands, Clacton. He complained to Hawkes of an 'anomalous state of affairs' which made it obvious 'that height above or below O.D. is no criterion of the age of a river terrace, and [...] the old terrace classification: 100ft., 50ft., etc. will have to be scrapped. Sorry and all that'.[23]

Oakley was not the first to recognize the complexity of ancient river behaviour or to address the problems of terrace classification. But his perception of an 'anomalous state of affairs' was also driven by a perceived disparity between the geological sequence and the newly-defined industrial phases of the 1930s. After excavating at Jaywick Sands in 1934, Oakley had

---

[21] Zeuner 1937, 141. On Boswell's changing views of the Hoxne industries, compare Boswell 1932a, 70 to Boswell 1936, 156.

[22] Boswell 1936, 157; Zeuner 1937, 153.

[23] Oakley to Hawkes, 22 October 1935: BM(F): Misc. correspondence.

become concerned about the relationship between the Clacton Channel and the Thames Valley river sediments. Both areas yielded Clactonian industries, and Oakley's attempts to maintain the expected sequence of Clactonian phases (I-IIA-IIB-III) inspired what he described as an unconventional and unorthodox interpretation of the geological sequence in the Thames Valley.[24] He presented his unorthodox opinions in two papers, one written with King: 'The Pleistocene Succession in the Lower Part of the Thames Valley' (1936); the other written with Mary Leakey: 'Report on Excavations at Jaywick Sands, Essex' (1937).

Warren had established a strong geological connection between Clacton and the Thames Valley when he decided that the Clacton Channel was not a tributary after all, but 'the actual channel of the Thames'.[25] Oakley now claimed that the Clacton Channel did not date to the earlier part of the 50-ft terrace, as Warren had suggested in the 1920s, but had to be matched instead to the middle of the 100-ft terrace.[26] This was his unorthodox view. Warren and Oakley saw different signs of geological time in river terraces, fauna, and industries, and they each described a different relationship between the Clacton Channel and the Swanscombe Middle Gravel. Neither believed the division between the 100-ft terrace and the 50-ft terrace to be as simple as some had assumed.

Warren's opinion that the Clacton Channel post-dated the Swanscombe Middle Gravel was influenced by the views of Hinton and Kennard, with whom he had often collaborated. The two palaeontologists had dated the Clacton Channel to the time of the early 50-ft terrace. As will be recalled from Chapter 7, Hinton and Kennard saw much similarity between the species in the early 50-ft terrace and those in the 100-ft terrace.[27] Warren agreed that the two series of deposits were similar in age, a view he derived from the character of the industries as well as the fauna, and from his consideration of the complicated behaviour of ancient rivers. These observations led Warren to group the early 50-ft and late 100-ft sediments together as an intermediate 'Furze Platt' stage. He correlated this Furze Platt stage to the Swanscombe Middle Gravel, and believed that the Clacton Channel was later in date.[28] Warren later reflected to Donald Baden-Powell (1897–1973), the Oxford

---

[24] Oakley and Leakey 1937, 256; Bull 1942, 31.

[25] Oakley and Leakey 1937, 217. On Warren's changing ideas about the Clacton Channel, compare Warren 1922b, 597 to Warren 1932a, 16.

[26] Warren 1922b, 597; 1923d, 618; 1932a, 24; Oakley and Leakey 1937, 217–218.

[27] Hinton 1910, 493–494, 503; 1926a, 337–338; Kennard 1916, 250, 255–257.

[28] Warren 1922b, 602; 1923d, 607; 1926, 43, footnote; 1932a, 5, 23–24. Treacher 1904, 18–19; 1910, 198–199 had also observed that the 100-ft and 50-ft terraces were accompanied by an intermediate terrace, and Burchell 1934 had argued that the 50-ft terrace ought to be split into two.

Figure 10.1. Mr and Mrs Warren at Lion Point, Clacton, in 1937 (UCL: Zeuner Diary 2, p. 12).

geologist: 'When Hinton concluded that the Early Middle [50-ft] Terrace was very near in date to Swanscombe (where most of the fossils had come from the *Lower* Gravel) I felt that this gave vitally important support to my own view'.[29]

Oakley reasoned differently. He trusted in the industrial phases, which seemed to provide a guide through the complexities of river-sediment stratigraphy. The tools from the Clacton Channel had been assigned to an earlier phase in the Palaeolithic sequence than those from the Middle Gravel. Oakley's argument went as follows. The Swanscombe Middle Gravel contained handaxes of Acheulian III-IV type. Breuil himself had assigned them to that phase of the Acheulian: a phase that was contemporary with Clactonian III. These Acheulian and Clactonian phases had both been found on Worthington Smith's old Floor at Stoke Newington, and Oakley reported in 1937 that Clactonian III tools were also present in the Swanscombe Middle Gravel. He concluded that the Middle Gravel had to be *later* than the Clacton Channel, because the Channel contained Clactonian IIB tools.[30] In 1938, Baden-Powell asked Oakley: 'In correlating the Clacton Channel with the time between

---

[29] Warren to Baden-Powell, 8 December 1952: UMO: DB-P: K147.
[30] King and Oakley 1936, 60; Oakley and Leakey 1937, 240–242. The connection between Oakley's view of the Palaeolithic sequence and his interpretations of the stratigraphic sequence is also noted by Bridgland 1994, 341.

Table 10.3. The sequences for Clactonian tool-bearing deposits suggested by Oakley and by Warren in the 1930s (after Warren 1932a, 24–26; 1932b; King and Oakley 1936; Oakley and Leakey 1937).

Oakley's sequence	Warren's sequence
	The High Lodge industry (Clacton technique and Mousterian finish)
Clactonian III: High Lodge; Stoke Newington; Swanscombe Middle Gravel	The Clactonian industry from the Clacton Channel (Oakley's Clactonian IIB)
Clactonian IIB: Clacton Channel; Grays (Little) Thurrock Channel	Swanscombe Middle Gravel
Clactonian IIA: Swanscombe Lower Gravel	The Swanscombe Clactonian: Swanscombe Lower Gravel (Oakley's Clactonian IIA)
Clactonian I: Found in derived condition in the Swanscombe Lower Gravel	The Swanscombe Clactonian I: (possibly earlier than Clactonian IIA, but Warren is cautious)

the Middle and Lower Gravels at Swanscombe, are you relying chiefly on the implements or on the fauna?' Oakley replied: 'Yes, it was mainly on the implements that I make it a shade later than the L. Gravels'.[31] The differences between the sequences developed by Oakley and by Warren are summarized in Table 10.3.

Oakley admitted that there was one great obstacle to the stratigraphic sequence suggested by the tools: the relative heights of these deposits. The Clacton Channel seemed to lie at too low a level for its sediments to have been laid down between the time of the Swanscombe Lower Gravel and Middle Gravel. Chandler had also found it difficult to explain the difference in level between the Clacton Channel and Swanscombe.[32] This was why Oakley had written to Hawkes about anomalies and the need to scrap the old terrace classification. But he developed an alternative geological classification with King (Fig. 10.2) that solved the Palaeolithic problem.

King, a Yorkshireman, had been appointed to the Chair of Geology at University College London in 1931. Before gaining his Professorship, he had trained in the Geological Survey, served as a military geologist during the Great War, and assisted Marr at Cambridge: giving lectures and accompanying him on his 'gravel grobs'.[33] King agreed with Oakley about the inadequacy of existing geological classifications to describe the sedimentary history of

[31] Baden-Powell to Oakley, 26 March 1938 and Oakley to Baden-Powell, 29 March 1938: UMO: DB-P: K99.
[32] Chandler 1930, 90 footnote 4; Oakley and Leakey 1937, 253.
[33] Nicholas 1963, 151–152; Shotton 1963, 171–173.

Figure 10.2. William B. R. King (1889–1963), geologist, and Oakley's co-author (GSL: P53/67).

the Thames Valley. They decided that the old system of terraces would have to be replaced with a longer series of more precise stages. These stages were based on cycles of erosion (down-cutting) and aggradation (accumulation of deposits); coarser time-markers were supplied by the glacial sequence.[34]

King and Oakley published their stages in 1936. Their paper was welcomed: few attempts had been made to synthesize such a vast amount of stratigraphical, palaeontological, and archaeological information from the Thames Valley since the efforts of Hinton and Kennard back in 1905 (see Chapter 7). Table 10.4 summarises the stages they established between the two great glacial markers provided by the pre-100-ft boulder clay and the Coombe Rock: the time-span of the problematic Palaeolithic correlations. King and Oakley avoided linking these glacial deposits to the Alpine glacial sequence, perhaps daunted by the kind of difficulties encountered by Boswell. They did, however, adopt the Palaeolithic terminology introduced by Breuil.

Oakley admitted: 'the low level of the Clacton Channel deposits requires some special explanation, if, as the evidence indubitably suggests, they post-date the Lower Gravel, but pre-date the Middle Gravel of the 100-ft. terrace'.[35] King and Oakley resolved the difficulty by invoking a period of down-cutting that they called the 'Inter-Boyn Hill Erosion Stage'.[36] (Boyn Hill was one of the

---

[34] King and Oakley 1936, 53–54.   [35] Oakley and Leakey 1937, 253.

[36] King and Oakley 1936, 57, footnote 1. On the detailed geological correlations of King and Oakley and the connections to the Little Thurrock fauna, see Bridgland 1994, 233–234.

Table 10.4. A summary of the Thames Valley stages developed by King and Oakley (1936) for the time-span of the problematic Palaeolithic correlations, with the industries and some of the sites associated with each stage.

Stages (King and Oakley)	Sites	Industries
Taplow	Taplow Station pit	Derived Early Levalloisian; Levalloisian III (?)
Baker's Hole (Main Coombe Rock)	Baker's Hole, Northfleet	(Coombe Rock lies above Levalloisian I–II)
Pre-Coombe Rock Erosion	Wansunt Channel; Globe pit, Greenhithe; Swanscombe Upper Loam	Late Acheulian; Early Levalloisian (I–II)
Middle Barnfield (Late Boyn Hill)	Final stage of aggradation: Swanscombe Middle Gravel; Stoke Newington	Middle Acheulian (III–IV); Clactonian III
Ilford	Uphall pit; Cauliflower pit, Ilford	
Clacton-on-Sea	Aggradation: deposits laid down in the channels at Clacton and Little Thurrock	Clactonian IIB
Inter-Boyn Hill Erosion	Uplift: cutting of the channels of Little Thurrock and Clacton	
Lower Barnfield (Early Boyn Hill)	Swanscombe Lower Gravel and Lower Loam	Clactonian IIA; Derived Clactonian I; Derived Abbevillian
Pre-Boyn Hill Erosion		
Great Eastern Glacier	Chalky Jurassic (Lower Chalky) Boulder Clay at Hornchurch	

names given to the 100-ft terrace.) They suggested that part way through the aggradation of the 100-ft terrace, between the time of the Lower and the Middle Gravel, the ancient river had cut down rapidly through its old bed. It was then that the channels at Clacton and Little Thurrock had been excavated. Afterwards, the river had returned to the level of the 100-ft terrace (see Table 10.4).

This Inter-Boyn Hill Erosion stage dated the Clacton Channel and its industry to the expected place on the Thames sequence, and enabled Oakley to claim the following year, in the Jaywick report: 'it is not unreasonable to expect to find deposits of approximately the age of the Clacton gravels at a lower level than the 100-ft Terrace in the region of Swanscombe. Such deposits do in fact occur at Grays (Little) Thurrock, in Essex, on the north side of the river opposite Swanscombe'.[37] In this report, Oakley maintained that his correlation of the Clacton Channel agreed with earlier observations: Hinton

[37] Oakley and Leakey 1937, 254.

had drawn a great distinction between the fauna from the Clacton Channel and the fauna from the later 50-ft terrace; Warren had suggested that the deposits of the early 50-ft terrace should be grouped with those from the late 100-ft terrace. Oakley also suggested that past interpretations might have been misled by the assumption that flake-bearing deposits had to be later in date than hand-axe-bearing deposits. Nonetheless, the strongest stimulus to his own correlation had been a belief in the new industrial phases.

The Jaywick report ended by noting a close correspondence between the physical history of the Somme Valley and the Thames Valley. King and Oakley stated early in their paper that recent research had established the practical identity of the British Palaeolithic sequence and Breuil's sequence in the Somme, and they praised his refinements to the industrial classification. Breuil returned the compliment the following year when he remarked on the renewal of British researches: King and Oakley were included in his list of commended researchers.[38]

Oakley's attempt to reconcile the geological and Palaeolithic sequences received a mixed response. In the early 1940s, Chandler, who knew the Swanscombe deposits well, argued that he saw no sign of Oakley's Inter-Boyn Hill Erosion stage in the section; he also rejected a slightly later period of channel cutting, part-way through the Middle Gravel, claimed by Marston.[39] Others were more convinced. Kennard wrote to Warren in November 1940 with a different view of the faunal sequence: he now dated the Clacton Channel deposits to a time before the deposition of the Swanscombe Middle Gravel. In 1942, Warren also adopted Oakley's view that the Clacton Channel, like the Little Thurrock Channel, had been cut between the deposition of the Lower and Middle Gravel of Swanscombe (although he still associated these channels with the early 50-ft rather than the 100-ft terrace).[40] Warren's change of mind was more closely connected to the industrial phases than to Kennard's fauna:

> The fossils prove a very close association between Grays, Clacton, and Swanscombe, but I think for more precise relative dating one must look to the human industries. In this connection it is noteworthy that the 'middle gravel' of Swanscombe yields a Clactonian that is intermediate between the Clacton II of the name site and Clacton III (= High Lodge).[41]

[38] King and Oakley 1936, 52–53; Breuil 1937b, 259; Oakley and Leakey 1937, 254, 256.

[39] Marston 1937, 351; Bull 1942, 29. McNabb 1996, 44–46 discusses how typology again 'set the agenda for stratigraphic interpretation' in this rejection of Marston's later channel, and describes Paterson's (1940a) alternative explanation of the stratigraphy and typology.

[40] Warren 1942, 173–174; 1951, 130. 'I still think that the Grays-Clacton-Barnham channels are cut through, & definitely severed from, the Boyn [100-ft] platform' (Warren to Baden-Powell, 23 November 1953: UMO: DB-P: K147). Warren enclosed notes of the changes in Kennard's private opinions about the Clacton-Thames correlations in a letter to Oakley dated 21 February 1951 (NHM: DF140/7).

[41] Warren 1942, 174.

British Quaternary research of the 1930s was pervaded by Breuil's Palaeo-lithic scheme. Sequences of boulder clays, river terraces, and tools were stretched to fit his long glacial chronology; the finer patterns between these glacial episodes were matched to his industrial phases. During this decade, a whisper that 'L'Abbé Breuil' approved an interpretation became an effective way to stifle opposition. Grahame Clark, Assistant Lecturer in Archaeology at Cambridge and an important figure in the Prehistoric Society, noted in 1938: 'in matters general the great French prehistorian has become something of an unofficial referee'.[42] But as these geological edifices were being constructed, Breuil confided to Oakley: 'I don't think the division in 7 stages of the Achelean is quite satisfactory; it was a trial essay'; he added: 'our knowledge of many details of levels and types is always in flux'.[43]

10.3 PATERSON'S PALAEOLITHIC AND THE DOUBTS OF THE 1940S

During the 1940s, flaws were recognized in the classifications of the previous decade. Palaeolithic patterns had been described on two different scales during the 1930s: broad divisions had been drawn between hand-axe (or core) industries and flake industries, and finer details had been identified in the many industrial phases that were sprinkled up these hand-axe and flake lines. But parallel cultures became overstretched as they were extended across the globe to Africa and East Asia, and faults also developed between local sequences of industrial phases. These changes in the perception of the Palaeo-lithic are evident in the efforts made by Thomas Thomson Paterson (1909–1994) to smooth over the cracks—and from the reactions to his solutions.

In 1933, Burkitt had encouraged Paterson to examine a small part of East Anglia known as the Breckland (between Cambridge and Norwich). There, he might establish an industrial–glacial sequence that would help correlate the tools and sediments of East Anglia with those of the Thames Valley. This popular topic of the 1930s supplied Paterson with material for his doctoral thesis on 'Lower Palaeolithic Man in the Cambridge District'.[44] When he became a Research Fellow at Trinity College, Cambridge, Paterson continued to work on East Anglian correlations, but he also began to study the Palaeolithic

---

[42] Clark 1938, 340. On Clark's influence on British prehistoric research, see Fagan 2001 *Grahame Clark. An Intellectual Biography of an Archaeologist.*
[43] Breuil to Oakley, 11 December 1936: NHM: DF140/6.
[44] Paterson 1942, 1.

of India. He remained at Cambridge as Curator of the Museum of Archaeology and Anthropology through much of the 1940s.[45] During this time, Paterson tried to adapt the Palaeolithic perspective introduced by Breuil in the 1930s to describe industrial variation on a regional and a global scale.

In his article 'On a World Correlation of the Pleistocene', published in 1941, Paterson suggested that the term 'Clactonian' be used as a 'family' name for *all* early Palaeolithic flake industries. He distinguished this flake family from his 'Acheulian' family of hand-axe industries. Arguing that Breuil's definition of the Clactonian was too generalized to apply to such specific industries as those from Swanscombe and Clacton, Paterson renamed the old Clactonian I and II of England and northern France, the 'Brecklandian'. He included the Brecklandian in his Clactonian family of tool-making traditions, alongside the Levalloisian, the Mousterian, and other flake industries from around the world.[46]

Paterson grouped the industries that produced *flakes* from cores within his Clactonian family; industries that produced *bifaces* (hand-axes) from cores or flakes were included within his Acheulian family.[47] Both industrial families were characterized by progressive improvement in the skill of tool-working techniques and in the quality of finished shapes.[48] Like Breuil, Paterson relied on tool-working techniques to define and connect industries within his flake family, and to maintain their distinction from his hand-axe-making traditions: 'In the so-called "flake culture", technique is the most important distinguishing criterion, whereas, in the "hand-axe culture", form is equally so'.[49]

The division between these tool-making families (or 'cultures') was reinforced by the gulf that Paterson placed between their makers: he linked the flake cultures to a Palaeoanthropoid line of hominids, and the hand-axe cultures to a separate Hominoid line. He was one of many prehistorians who had associated the Swanscombe skull with Acheulian hand-axes and our own ancestors.[50] Similar arguments had characterized earlier generalizations about the Palaeolithic sequence: the division between technique and tool-type, the connection between different industrial groups and different hominids, and the line of progress that allowed links to be made between the industries of distant areas in the absence of secure dates.

Although Paterson stretched these arguments to accommodate a global picture of the Palaeolithic, he also wanted to explain the local diversity of industries. He did not consider Breuil's solution adequate. Breuil had accounted for industrial diversity by creating his industrial phases—the

---

[45] Clark 1989, 60–61.   [46] Paterson 1941, 378–379; 1942, 184–185.
[47] Paterson and Fagg 1940, 6.   [48] Paterson 1941, 378, 380–383; McNabb 1998, 10.
[49] Paterson 1945, 5.   [50] Paterson 1940a, 167–169.

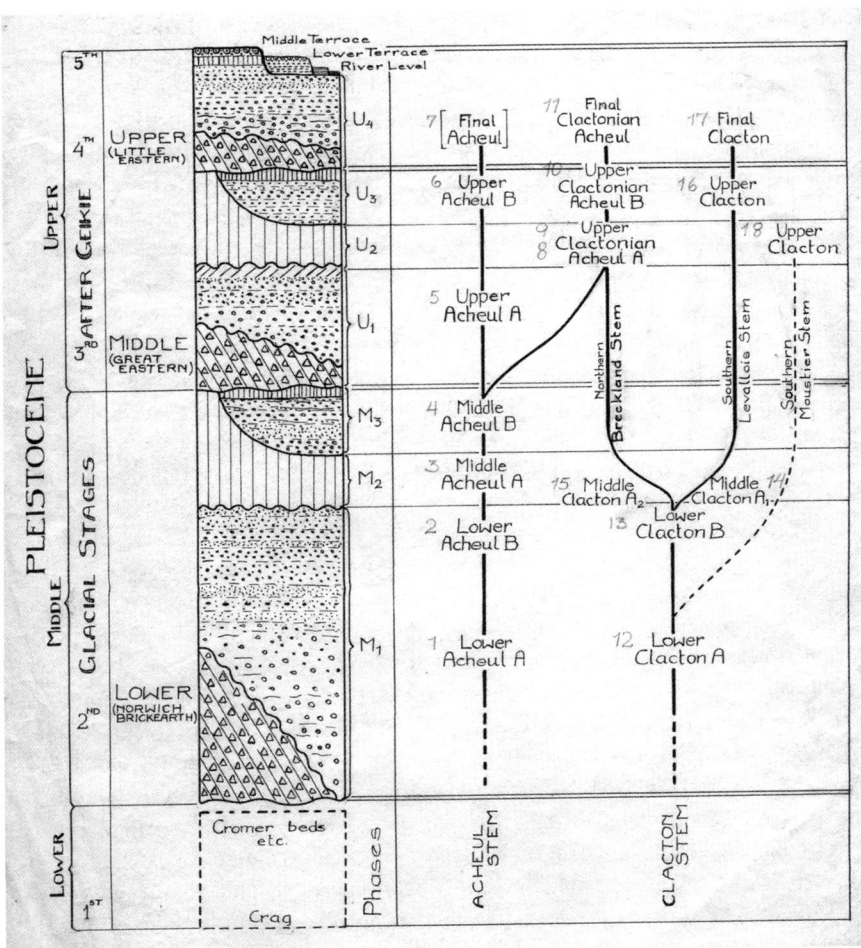

Figure 10.3. Paterson's diagram illustrating the development of the Acheulian and Clactonian families (Paterson papers, undated, *c.* 1945: AAM: W21/1/3). The industries were connected to the geological sequences in the Thames Valley and East Anglia, and to the five glaciations of James Geikie. The wavy lines in the section indicate periods of erosion (unconformities). The Lower terrace was also known as the 100-ft; the Middle terrace as the 50-ft. For the East Anglian glacial terminology, see Solomon 1932, in Table 10.2.

fine time-markers beloved by stratigraphers—but the variety of character displayed by these phases had been restricted by Breuil's insistence on defending a great distinction between hand-axe and flake industries. His belief in this distinction had stimulated the attacks on Warren's bifacial core-tools from Clacton in the 1930s (described in Chapter 9). Paterson, however, took a more flexible approach to his two great families: he encouraged them to hybridize.[51]

Like some of his contemporaries, Paterson used the idea of 'fusion, by contact, of separate cultural entities' to explain the presence of both Acheulian *and* Clactonian elements in the same assemblage of tools.[52] Over the 1940s, this idea of 'culture-contact' became a popular explanation for the growing numbers of assemblages that were too varied to fit neatly into Breuil's industrial scheme.[53] Industrial hybridization nurtured a proliferation of terms like 'Brecklandian Acheul': a description used by Paterson to describe a dominant (generic) Acheulian influence and a secondary (specific) Brecklandian influence on a particular industry.[54]

As culture-contact became used more frequently to describe industrial diversity, it became more difficult to accommodate the expanding variety of industries within a broader Palaeolithic pattern. Industrial terminology was losing its precision. Palaeolithic researchers began to question their definitions; they even reconsidered the aim of their researches. Palaeolithic classifications had a long history of ambiguity: the same term might refer to the age or to the character of an industry—or to both. But Breuil's popularization of parallel cultures, and the latest craze for hybridization, had caused the terms 'Acheulian' and 'Clactonian' to be applied so widely that the meaning of a distinct Palaeolithic 'culture' was now perceived to be under threat.

The labels 'Acheulian' and 'Clactonian' might describe many varied industries in western Europe, and had even been extended to industries in Africa and Asia. The term 'culture' had once implied an ethnological connection between the makers of Palaeolithic industries. Now, this old meaning was dissolving. Some complained that an industrial term might describe any chance similarity of appearance between two assemblages, regardless of the relationship, in time or space, between the tool-makers who had made those two industries. The term 'Clactonian,' for example, might describe the technique displayed by an industry with no chronological or cultural (i.e. ethnological) connection to other Clactonian industries.[55]

---

[51] Paterson 1940b, 50, 52.   [52] Paterson and Fagg 1940, 22.
[53] Oakley *et al.* 1948a, 22.   [54] Paterson 1940b, 49.
[55] Burkitt 1936a, 103; 1936b, 216; Paterson 1937, 135; Oakley *et al.* 1948a, 22–23; Movius 1953, 167, 188.

10.4 NEW PERSPECTIVES ON THE PALAEOLITHIC FROM AFRICA AND ASIA

Garrod had observed in 1928 that it was difficult to fit Palaeolithic discoveries from beyond western Europe into the existing industrial framework.[56] Back then she was referring to the system of de Mortillet, and had seen the problem as one of scope, rather than with the framework itself. By the 1940s and early 1950s, archaeologists working far outside this classic heartland of Palaeolithic research focused their attacks on the framework of the 1930s. They found it difficult to interpret new discoveries in terms of contemporary hand-axe-making and flake-making cultures. They questioned the existence of two industrial families or cultures, and criticized the concept of culture-contact as an explanation for regional variety.[57]

Clarence (Peter) van Riet Lowe (1894–1956), Professor of Archaeology at the University of Witwatersrand, and A. J. H. (John) Goodwin (1900–1959), at the University of Cape Town, had both been working on the South African sequence since the 1920s. They found the standard Palaeolithic terms and concepts inadequate to describe and explain their varied industries. Lowe and Goodwin had been cautious in their use of European industrial terminology, and employed a range of terms to describe South African industries: their major hand-axe industry, for example, was called the 'Stellenbosch'.[58] Lowe, however, still relied on European terms to describe similarities in technique, which he thought suggested an affinity between the two continents, but some of the differences made him uneasy.

The relationship between hand-axe and flake cultures in South Africa seemed different from the European pattern. In Europe, flake cultures (like the Levalloisian and the Clactonian) and hand-axe cultures (like the Abbe-villian and Acheulian) had developed independently and progressed in parallel, albeit with some culture-contact, particularly in the later stages. In South Africa, however, Clacton and Levallois techniques had been an *integral* part of the core (hand-axe) culture since early Palaeolithic times.[59] This made Lowe suspicious of the 'extraordinary two-stream development in Europe': he hoped for 'a final interpretation which [would] strike a less discordant note with the state of affairs we know existed in Africa during Old Palaeolithic times'.[60]

---

[56] Garrod 1928, 260.   [57] Childe 1944, 19; Movius 1953, 164.
[58] Goodwin and Lowe 1929, 9–11, 78, 96.
[59] Lowe 1936, 199; Lowe and Breuil 1945, 50–51.
[60] Lowe and Breuil 1945, 54.

Figure 10.4. Cartoon of Breuil (short) and Lowe (tall), drawn in 1944 (postcard from Lowe to Burkitt, 24 November 1950: ULC: 7959, Box 1).

During the 1940s, Breuil worked closely with Lowe on the South African sequence. They must have become a well-known pair: Lowe sent Burkitt a cartoon (illustrated above in Fig. 10.4) that had been drawn of them both. General Smuts invited Breuil to spend the Second World War in South Africa as a Research Officer on the Archaeological Survey. This move made him unpopular with those who had remained in Paris under German occupation.[61] It also reduced his influence on British Palaeolithic research.

[61] Mason 1965, 142; Leakey 1974, 195.

Breuil and Lowe had to use an unwieldy terminology to describe the South African industries, with labels like: 'core-cum-flake Stellenbosch I of Clacto-Abbevillian facies'.[62] The use of terms like 'Clacto-Abbevillian' to describe South African industries worried researchers who worked on British industries. Mary Leakey, who had excavated at Jaywick Sands, Clacton, before she turned to the Palaeolithic sequence of East Africa, argued that the Clactonian was a distinct cultural entity of western Europe. She believed that the simple flakes of these African industries had only a superficial, technical resemblance to the Clactonian 'culture' (in its discrete ethnological sense). In the late 1940s, she shared her concerns with Oakley: 'I'm so glad you agree about the Clacton in Africa, or rather the lack of it.' She added that Lowe 'for one hasn't the vaguest idea what we really mean by the Clacton & has been led off on red herrings by Breuil & Co'.[63]

Soon afterwards Leakey persuaded Lowe to read an offprint of the Jaywick Sands report. She wrote again to Oakley, reassuring him that this had finally convinced Lowe that a distinct Clactonian 'culture' really existed, and suggesting that Oakley take Edward James Wayland in hand about the East African sequence: 'and convince him that the Clacton is a thing on its own & not a waste product of other cultures'.[64] Wayland had been working on the Palaeolithic of Uganda since 1919, when he had arrived in the country to work for their Geological Survey.[65] British researchers were becoming concerned that the meaning of Palaeolithic 'cultures' as the products of distinct peoples was being lost as the terminology popularized by Breuil and elaborated by Paterson spread across the globe.

Meanwhile, the geographical centre for general, standard Palaeolithic sequences, traditionally located in western Europe, was shifting to East Africa. Mary and Louis Leakey had been working on the immense industrial sequence from Olduvai Gorge in the Serengeti Plain, Tanzania, from the 1930s. After long delays, caused partly by publication costs and the war, the results were published in 1951.[66] As the East African sequence rose in importance over the 1950s, the space previously devoted to the Eolithic period in British publications became occupied by the Oldowan: the earliest industry from Olduvai Gorge.[67] British eoliths and pre-palaeoliths became overwhelmed by African industries, which had once helped to boost the credibility of Moir's

---

[62] Lowe and Breuil 1945, 50.   [63] M. Leakey to Oakley (?1949): NHM: DF140/7.
[64] M. Leakey to Oakley (?1950): NHM: DF140/7.
[65] Wayland 1923, 101.
[66] Leakey 1951, xv.
[67] Watson 1950, 28–29; McBurney 1953, 127; Coles and Higgs 1969, 202. It was still not until the third (1968) edition of the British Museum publication, *Flint Implements*, that British eoliths were entirely replaced by African palaeoliths in the section on the earliest tools: compare Watson 1950, 27–28; 1956, 29–30 to Watson 1968, 42.

rostro-carinates. Their decline was hastened by the death of Moir in 1944; they lost their most fervent supporter and gradually faded from the literature.

A perspective of the Palaeolithic in Asia led Hallam L. Movius, Jr. (1907–1987), Professor in the Department of Anthropology at Harvard University, to question the distinction that had been drawn in Europe between hand-axe (core) cultures and flake cultures. He observed that Commont and Obermaier had found hand-axes in intimate association with flake industries in the Somme Valley, and argued that observations made more recently in Africa, the Middle East, and India cast doubt on the existence of parallel cultures. Movius included both flake and core industries from western Europe within his 'Great Hand-Axe Complex' (although he believed that the Clactonian industry of England was a distinct entity).[68]

This Hand-Axe Complex (or 'tradition') stretched eastwards from Europe, through Africa, and across Asia as far as the geographical boundary known as the 'Movius line'; on the other side, in southern and eastern Asia, the hand-axe innovation had apparently failed to penetrate. Here, Movius described a Chopper/Chopping-Tool Complex. Paterson had placed the early Asian choppers and chopping-tools at the base of the Clactonian family line of flake cultures, where they remained distinct from the Acheulian line. Movius, however, chose to interpret them as a more widespread 'basic cultural substratum' to *both* industrial traditions, with the choppers later becoming less important in areas where the hand-axe had been developed.[69]

Movius also disagreed with those, like Paterson, who had divided core and flake industries between two hominid lines. The core (hand-axe) implements were usually assigned to our own ancestors, but Movius cited a discovery made in 1947 at the cave of Fontéchevade in the Charente, Central France, where the remains of modern humans were associated with a pre-Mousterian flake industry known as the Tayacian.[70] Back in the 1930s, Breuil had placed the Tayacian mid-way between the Clactonian and the Mousterian on his line of flake cultures; Louis Leakey, following similar reasoning, had preferred to call the Tayacian 'Clactonian IV'.[71]

As Movius, Lowe, Goodwin, Garrod, and others added their criticisms to the core/flake dichotomy, interpretations of the Palaeolithic changed. The sequence from western Europe was regarded doubtfully in the late 1940s and early 1950s. The patterns from this Continental cul-de-sac, which had previously led Palaeolithic classifications, were now starting to look like a tiny part

[68] Movius 1953, 164.
[69] Paterson 1941, 379; Movius 1948, 350, 409–410; 1953, 181–182.
[70] Movius 1948, 367; 1949, 1447–1448.
[71] Breuil 1932a, 571–572; Leakey 1934, 128–129.

of a much broader picture: at best atypical; at worse suspect.[72] Hawkes asserted in 1951: 'It is not enough, to-day, to have out-grown the one-track Palaeolithic scheme of before 1925: the two track scheme—"core" and "flake" of the 1930s has been transcended too'.[73]

In the reaction against this two-track scheme, different approaches to Palaeolithic research rose in reputation and popularity. Garrod, Goodwin, and Movius urged Palaeolithic researchers to move beyond a restricted interest in classifications and chronologies and try, instead, to interpret the diversity of former hominid activity. They suggested that the task of building chronological sequences ought to be the province of geologists, not archaeologists.[74] The role of the archaeologist was to explain change in tool-types and techniques in terms of the response of tool-makers to their environment; and to use the tools to try and understand the activities and relationships of their makers.[75]

In an attempt to distance themselves from past research, which was thought to have been overly reliant on an intuitive selection of Palaeolithic type-fossils, archaeologists tried to draw on the entire industrial assemblage in their interpretations (an 'assemblage' being a group of artefacts found together in a single area at a given site, discarded over a restricted period of time). Technological and statistical analyses became widespread through the 1950s. Such techniques seemed to be less subjective than earlier methods of extracting Palaeolithic patterns from stone tools; many archaeologists now hoped to measure the truth out of them.[76] Although the broader conception of industrial families and traditions persisted, this was now articulated in terms of a pebble-tool/chopper-tool tradition and a hand-axe-making tradition. References to core and flake cultures dwindled, and the term 'culture' began to return to one of its old meanings: the material reflection of a particular ethnological group.[77]

Inferences of time based on tool-types or techniques did not disappear, but they became seen as old-fashioned. Equally, an interest in the ancient behaviours that had produced these stone tools was not new, but such approaches attracted more attention once they had been characterized as a favourable alternative to chronology-building and 'the overworked conception of culture-contact to account for modifications and new introductions

---

[72] Childe 1944, 19; Lowe and Breuil 1945, 54; Caton-Thompson 1946a, 87; Movius 1948, 409.
[73] Hawkes 1951, 7.
[74] Goodwin 1946, 100; Movius 1949, 1446–1447.
[75] Childe 1944; Caton Thompson 1946a 88; Garrod 1946, 8–21; Goodwin 1946, 91; Movius 1948, 330–331, 349; McBurney 1950; Bordes and Bourgon 1951, 22.
[76] Oakley *et al.* 1948b, 81–82; Bordes 1950b; Bordes and Bourgon 1951; Spaulding 1953.
[77] Movius 1948, 410; Warren 1951, 130–132.

within a given stone industry'. This estimation of previous work was given by Gertrude Caton-Thompson (1888–1985), Fellow of Newnham College, Cambridge, and President of the Prehistoric Society during the War years, who had encountered her own difficulties with the Palaeolithic framework whilst working in North Africa and Egypt.[78]

Back in Britain, the Clactonian industry was looking rather different in light of these perspectives developed in Africa and Asia. The identification of 'Clactonian' industries around the world had driven Mary Leakey to protest that the Clactonian was a distinct regional entity. Despite his attacks on the core/flake distinction, Movius asserted that the Clactonian flake industry, which was mainly restricted to England, was separate and distinct from his Great Hand-Axe Complex.[79] Caton-Thompson, in her reaction against the idea of the Clactonian as a generic flake 'culture', suggested that, like many other industries, it might simply reflect 'environmental conditions, and the local need at that remote moment for one sort of artifact rather than the other'.[80] Nonetheless, the old conflict between interpretations of specific regional industries and general Palaeolithic patterns would continue, as the Clactonian was connected to the pebble-tool and chopping-tool traditions of Africa and Asia.

Warren's Clactonian core-tools, which had once sat uneasily between the lines of core (hand-axe) and flake industries, lost much of their strangeness in the late 1940s and early 1950s. After working in Asia, Movius recognized the importance of core-tools alongside the flakes of Warren's Clactonian. He had described the Chopper/Chopping-Tool Complex from Southern and Eastern Asia as a *core-tool* complex, despite the presence of flake-tools in these industries. The Western core-tool tradition, characterized by hand-axes, was contemporary with this Eastern complex. In the 1950s, Warren presented his Clactonian industry as an offshoot from the primitive African and Asian pebble-tool and flake-tool industries.[81] Oakley, who had once taken Breuil's part in the attack on Warren's side-choppers, conceded to Louis Leakey in 1950 that 'many of the Clactonian cores were chopper-tools'.[82]

Meanwhile, the varied hand-axe industries of Europe, formerly divided by Breuil into seven different phases, were becoming subsumed under a giant hand-axe tradition: initially termed 'Abbeville-Acheulian' and later, simply 'Acheulian'.[83] Oakley even suggested in the 1960s: 'the Clactonian flake-industries

---

[78] Caton-Thompson 1946b, 57.
[79] M. Leakey to Oakley (?1949 and ?1950): NHM: DF140/7; Movius 1953, 164.
[80] Caton-Thompson 1946a, 87.
[81] Movius 1948, 410–411; 1953, 166; Warren 1951, 109, 132; 1958, 128.
[82] Oakley to L. Leakey, 20 February 1950: NHM: DF140/7.
[83] Watson 1950, 28–29; 1956, 29–30, 1968, 42; Mason 1967, 766; Coles and Higgs 1969, 62, 85; Butzer 1972, 436, 448.

in all situations should perhaps be regarded rather as aspects or facies of Abbevillian-Acheulian culture'.[84] New generalizations persisted alongside attempts to rescue specific ethnological cultures from the old simplistic core/flake dichotomy; the old battles continued, though they wore new names.

10.5 CRITICISM OF THE BRITISH PALAEOLITHIC SEQUENCE BY GEOLOGISTS AND THE RISE OF NEW DATING TECHNIQUES

In the late 1940s, when archaeologists were turning away from former Palaeolithic classifications and handing the task of dating to geologists, the geologists were trying to detach their chronological conclusions from archaeological sequences. As the core/flake dichotomy was being criticized by Movius, Goodwin, Garrod, and others, the fine, time-specific industrial phases had come under attack in geological literature. In the 1940s and 1950s, geologists looked with suspicion at dates and interpretations that had been built on Palaeolithic foundations.[85]

Geologists often used the site of Baker's Hole, at Northfleet in the Thames Valley, as a cautionary warning about the dubious chronological value of Palaeolithic industries.[86] Spurrell had remarked on the presence of 'turtle-backed' flakes at Northfleet back in the nineteenth century (see Chapter 3). These had become known as Levallois flakes. In 1931, Breuil had made a more specific identification of the Baker's Hole tools: Levalloisian I–II.[87] This industrial phase had linked the Coombe Rock to the Riss glaciation rather than the Würm, and had helped to push the Thames Valley into line with Breuil's industrial–glacial framework (see Table 10.2).

In 1943, when Oakley turned out one of Spurrell's old assemblages from Baker's Hole at the British Museum (Natural History), he was shocked to find hand-axes amongst the Levallois flakes. Until then, everyone had taken Breuil's word that this was an early Levalloisian industry, but these hand-axes indicated a later date. They were a distinctive feature of Breuil's Levalloisian V in the Somme Valley, an industrial stage that was thought to reflect culture-contact between Levalloisian and Acheulian populations. Oakley explained the geological repercussions. The Coombe Rock now had to

---

[84] Oakley 1964b, 140; see also Oakley 1969, 226, 274, Note 114.
[85] Oakley *et al.* 1948b, 80; Movius 1949, 1446–1447.
[86] Zeuner 1944, 19; Oakley *et al.* 1948a, 1948b; Movius 1949, 1447; Bridgland 1994, 272.
[87] Spurrell 1884, 113; Breuil to Sollas, 24 August 1931: BGS: GSM1/445.

represent a later glaciation than the Riss: probably Würm I.[88] This threw the correlations of the Thames Valley and East Anglia into confusion.

There was, however, an alternative explanation that retained the original geological age of the deposit. Breuil defended the Rissian date of the Coombe Rock in 1947, hypothesizing that culture-contact had occurred *earlier* in England than in France: this newly-identified Levalloisian V stage in the Thames Valley was contemporaneous with the Levalloisian I–II stage in the Somme Valley. He added that such regional differences were to be expected with climate change and migration: an old solution for shoe-horning sequences into alignment.[89] But as Oakley observed: 'Such possibilities make many geologists distrust the use of archaeological evidence'.[90]

Doubts were also gathering over Breuil's industrial phases.[91] Oakley admitted in 1947 'that he was less confident than he [had been] in 1936 about the value of palaeoliths for close dating, in view of the cultural complexities within the Lower Palaeolithic, which recent studies had revealed'.[92] Oakley still had some faith in the chronological value of Palaeolithic industries, as he showed in his arguments about the date of the Baker's Hole Coombe Rock, but he reduced Breuil's seven Somme-based Levalloisian phases to only three British divisions: Early, Middle, and Late.[93] Zeuner had also turned away from previous 'unhappy attempts to base a chronology in part on conjectural views of human cultural evolution', as a reviewer observed with satisfaction in his report of Zeuner's latest book: *The Pleistocene Period, its Climate, Chronology and Faunal Successions* (1945).[94] The following year, in *Dating the Past: An Introduction to Geochronology* (1946), Zeuner warned: 'If we are to obtain a clear idea of the sequence, overlap, alternation and duration of the industries of the Palaeolithic, it is absolutely essential to keep apart the geological (and palaeontological) evidence for the climatic chronology from the typological classification of the industries of early man'.[95]

Oakley, suspicious of the Clactonian phases, had to unpick some parts of his old geological interpretation of the Thames Valley and its connections to the Clacton Channel. In 1951, he was in correspondence with Louis Leakey, who was working on a new edition of his book: *Adam's Ancestors*. Leakey suggested to Oakley that there was no good evidence for a discrete Clactonian

[88] Oakley to Arkell, 9 May 1943: UMO: WJA: C132; Oakley 1943, 31; Oakley and King 1945, 51–52; Oakley *et al.* 1948a, 23; Bridgland 1994, 272.

[89] Breuil 1947, 831; Breuil to Arkell, 1947: OUM: WJA: C128. Oakley had pondered a similar solution in 1943 (Oakley to Arkell, 23 June 1943: OUM: WJA C132).

[90] Oakley 1943, 31.

[91] Kelley to Oakley, 11 March 1949: BM(F): misc. correspondence; Oakley *et al.* 1948a, 23; Oakley 1952, 285; Movius 1953, 165; West and McBurney 1954, 145; Bordes 1956, 1.

[92] Hare 1947, 337.   [93] Oakley *et al.* 1948a, 23.   [94] Hollingworth 1947, 187.

[95] Zeuner 1946, 146.

I phase: the flakes were made by a technique common in Abbevillian industries, and no 'genuine Clacton tool types' seemed to be present.[96] In reply, Oakley agreed that there was no good evidence for Clactonian I, and wrote: 'I would go further, and say that it is rather doubtful if Clacton III is any more than a local facies of Late Middle or Upper Acheulian. [...] At any rate, I see no justification for recognizing two cultures in the Middle Gravels.' Oakley described the position to Leakey in summary form:

Clacton II	Acheulian III
Clacton II	Acheulian II
Clacton II	Acheulian I.[97]

This was a significant shift from Oakley's original interpretation. In the same format, the version he formulated with King back in the 1930s would have looked like this:

Clactonian III	Acheulian III–IV
Clactonian II	
Clactonian I	Acheulian I–II, Abbevillian.[98]

The interpretation of the geological history of the Thames Valley would also have to be changed. In the 1930s, Oakley had placed the sediments and industries in the following order: Swanscombe Lower Gravel (Clactonian IIA), Channels at Clacton and Grays Thurrock (Clactonian IIB), Swanscombe Middle Gravel (Clactonian III). A down-cutting event—the 'Inter-Boyn Hill Erosion Stage'—had helped him to explain how the Swanscombe Middle Gravel could be later in date than the Clacton Channel deposits, despite the lower level of the Channel.[99] Warren had come to accept Oakley's erosion stage (although he placed it in the 50-ft terrace rather than the 100-ft (Boyn Hill) terrace), but some geologists were less sure of this sequence.[100] In 1951, however, Oakley told Louis Leakey that he had 'given up the idea of the Clacton Channel representing a low sea level in the middle of the Swanscombe aggradation':

and, moreover, since the typological advancement of any particular 'Clacton II' industry appears to vary horizontally (from place to place) as much as it does vertically, I would not be prepared to say on typological grounds whether the Clacton Channel represents the low sea-level at the end of the Mindel glaciation or the low sea-level at the initiation of the Riss glaciation.

96 L. Leakey to Oakley, 17 January 1951: NHM: DF140/7.
97 Oakley to L. Leakey, 25 January 1951: NHM: DF140/7.
98 After King and Oakley 1936, 55, 60.
99 King and Oakley 1936, 57–60; Oakley and Leakey 1937, 240–242, 253–254.
100 Chandler in Bull 1942, 29; Warren 1942, 173–174; 1951, 130; Hare 1947, 337–338.

He closed his letter with a hopeful thought:

I think the question will eventually be settled by the pollen-spectrum at Clacton. The sequence of tree-pollen at the beginning of an interglacial is different from that in the closing stages. In September Warren, Godwin and I were engaged on boring into the Clacton beds with this research in view. The results will not be available for several months. I only hope they will be conclusive.[101]

Warren had been visiting the Clacton exposures regularly for several decades now, spending one fortnight there in June, and another in September.[102] He had initiated this project of pollen identification at Clacton in 1950 with Oakley and Harry Godwin (1901–1985), Reader in Quaternary Research at the University of Cambridge, and Godwin's research student, Kathleen Pike. Godwin had graduated from Cambridge in 1922 with a degree in Botany and did much to further this subject at the university. In 1948, he became the first Director of the Sub-Department of Quaternary Research, which he had helped to establish in the hope of encouraging collaboration between geologists, archaeologists, and palaeobotanists.[103]

British geologists had used the remains of plants to date and correlate Quaternary deposits since the nineteenth century. Reid, an enthusiastic palaeobotanist, had relied on plant fragments to identify climatic waves in the Hoxne sediments during the 1890s.[104] Pollen preserved another reflection of climate change. Different types of forest had spread and declined over Europe for millennia, as the climate grew warm and then cooled. The wind blew their pollen across the land, and it survived in waterlogged deposits with the potential to reveal these old forest patterns. Each tree species produced a distinctive shape of pollen grain, which could be distinguished under the microscope. In the early twentieth century, however, the study of pollen was restricted largely to the post-glacial period.

Godwin felt that pollen-analysis offered a solution to many of the problems of correlation faced by geologists. In 1941 he urged British researchers to follow the example of their Continental peers and apply this technique to older, interglacial sediments.[105] Palaeobotanists on the Continent had produced detailed vegetational histories from pollen in the 1920s, stretching back through several interglacials. The most influential of these early schemes was developed in 1928 by Jessen and Milthers in Denmark. They used differences

---

[101] Oakley to L. Leakey, 25 January 1951: NHM: DF140/7.
[102] Warren to Baden-Powell, 24 April 1951: UMO: DB-P: K146.
[103] Godwin, 'The Development of Quaternary Research in the University' 1943: BLO: CH, 95; Edwards, Clark *et al.* 1986, 301; Turner and Gibbard 1996, 378. On Godwin's life and research, see West (1988).
[104] Read 1896, 406–411.   [105] Godwin 1941, 329.

in the abundance of tree species to distinguish between different interglacials in northern Europe.[106]

In 1930, Godwin and his wife Margaret had started to apply the pollen-analysis techniques developed by Lennart von Post in Sweden to British deposits, encouraged by Albert Charles Seward (1863–1941), Professor of Botany at Cambridge. By 1950, it seemed from the work of Godwin, Paul Woldstedt, and others that the different interglacials in East Anglia had different vegetational signatures, which could indeed help to correlate different deposits. This was the background to the pollen-sampling project at Clacton.[107]

One of the main reasons for retrieving pollen from Clacton in 1950 was to give the Clacton Channel deposits a more secure and precise position within its interglacial episode. Louis Leakey struck a fortunate moment when he sought Oakley's opinion about the correlation of industries and glaciation in 1951; Oakley and Warren had both been considering their views about the correlation of the Clacton Channel in preparation for Godwin's publication of the pollen spectrum.[108] In one of his letters, Leakey queried the likelihood of a down-cutting event in the middle of an interglacial. Warren wrote to reassure Oakley: 'I still think your original suggestion for the position of the Clacton bed is the most probable,' but received the following reply:[109]

I am inclined to leave the problem of the relation of the channel to the Swanscombe terrace as an open one. I have not abandoned my original theory to the extent of suggesting that it can't be right; but I want to emphasize that it is by no means proved, and that it is even possible that the C. channel is later than the Middle Gravels.[110]

But a few months later, when Godwin published his interpretation of the pollen profile at Clacton, Oakley's original, unorthodox view of channel-cutting seemed to be confirmed. Pike and Godwin read their report to the Geological Society at the end of 1951. They dated the Clacton Channel to the middle of the Mindel-Riss interglacial.[111] Regardless of this outcome, Oakley had nonetheless been shaken in his geological opinion by uncertainties about the Palaeolithic sequence; his support for Godwin's pollen work had been strengthened by a desire for an independent criterion of geological time.

---

[106] Hollingworth *et al.* 1950, 503; West 1981, 129–130.
[107] Godwin 1941, 354, 358; Hollingworth *et al.* 1950, 503–506; Woldstedt 1950, 1002–1003; West 1981, 130; 1988, 271–278, 282–284.
[108] Warren to Oakley, 28 January 1951: NHM: DF140/7; Pike and Godwin 1952, 262, 267–268.
[109] Leakey to Oakley, 17 January 1951: NHM: DF140/7; Warren to Oakley, 3 February 1951: NHM: DF140/7.
[110] Oakley to Warren, 12 February 1951: NHM: DF140/7.
[111] Pike and Godwin 1952, 268–269. See also the correspondence between Baden-Powell, Warren, and Godwin in UMO: DB-P: K146.

According to Kennard, even Breuil had changed his old views about the Palaeolithic. In 1948, he wrote to a fellow palaeontologist and shell expert: 'I am afraid that I have some bomb-shells for you. The first is that the Abbé Breuil has given up his scheme of typology. He told Marston "Geologists must tell us the age of the beds" if he had said Palaeontologists he would have been nearer the truth'.[112] But Breuil would have no qualms about re-publishing descriptions of culture-contact between his old industrial phases in the late 1950s and early 1960s.[113]

A coarse typology of tools continued to be used by some British geologists to date deposits. Oakley, for example, still believed that Acheulian hand-axes had become more refined in their workmanship through time, although when he used these ideas to date a tool-bearing channel at Caversham, he was criticized by fellow geologists for 'incessant wavering of typological judgement'.[114] But the finer industrial phases sank back into obscurity, and geologists preferred to use other timescales to correlate the sediments of different sites. Richard West, who joined Godwin at Cambridge as a research student in 1951, built on Godwin's early pollen work at Clacton: identifying interglacial stages in East Anglia and developing a sequence of pollen zones from sites like Hoxne and Marks Tey.[115] Shells and bones continued to play an important role in correlation, particularly for sites without reliable pollen-profiles.

The names of the Alpine glaciations, which had been strongly associated with Breuil's industrial–glacial scheme of the 1930s, continued in use into the 1960s.[116] They were, however, becoming rare; tired of endless arguments about Alpine correlations, geologists replaced them with various other terms. After Woldstedt demonstrated a connection between the glacial sequences of East Anglia and north Germany (assisted by pollen-analysis), glacial episodes were given the names of glaciations from north-west Europe: Elster, Saale, and Weichsel.[117]

William Joscelyn Arkell (1904–1958), a geologist best known for his work on the Jurassic, wanted to assign important local areas their own glacial nomenclature until a single scheme could be developed. 'Nothing but harm can come of attempts to twist local evidence in aid of a general hypothesis', he wrote in 1951.[118] For the main drift sheets in the Thames Valley and the

---

[112] Kennard to Jackson, 26 January 1948: BMD: JWJ, Box 48. On Kennard's continued defence of a single glaciation, late in the Quaternary, see Kennard 1944, 160, 168.

[113] Breuil and Lantier 1965, 112–115.

[114] Wooldridge 1957, 8. See also Treacher, Arkell *et al.* 1948, 153; and West and Donner 1956, 89.

[115] Pike and Godwin 1952; West 1954; 1956, 302–340; Turner 1970.

[116] Oakley 1964b, 24, 28; Lacaille 1964, 109.

[117] Woldstedt 1950; West 1956, 339. See the correspondence between Woldstedt and Baden-Powell, 1948–1951: UMO: DB-P: K156.

[118] Arkell 1951, 19.

Midlands, Arkell chose the names 'Berrocian', 'Catuvellaunian', 'Cornovian', and 'Cymrian', inspired by the strange similarity between the area occupied by the Eastern Drift and the Catuvellauni: the tribe discovered north of the Thames by the Romans. He named his interglacials after hand-axe industries: 'Abbevillian', 'Middle Acheulian', and 'Micoquian'. Although Oakley thought this an excellent idea, the terms did not become widely used.[119]

Arkell thought that East Anglia also ought to have its own glacial terminology. He discussed the problem with Baden-Powell, who had been using the direction of ice-flow to confirm the sequence in this area. Baden-Powell hoped that definite geological correlations would allow a return to a standard scheme, and tried to avoid new terms. Reluctantly, however, he gave the names 'Lowestoft' and 'Gipping' to the Lower and Upper Chalky Boulder Clays from these famous localities. The great interglacial episode between the two, which was connected with the implements at Hoxne, the Clacton Channel, and the 100-ft terrace—the interglacial that had been matched to the 'Riss-Würm' in the 1920s and the 'Mindel-Riss' in the 1930s—became described as the 'Hoxnian'.[120]

Then there were other ways to describe climatic change that avoided reference to terrestrial glaciations. A marine sequence of oxygen isotope stages soon overtook the old Alpine sequence of glaciations. The discovery of these stages is explained clearly by John and Katherine Palmer Imbrie in their book: *Ice Ages: Solving the Mystery* (1979). The sediments that gathered on the sea floor preserved an indication of climatic change in tiny shells of foraminifera. As these small creatures swam on the surface waters of the sea, they converted the oxygen in the water to construct their calcareous shells. When they died, they sank to the ocean floor and were incorporated in growing layers of sediment. In December 1946, Harold C. Urey, from the Institute of Nuclear Studies at the University of Chicago, demonstrated that the ratio of two oxygen isotopes—O^{16} and O^{18}—varied with the temperature of the water. He thought his calculations might be of interest to geologists. If these proportions could be extracted from the sequence of foraminifera, they would reveal a pattern of ancient climatic fluctuations.[121]

---

[119] Arkell 1943, 150–153; West 1981, 129. See also Arkell to Oakley, 15 June 1943 and Oakley to Arkell, 16 June 1943: UMO: WJA: C132; Arkell to Baden-Powell, 5 June 1944 and 3 November 1947: UMO: DB-P: K4.

[120] Baden-Powell 1948, 280; West 1954; 1956, 270, 339; 1963, 165; King 1955, 206–207; Clark Howell 1965, 1044; Coles and Higgs 1969, 205, 213–214. See also Baden-Powell to Arkell, 10 March 1944: UMO: DB-P: K4; Baden-Powell to West, 21 March 1954 and 3 November 1954: UMO: DB-P: K152; Baden-Powell to Warren, 20 February 1955: UMO: DB-P: K149.

[121] Urey 1947, 578–581; Oakley 1964b, 41–45; Imbrie and Imbrie 1979, 135. On the development of Quaternary dating methods, see also Oldroyd (2006).

Urey's isotopic work coincided with the development of new deep-sea boring techniques. Cores had been collected from the ocean floor for decades, most famously by the *Challenger* Expedition in the 1870s, but their equipment could only penetrate a few feet into the sea bed. In 1947 the problem of internal friction was solved by a Swedish invention: the Kullenberg piston core sampler. Piston corers could soon extract fifteen to thirty metres of sediment from the sea floor, reaching down to the older foraminifera of early Palaeolithic times and beyond to the Pliocene.[122]

As cores were collected and analyzed around the world, a scheme of climatic stages emerged based on oxygen isotopes, which was founded on the numbered core stages developed by Cesare Emiliani in 1955. Soon, however, it was realized that oxygen isotope studies were revealing more than palaeotemperatures. They reflected the volume of glacial ice, and provided an index of past glacial episodes. When ice-sheets had gathered on the land they had locked up a high proportion of the lighter isotope O^{16}, and the concentration of O^{18} had risen in the oceans. This glacial record from the sea offered a welcome alternative to the scattered terrestrial deposits: oxygen-isotope stages gave a global perspective on climatic change.[123]

Two problems still remained: how to correlate cores over long distances, and how to date their glacial episodes. In the 1960s, the geomagnetic time-scale provided a solution to the first problem. Bernard Brunhes, the French geophysicist, had discovered back in 1906 that the iron-rich minerals in cooling lava aligned themselves to the earth's magnetic field. He realized that ancient lava would preserve the direction of the earth's magnetism at the time when it had solidified. After studying several lava flows, he found that the magnetic field had been reversed in the past. Motonori Matuyama reported several such events from the lava flows of Japan and Korea, but it was not until 1963 that the hypothesis of magnetic field reversal was finally accepted. The deep-sea cores were also found to contain minerals that retained magnetic signatures.

Meanwhile, a new technique of dating lavas began to be applied in the late 1950s: the potassium-argon (K/Ar) dating method. This was a radioactive clock that started when lava solidified. It ticked at a constant rate as K^{40} decayed to A^{40}, which became trapped in the rock. K^{40} had a very long half-life, and K/Ar dates could reach back for millions of years. Researchers working on the lava-rich Palaeolithic sequence in Africa welcomed this new radiometric dating method, which allowed them to date their sediments directly. In the early 1960s, the magnetic-reversal sequence was supplied

---

[122] Wiseman and Ovey 1950, 49–50; Shackleton 1975, 2; Imbrie and Imbrie 1979, 123–126.
[123] Emiliani 1955, 565, Table 15; Shackleton 1967; 1975, 4–6; Imbrie and Imbrie 1979, 135–140.

with a K/Ar timescale, and these dates were transferred to the glacial episodes recorded in the deep-sea cores.[124] From the mid-1960s, oxygen-isotope stages began to replace the old master-sequences of glacial–interglacial cycles. Terms like 'stage 5' or 'sub-stage 5e' appeared in publications instead of the old glacial terminology. This development did not, however, end controversy about correlation.[125]

The downfall of Breuil's scheme has been linked by Robin Dennell to the 1960s, and to the advent of absolute dating techniques. But criticisms of the core/flake dichotomy and Breuil's detailed sequence of industrial phases were being made in the late 1940s and early 1950s. This was some time before K/Ar, palaeomagnetic, and radiocarbon dating became widely used.[126] The shift in interpretations of the Palaeolithic did not follow directly from changes in the way geologists dated their deposits; geologists and Palaeolithic archaeologists seized these approaches of dating and interpreting the British Palaeolithic as they searched for alternatives to an outmoded typological classification.

10.6 DIVISIONS IN PALAEOLITHIC RESEARCH: THE END OF AN ERA?

This account of the search for chronological patterns in the British Palaeolithic ends here, in the late 1950s: a good point at which to stop. The rise of these different dating techniques would alter the terminology and foundation of arguments about Palaeolithic time. Their tone had already changed after archaeologists cast the problem of dating into the hands of geologists. After a longstanding mutual interest in classifications and chronological sequences became clouded with doubt in the late 1940s and early 1950s, geologists distanced themselves from the use of industries as time-markers and relied on other guides to age. Archaeologists self-consciously avoided classifications and chronologies, and emphasized the importance of interpreting the behaviour behind the tools and the reactions of tool-makers to their environment.

This partitioning of roles was stimulated in part by criticism of previous Palaeolithic research, but it also reflected the emergence of a distinct identity

---

[124] Dalrymple and Lanphere 1969, 43–51, 209–213; Shackleton 1975, 6–21; Imbrie and Imbrie 1979, 147–152; Berry 1987, 161–165.

[125] Coles and Higgs 1969, 25–26; Isaac 1975, 876. The controversies about correlation between marine and terrestrial sequences are mentioned by Bridgland 1994, 194.

[126] Oakley 1964b, 8; Dennell 1990, 555; Sutcliffe 1998, 130. On the role of Arthur Holmes in the early development of radiometric dating (though for periods far more ancient than Quaternary times), see Lewis, *The Dating Game. One Man's Search for the Age of the Earth* (2000).

amongst archaeologists and the growing specialization of geological research. In the past, Palaeolithic researchers in Britain had emerged from a broad spectrum of backgrounds with wide-ranging interests: many might, like Warren, have described themselves as geologists. By the mid twentieth century there were more opportunities for professional training and academic careers in prehistoric archaeology at the universities. Meanwhile, the research interests of geologists were contracting. Their discipline had gone through a similar process back in the 1870s, and was dominated by professionals by the end of the First World War.[127]

Many of the Palaeolithic researchers and geologists introduced in previous chapters were self-taught and (for much of their lives) had no paid employment in the subject. They included Hinton, Kennard, Burchell, Chandler, Haward, Warren, Sainty, Moir, Sturge, Treacher, and Peake. Their most prominent publications clustered in the earlier decades of the twentieth century. Boswell, Solomon, Oakley, and King, however, all studied geology as undergraduates and entered paid geological employment.

The growth of specialization in Quaternary research is seen in the rise of interdisciplinary centres that sought to re-integrate now-disparate subjects. The Sub-Department of Quaternary Research at Cambridge, for example, had opened in 1948 under Godwin's directorship. In 1966, he was succeeded by West, who had studied botany under Godwin as an undergraduate and later lectured in the Botany Department. By then, West was considered to be exceptionally broad in his interests, which ranged from botany to glacial geology.[128]

Meanwhile, the University of Cambridge had emerged as a major centre for training and research in prehistoric archaeology.[129] Many of the Palaeolithic researchers who regarded themselves as 'archaeologists' had a connection to Cambridge. Burkitt and Garrod attended Cambridge as undergraduates (Burkitt took Natural Sciences; Garrod studied history) and returned to teach prehistory after experiencing the tutorship of Breuil abroad. Louis Leakey took archaeology courses as an undergraduate at St John's College and continued his Palaeolithic work as a Research Fellow. Paterson followed a similar route at Trinity College, and Caton-Thompson carried out much of her research as a Fellow of Newnham College. These Cambridge archaeologists had also made a sizeable contribution to the attacks on Palaeolithic classifications in the 1940s and 1950s. Leakey, Paterson, Garrod, and Caton-Thompson all criticized Breuil's Palaeolithic scheme, and called for a different approach to the past. Another Cambridge researcher, Charles McBurney

---

[127] O'Connor and Meadowes 1976, 78–79, 87; Porter 1978, 810, 835–836; O'Connor 2005.
[128] Turner and Gibbard 1996, 376.   [129] Clark 1989; Smith 2004.

(1914–1979), promoted new geographical methods of interpreting the Palaeolithic: he had read archaeology at Cambridge as a third-year under-graduate in 1935.[130]

The days of Warren and Burchell, of Hinton and Kennard, and others, were coming to an end. The career structure of geology and, increasingly, of archaeology, encouraged practitioners to focus their interests in specific, rigid directions. Specialist academic journals multiplied as professional research narrowed. Collaborations between specialists had characterized geological and Palaeolithic research since the days of Brixham Cave, when Falconer was intending to take the bones in hand, and Prestwich the gravels, whilst the shells and flint tools were handed out to other experts.[131] But when specialist professionals turned their efforts towards specific research projects, they became less interested in the kind of detailed local studies carried out constantly over the course of several years by researchers like Pengelly and Warren. Changes in gravel-working techniques offered less potential for Palaeolithic discoveries. The constant observations of building developments, drainage works, coastal erosion, and other temporary exposures of Palaeolithic deposits in a single region—once regarded as a standard part of Palaeolithic research—became the province of 'amateurs', the term acquired its present-day stigma, and its representatives became marginalized from university-based research. In 1949, Baden-Powell observed: 'Today my impression of Norfolk and Suffolk, not to mention other areas, is that this amateur race is nearing extinction. I do not know of more than half a dozen people in Britain who are collecting palaeoliths systematically, and large numbers of pits remain unvisited for years at a time.'[132]

As Palaeolithic archaeologists and geologists distanced themselves from Palaeolithic timescales in the 1940s, some of the old polymathic Palaeolithic researchers eyed the new breed of professionals with annoyance. Kennard, though in his seventies, was still spending long hours sieving material at Swanscombe. The tone of his complaints, though coloured by irritation, recall a style of working that had characterized Palaeolithic research since the 1860s, and which he feared was now in decline: 'I don't see the young 'uns coming on. [...] What we want is local observers & the landed folk that used to do this sort of thing don't do it now & the trained men in office want the stuff brought to them & do not work [...] They know everything & talk like God Almighty to a blackbeetle'.[133]

---

[130] McBurney 1950, 163; Roe 2004b.
[131] Pengelly to Prestwich, 21 July 1865: GSL: 8, p.4.
[132] Baden-Powell to Burkitt, 15 September 1949: UMO: DB-P: K20. On the contributions of amateurs to Palaeolithic archaeology, see Roe 1981.
[133] Kennard to Jackson, 10 (?September) 1942: BMD: JWJ, Box 48.

Conclusion

'I believe [...] in the fundamental interconnectedness of all things'
(Dirk Gently, in Adams 1993, 118).

Ideas about the British Palaeolithic and its connections to geological time changed enormously between the days of the early eighteenth century, when Bagford wondered whether the implement found by Conyers at Gray's Inn Lane had been left by an Ancient Briton near the bones of a Roman elephant, and the century covered in this book. In the hundred years that followed the acceptance of human antiquity—between c.1860 and c.1960—similar tools were scrutinized by many other interested eyes; they were labelled and classified, and their age and meaning were vigorously debated. In the present study, I have provided a picture of changing ideas about British Palaeolithic tools and their place in geological time, and have also tried to recover the excitement of the arguments that swept through this century of geological research and its little-known relations with archaeology. Views of the past were not built up by dispassionate authorities, coolly observing the range of available and expanding data; the gaze of each individual was restricted by different questions and expectations that encouraged them to describe, interpret, and defend different patterns in the ancient stone tools of Britain.

It is now time, before closing this chapter on the history of British Palaeolithic research, to stand back and take a broader look at some of the reasons for these differing beliefs and for their varying success. But first, a recapitulation is offered of the major developments. In this summary, presented below, the Gray's Inn Lane implement is followed through time to highlight the changes in perception of the Palaeolithic. During the latter part of the tale, this pear-shaped 'hand-axe' found by Conyers is accompanied by the Clactonian industry, which has supplied a more familiar anchor point for the shifting interpretations described in previous chapters.

A CENTURY OF RESEARCH ON THE BRITISH
PALAEOLITHIC, *C*.1860–*C*.1960

Human antiquity was widely accepted in learned circles after they heard the famous papers of 1859. But it was a few years more before the hand-axe from Gray's Inn Lane became described as the contemporary of many extinct prehistoric animals and assigned to post-submergence, post-glacial times. The work of geologists was central to the task of placing such implements more precisely in Britain's distant past. Chapters 1 and 2 described how geologists had been attracted in increasing numbers to the once-unpopular drifts and the bones and tools that they preserved. Officers of the Geological Survey played an important part in refining the drift sequence: a task that was not necessarily part of their official duties or of any particular interest to their superiors. The study of Quaternary drifts was also carried out in the spare time of researchers who had no paid geological employment, such as Wood or Prestwich (in his earlier decades). As the century drew on, more young geologists entered the field with university experience, and the number of geological posts in universities increased steadily.

Chapters 3 and 4 took up the tale from a Palaeolithic perspective. Through the latter decades of the nineteenth century, more hand-axes and other implements were discovered and discussed by a varied group of interested individuals, fighting off forgers and curio-collectors in the busy gravel pits of southern England. These artefacts were distinguished from each other in a range of early classifications. The Gray's Inn Lane implement seemed to belong to the River-Drift people. Lubbock was one of many to assign such tools to the Palaeolithic period (in its restricted, nineteenth-century sense), and to distinguish them from those made in the later Reindeer period (which tended to be associated with the Cave peoples of southern France described by Lartet and Christy).

Other researchers made more precise, though less well-known, estimates of the age and character of the Gray's Inn Lane implement. Brown thought that this hand-axe was one of the older river-drift palaeoliths; Evans placed it in his class of pointed weapons, where it was distinguished from the later ovate hand-axes and flint flakes; and it might have been included in the second class of the classification developed by Worthington Smith. In Chapters 5 and 6, it became apparent that supporters of Eolithic and Pre-Palaeolithic stages also managed to divide their finds into distinctive types and arrange them along progressive sequences—finds which are now generally accepted to have been produced by the forces of nature rather than by the hands of ancient tool-makers—an achievement that illustrated the power of expectation and clarified the major principles of Palaeolithic classification.

Chapters 7 and 8 described the emergence of a standard industrial terminology for British implements in the early twentieth century, when classifications from France and Belgium were adopted but also adapted. Despite the prominence of Gabriel de Mortillet in histories of prehistoric research in the nineteenth century, it was only in the twentieth century that his Palaeolithic labels ('Chellean', 'Acheulian', and 'Mousterian') began to pepper British works. Researchers like Hinton and Kennard, who worked on the tools and terraces of the Thames Valley, helped to strengthen the connections between the age and the character of industries; such links encouraged the acceptance of these generalized, widely applicable terms. The work of Commont in the Somme Valley gave stronger confirmation to de Mortillet's scheme, and the acceptance of a standard industrial sequence in Britain was reinforced by the work of Reginald Smith and Dewey on the classic Swanscombe section in 1912 and 1913. They made enthusiastic reference to Commont's work in their conclusions. Tools like the Gray's Inn Lane implement, when found at Swanscombe and at other sites, now became known as 'Chellean'.

The Swanscombe collaboration also reflected the scattered position of Palaeolithic research in the early twentieth century. Smith and Dewey were prepared to take a holiday to excavate at Swanscombe if their respective institutions (the British Museum and the Geological Survey) proved uninterested in the project. Prehistoric studies had gained a base at the PSEA, but this was a local, eolith-centred society in its earlier years. Palaeolithic research was still carried out by range of researchers and shared in a variety of forums, as in times past. The close connections to geological research would also continue for a few more decades.

One of the most absorbing developments in Palaeolithic research occurred when the tools themselves became another chronological guide. The President of the Geological Society had asked, in 1913, whether a sequence of implement types could be established to enable geologists to use implements as zone-fossils.[1] The grouping of tools into sets (or industries) with standard names, the connection between their age and character, and the sequence found at Swanscombe all contributed to the answer, 'Yes'. Interpretations of geology and of the Palaeolithic were now tied together firmly. But reports of anomalous Palaeolithic industries multiplied in the wake of these expectations: some fell in the wrong place in the expected sequence; others failed to exhibit the expected character for their age. Warren's industry from Clacton offered one example of such an anomaly.

---

[1] Strahan in Anon 1914, ii.

By the early 1920s, Palaeolithic tools were being connected to a more complicated glacial chronology. The Chalky Boulder Clay, once such an important division between pre- and post-glacial times, was no longer regarded as the last reflection of glaciation in East Anglia: tools had also been recovered from older drifts. The four-fold Alpine scheme, developed by Penck and Brückner in the first decade of the twentieth century, had become widely used on the Continent for industrial correlations. The British river-drift tools, likewise, were assigned a position in the Alpine glacial sequence and became dispersed through several glaciations. The Gray's Inn Lane implement, once regarded as 'Chellean' and post-glacial, was now more commonly referred to as 'Acheulian' and 'Riss-Würm.' Towards the end of the 1920s, however, suspicions were growing that industries might not serve as reliable zone-fossils for geological correlation if, as Peake noted, 'we have evidence of successive stages of two industries, a core industry and a flake industry'.[2]

Some suggested that the anomalous flake industries, which dated to the time of the hand-axes (or 'core' implements), might be swept up into their own cultural sequence and placed *alongside* the hand-axe cultures as their contemporaries. This suggestion was famously articulated by Breuil in the 1930s. Breuil also offered a solution to Peake's problem. Chapter 9 explained how the finer industrial sub-divisions (or phases) of Breuil's Somme-based industrial–glacial scheme managed to retain the chronological value of industries. The Gray's Inn Lane implement had then become known as Acheulian 'I' or 'II,' and had been given a clear position in time—both in relation to the industries on other cultural lines such as the Clactonian, and in relation to the glacial–interglacial cycle.[3]

Following the example of Breuil, British industries were classified according to these new standard labels. Boswell urged researchers in 1936 to look again at old definitions of flake industries, previously dominated by the umbrella term 'Mousterian,' and to ascertain their stage more precisely if they were to be used for correlation or dating. The 'Strépyan' industry from Swanscombe, Warren's 'Mesvinian' from Clacton, and other flake industries were soon described and classified as discrete phases of the 'Clactonian' industry. This new diversity of industries spread further back along the sequence of Alpine glaciations, and the Gray's Inn Lane hand-axe became widely associated with an earlier interglacial period: the Mindel-Riss.

Chapter 10 described how the varied sediments and industries of southern England were incorporated into this new industrial–glacial scheme during the 1930s. The sequences of deposits in the Thames Valley and East Anglia received

---

[2] Peake 1930, 382.   [3] Warren 1924b, 279, footnote; 1926, 41; 1932a, 7.

particularly close attention as old correlations were re-shuffled. Oakley even included an 'Inter-Boyn Hill Erosion Stage' in his account of the geological history of the Thames Valley, to maintain the expected sequence of Clactonian industrial phases. In the 1940s, however, doubts began to gather about the patterns in age and character described for the Palaeolithic of Britain and, indeed, for the world.

Many of these problems were highlighted when Paterson tried to adapt the old industrial scheme to incorporate regional variation as well as generalized global patterns. Like many others, Paterson used the idea of industrial hybridization or 'culture-contact' to accommodate local variety, but this explanation attracted criticism. Some pointed to differences between the Palaeolithic sequence from the Somme Valley and the Thames Valley: differences that suggested that the fine sequence described for the Somme had a far simpler reflection in Britain. Others, working in Africa, Asia, and the Middle East, complained that the two contemporaneous flake and hand-axe families might be visible in Europe, but it was difficult to discern or impose the same pattern elsewhere in the world. There was dejection and confusion about the meaning of Palaeolithic labels: were they reflections of different biological peoples or ethnological groups, or were they merely chance similarities of technique? Above these questions loomed greater uncertainties and doubts about the very aims of Palaeolithic research. Suspicious of the complicated knot that had developed between the character and the date of an industry, researchers asked different questions of the geological and Palaeolithic pasts.

The burden of dating now began to fall squarely on the geologists, who had also become more cynical about the chronological value of industrial phases. Oakley looked again at his interpretations of the geological sequences in the Thames Valley and at Clacton, and turned to the new pollen work to verify his correlations. The time-scale provided by oxygen-isotope stages from marine-cores also rose in popularity, despite controversy about the connections between these marine-based climatic cycles and terrestrial sequences of glacial cycles, river sediments, pollen, and fauna. The old industrial labels for the British Palaeolithic became simpler. The use of tools as time-markers did not die away, but localized industrial variety tended to be explained in terms of the environment in which tool-makers lived, rather than as culture-contact between groups. Industrial patterns from around the world were grouped together in techno-complexes and, by the 1960s, the Gray's Inn Lane hand-axe was swept up with representatives of other, once-specific, industries as part of a generalized Acheulian 'tradition'. Chapter 10 ended by describing how self-proclaimed archaeologists, many of whom now had professional training and posts in universities, were self-consciously turning away from attempts to order and classify tools and retreating from the style of research carried out by the old geologist/Palaeolithic polymaths like Warren and Kennard.

These, then, were the major developments in the classifications and chronological sequences that were suggested for British Palaeolithic tools between 1860 and 1960. Of course, behind this simple narrative of the rise and fall of different sequences there lay a more complicated maze of different beliefs, interests, and questions. Although the emphasis of this book has rested on a detailed history of the period rather than the philosophy of science, it is important to look now, albeit briefly, at *why* knowledge developed in that way, and to try and extract some patterns of knowledge formation and scientific change from this detail. Before closing, this concluding chapter will explore some of the reasons behind the different approaches to and beliefs about the Palaeolithic, ask why some opinions achieved wider recognition than others, and unpick a few of the complicated connections between explanations and expectations of Palaeolithic classifications.

THE VARIETY OF SPECIALIST INTERESTS
BROUGHT TO PALAEOLITHIC RESEARCH

One obvious source of different beliefs about and interests in the British Palaeolithic lies in the diversity of individuals who converged on the subject. They arrived from a variety of social and institutional backgrounds, bringing with them a range of specialist interests. The task of developing classifications and chronological sequences for these stone tools, though a central theme of this book, was not necessarily of central concern to those who influenced the answers to this problem. They approached the subject in different ways and saw different patterns in the Palaeolithic, but such differences did not always lead to conflict or to a change in ideas. Evans, for example, was better known for his vast catalogue of finds from around Britain than for his few brief comments on classification; Worthington Smith's work was inspired by discoveries in the more restricted area of London; whilst Spurrell was famous for reconstructing the fine detail of specific tool-making episodes. All three described different patterns, but rarely attacked each other on that account.

Some of the strongest disagreements about the sequence of Palaeolithic tools can be traced to the work of geologists, many of whom contributed to Palaeolithic research through their desire to understand order in the bones, river drifts, and glacial deposits of the Quaternary period. Driven by beliefs in different geological patterns and interests in different geological questions, they defended conflicting interpretations of the Palaeolithic. An understanding of those beliefs has, in some cases, helped to explain why Palaeolithic research took the direction it did. The attention devoted to geological researches in these chapters might seem surprising to readers who are more used to classing the Palaeolithic with 'archaeology' than with 'geology,' but the histories of the two disciplines were intertwined.

Many interpretations examined in these chapters would look strange if they were removed from the context of geological debate. To take one example, the account of Cave and River-Drift peoples outlined by Dawkins in *Cave Hunting* (1874), described in Chapter 4, appears unclear and confused until viewed in light of his work on the classification of Quaternary mammals. And, just as Dawkins had brought the movements of tool-making groups into his faunal arguments, Geikie linked them to his views about glacial cycles in *Prehistoric Europe* (1881). The behaviour of ancient tool-makers presented by these two antagonists was strongly coloured by their opposing geological views.

Ever since the late nineteenth century, research on the British Palaeolithic has been compared and contrasted to work in France, usually to the detriment of the former. Often has it been said that France dominated Palaeolithic research in the late nineteenth and early twentieth centuries, with its prolific caves and the Somme Valley sequence. But this ignores the detailed scale and geological context of much British work. The British record was not being ignored as the French prehistorians composed their famous syntheses; researchers like Worthington Smith were merely writing more detailed articles. Meanwhile, the Palaeolithic was also attracting the attention of geologists as they examined the complicated drifts of southern England. For these geologists, one of the most interesting outcomes of the Swanscombe excavations was to suggest that stone tools could be used as zone-fossils to date geological sediments. This had important implications for the direction of future research in geology as well as Palaeolithic archaeology.

Even French researchers held the British boulder-clays and periglacial deposits in some esteem; Breuil was not the first to lean on the observations of British geologists when developing the chronological key for his famous industrial–glacial sequences. One reason for the great interest that Breuil's scheme attracted in Britain lies in the despair of certain geologists who, steeped so long in problems of correlation, hoped that these finely dated industrial phases might guide them through their maze of geological difficulties. The close connections between geology and Palaeolithic archaeology are particularly evident in the remarkable attempts made during the 1930s to reconcile British geological patterns to Breuil's industrial–glacial scheme. As in previous decades, numerous opinions about the British Palaeolithic could be traced to beliefs and interests cultivated through research on the sediments and fauna of Quaternary Britain.

THE SOCIAL CONTEXTS OF PALAEOLITHIC RESEARCH

Another important and obvious influence on views about the British Palaeolithic lay in the social contexts within which ideas were conceived, delivered, and judged. Ever since 1859, when the case for human antiquity was presented to the leading British societies, it had been evident that tone and rhetoric could be as important as content in advancing a position. The selection of particular forums for debate, journals for publication, specimens for display, and individuals for persuasion was rarely accidental. The strategies used by Moir and Lankester as they tried to gain a favourable hearing for their pre-palaeoliths, though perhaps exceptionally blatant, were not unusual. The attention paid to Breuil's pronouncements on this topic, and his later hold over interpretations of the Palaeolithic, owed much to his respectable position in learned societies and institutions. Many of the tools identified in Britain and placed in their expected position in Breuil's sequence were identified *by Breuil himself*—and the labels that he assigned to them were valued the more for this.

The institutions that hosted research could also curtail the range of questions asked of the Palaeolithic. Researchers directed their stances to match the interest and scope of specific learned societies, and chose those societies to compliment and reinforce their stances. Some leaned more towards geology; others towards archaeology or palaeoanthropology; some embraced research on a national scale, others worked on local details. The interests of individuals could conflict with the aims of the institutions where he or she worked. The Geological Survey, for example, employed several of the researchers introduced in these chapters, but did not always encourage their work on the superficial drifts or the Palaeolithic. Long after drifts had joined solid geology as a legitimate field for mapping, Dewey would be instructed to stop working at Swanscombe in his official Survey time. As the twentieth century drew on, the questions asked by expanding numbers of professional archaeologists in university departments would also be directed down the increasingly specialist channels of approved academic research.

The variety of specialist interests that had been brought to Palaeolithic research from its earliest decades could be bridged in a number of ways. Individuals sometimes contributed their different, but complementary, interests to a particular problem, gathering in large numbers around topics like the eolith debates. But more often they gathered on a very small scale, too small, perhaps, for analogies to be drawn to the 'core-sets' of Collins.[4] The problems

---

[4] Collins 1981, uses the term 'core-set' to describe the small and ephemeral sets of scientists who gathered around a particular controversy and contributed to its outcome, and he uses their

in these cases often involved the explanation of different aspects revealed by a site: fauna, shells, sediments, implements, and so on. These specialists had usually been drawn together by the director of some project or excavation; they were often less concerned about controversy and more interested in the orderly production of their allotted part of the report.

More informal partnerships developed between researchers with complementary skills and compatible personalities, and these became centres for vigorous discussions of broader questions. Very often, researchers worked together in pairs. Prestwich, for example, often worked with Falconer; Hinton with Kennard; Smith with Dewey; and Boswell with Moir. Though small in scale, these partnerships were important in generating knowledge. Interpretations were reached through private negotiation, and inconsistencies were polished away before conclusions were presented in public. Despite the small scale of these partnerships, their views could reach a large proportion of their peers because each partner tended to have links with a different range of societies and journals.

Though individuals might not always agree with their partners, they usually took care to avoid public confrontation within the partnership; together, they could mount a stronger attack on opposing views held by other individuals, who might themselves be members of other partnerships. The security experienced by members of these tiny groups, their conviction strengthened by personal standing in one of the many local or national societies, encouraged researchers to maintain ideas in the face of opposition. Palaeolithic research was remarkable for the variety of views that coincided relatively amicably, perhaps erupting occasionally but not necessarily attracting many adherents or ending in general consensus.

INTO THE FIELD: THE ROLE OF CLASSIC SITES

The standard sequences, described in the summary above, enjoyed broad consensus; but they, too, were forged through an interplay of social processes and the material world. This interplay is illustrated by the role of classic sites in the development and defence of such sequences. A small number of sites became seen as classic guides to different aspects of Britain's past, and each tended to be viewed as the personal territory of a particular researcher (or research partnership). The pits around Swanscombe, for example, provided

activities to show that the production of scientific conclusions is a social process. See also Rudwick 1985, 426–428.

Smith and Dewey with an unusually long sequence of implement-bearing deposits and became a classic stratigraphical sequence for the British Palaeolithic. The deposits of Clacton, however, gave Warren the prototypes for a discrete industry of specific character and date. The names of other sites became familiar as examples of particular time periods: the Thames Valley is full of obscure districts that have become famous through their connections to a particular Quaternary stage or Palaeolithic industry.

These classic sites, where investigators looked down into the ancient sediments and saw the stone tools which they had entombed, offered flashes of light in the darkness of the geological past: a way to link scattered industries, sediments, and bones together within a detailed framework. They helped researchers to build up Palaeolithic sequences, tighten the bonds between the character and date of tools, and formalize the mutual reliance between Palaeolithic and geological research. Sites rose to fame partly through their Palaeolithic and geological wealth, and partly through successful efforts of their discoverers to promote their value; association with, access to, and knowledge of a classic site could boost personal power and prestige.

But the empirical evidence gathered on journeys to gravel pits, railway cuttings, or exposures cut away by the sea was not just informed by what was seen; this evidence was also influenced by expectations of what was likely be seen. In the early 1910s, when Reginald Smith looked into Barnfield pit, Swanscombe, his view of these London implements and sediments was coloured by his awareness of the splendid Palaeolithic sequences recorded recently by Commont in the Somme Valley. Researchers looking into the same pit during the 1930s, when different patterns seemed more likely, saw the stone tools in a different light; they described a wealth of additional industries and Oakley re-interpreted the geological sequence.

Exposure to material evidence in the field added value to an opinion, and offered a powerful chance for guardians of a 'territory' to persuade others to accept their point of view. Moir and Lankester took care to plan the visit of Boule and Breuil to the Pre-Palaeolithic sites of East Anglia; and Warren, as an authority on the Clacton deposits, was often approached by interested peers who wanted a tour. Researchers would take pains to visit a site in company with its unofficial guardians to bolster old ideas, embrace new discoveries, or hear different points of view. Baden-Powell remarked to Oakley during the Second World War:

I think that informal visits to problems in the field are of special importance. Excursions like those of the G.A. [Geologists' Association] are good in their way, but I think one person showing two or three other specialists round what he is doing in the field is the most useful of all. Before the war I remember getting far more from one day in the field with Patterson [Paterson] in Breckland than I had got from all his publications and correspondence, and I hope the same was true of Patterson, Zeuner and Day Kimball, who all came and visited Moir and myself when we were working round Cromer.[5]

When a site gained classic status, it also gained mystique. Hoxne, for example—the site used by Prestwich in his post-glacial arguments of 1859—was re-visited at times of uncertainty about the connections between industries and climatic stages, and used by researchers to defend and justify their changing expectations. Reid travelled there in the late nineteenth century, when his colleagues were disconcerted by Geikie's interglacial theories. Moir dug there in the mid-1920s, when the interglacial Palaeolithic was more acceptable, and reported tools sandwiched between two boulder clays. Boswell adapted the date of the Hoxne industries to match the 'long' Alpine glacial chronology in the late 1930s. West worked on the rich Hoxne pollen profiles in the 1950s with Godwin when the timescale offered by Palaeolithic industries was looking more dubious.

The attention paid to these rich Palaeolithic sites was not surprising—they were, after all, not very common; they were also being extinguished rapidly in the quest for gravel and building materials. Sites were mines of information, but the gathering of fresh material was mingled with a deeper desire to exorcise old interpretations in the field, particularly at the classic sites that gripped the imagination of researchers. Their wealth was seen in the light of expectation; interpretations had to run a social gauntlet before consensus could be achieved; but the weight of an argument could be increased by invoking the name of a famous site and by arguing a case in the field.

EXPECTATIONS DERIVED FROM CLASSIC SEQUENCES

But what of the explanations that were given to account for Palaeolithic patterns? Why did they take some forms rather than others? One major restriction to explanation, one that comes across clearly in this examination of classifications and chronologies, lies in the expectations derived from classic sequences, geological as well as Palaeolithic. Explanations also took much of their colour from the interplay between interpretations of geological sequences and Palaeolithic behaviour, which could be combined in a remarkable variety of ways: some corresponded with expectations of broad, standard sequences; others matched more specific beliefs.

---

[5] Baden-Powell to Oakley, 19 March 1944: BLO: CH, temp. box 95.

There was a close connection between interpretations of the British Palaeolithic and the geological deposits in which these tools had been found. In the nineteenth century, for example, Dawkins referred to groups of River-Drift and Cave men in justification of his mammal classification, and Geikie pointed to the presence of tools beneath boulder clays in support of his interglacial hypothesis. But detailed, standard industrial sequences became more widely used in the twentieth century, when tools were positioned more precisely along the sequences of glaciations, fauna, and river drifts. The character of a stone tool was then bound closer to a particular moment in geological time, associated with a specific position in a standard Palaeolithic sequence, and assigned a familiar label. Interpretations were restricted by these expectations, and were often directed towards the explanation of anomalies. Each Palaeolithic industry had long had a dual meaning—as a marker of Palaeolithic behaviour and as a marker of geological time—but the connections between the two had been tightened.

Industrial sequences offered geologists another guess at the date of geological deposits—but interpretations of the age of those deposits could also be modified to justify expectations of the same industrial sequences. Industrial patterns offered a glimpse of the movements, cultural affinities, and reactions to the environment displayed by ancient tool-makers—but this same range of interpretations could also be used to restrict industrial patterns within the bounds of expectation, to defend long-standing sequences and classifications. The distinction between expectation and interpretation was hazy; interpretations were often restricted by a belief in a particular Palaeolithic sequence. This was seen vividly in the interpretation of the Swanscombe sequence offered by Reginald Smith in the 1910s (influenced by the Somme sequence), the reaction to flake industries like the Clactonian in the 1910s and 1920s (influenced by the standard linear sequence), and the interpretations of river and glacial sequences in the 1930s (influenced by the industrial–glacial sequence popularized by Breuil).

The 1930s witnessed the remarkable development, described in Chapter 10, whereby the geological framework of glacial cycles and river sediments, from East Anglia and the Thames Valley, was set to match expectations derived from the Palaeolithic framework. John McNabb has explained this development in terms of a 'conceptual lock' on interpretation, describing how the belief in this industrial sequence and in its value for dating geological layers acted to funnel interpretations and restrict new ideas in the mid twentieth century.[6] But the lock applied to interpretations of the character as well as the date of stone tools. Industries were not just time-markers: they were also

[6] McNabb 1996, 47.

the cultural reflection of mobile, tool-making groups, reacting to different circumstances. Industrial sequences could be defended by re-interpreting geological deposits but also by varying the habits, movements, or race of the tool-makers supposed to have produced the tools. Since the nineteenth century, overlap in time or space had been explained in terms of the movement of different tool-making groups, and such reasoning became integral to the detailed Palaeolithic classifications of the 1930s and 1940s. Different industrial branches were associated with different peoples, or culture-groups, each with a different geographical heartland; and local industrial variation could be accommodated to expectation by invoking culture-contact and industrial hybridization.

The complicated interplay between interpretations of geology and of Palaeolithic behaviour, and their connections to expectations derived from standard sequences (such as Breuil's scheme of the 1930s or the Swanscombe sequence of the 1910s) might be compared to the 'normal science' and 'paradigms' of Thomas Kuhn. Kuhn's paradigms were the classic observations, experiments, models, or examples on which expectations and future researches were based; but the term was also applied in a sociological sense to the community of researchers who shared those beliefs.[7] The kind of research carried out by those communities under the influence of a paradigm was described by Kuhn as 'normal science', observations being shoe-horned into conceptual boxes. The rise of anomalies might eventually urge researchers to adopt a new paradigm and a new set of expectations in one of Kuhn's 'scientific revolutions'.[8] The process described by McNabb as a conceptual lock is similar to what Kuhn called the 'mopping-up operations' of normal science.[9]

Belief in the reliability and usefulness of these classic sequences certainly had a powerful influence on the direction of Palaeolithic research. Interpretations cannot be separated neatly from expectations, geology from archaeology, or chronology from culture. But the reasons why those sequences gained credibility lies in the historical and social context of research. Although the grand schemes and sequences tend to be remembered most vividly today, not all arguments were geared towards their maintenance and interpretation: not all were mopping in the same direction.

Kuhn is not the only philosopher of science to have studied how ideas emerged, changed, and gained acceptance—although he is still the best known to those outside his field. Susan Leigh Star and James R. Griesemer, for example, explore the tension between varied viewpoints and the desire to

---

[7] Kuhn 1996, 10, 23, 175.   [8] Kuhn 1996, 5–6, 24, 175.
[9] Kuhn 1996, 24.

develop a synthesis in their study of how the aims of people from different social worlds are reconciled in the scientific workplace (specifically, Berkeley's Museum of Vertebrate Zoology between 1907 and 1939). Thomas F. Gieryn uses the concept of 'boundary-work' to examine how scientists promote their authority through ideology and rhetoric by distinguishing science from pseudo-science.[10] Although their ideas are subtle and attractive, they lose power when separated from the case studies on which they are based. It is difficult to apply their philosophical models to the history described in this book without twisting a complicated tale into an awkward shape. Kuhn might be unpopular amongst today's philosophers of science, but his ideas mesh neatly with this history of Palaeolithic research.

FINAL COMMENTS

In the late nineteenth century, Whitaker drew attention to the 'tendency to dogmatise' on one's own subject, 'and to pass over' the lines of evidence and findings of others, 'as things of small importance'.[11] He was denouncing those like Dawkins who seemed overly concerned with fauna, and less interested in Whitaker's stratigraphy. But this picture of blinkered researchers, their eyes fixed on the answers they required for specific questions and blinded to the views of others, is applicable to many other situations. Take the belief in standard sequences above, for example, or the eolith debates where arguments were built on different foundations, leading protagonists to fire past each other and fail to score direct hits.

An analogy might be drawn between the aims of the Palaeolithic researchers who, like Breuil or Smith and Dewey, were trying to discover a broad, general pattern in the stone tools, and the attempts to identify patterns of knowledge formation and scientific change in the historical detail of this century of research. Broad patterns, syntheses, and consensus can be extracted, such as the models of philosophers of science, or the 'conceptual lock' of McNabb. But answers change according to the questions asked and the scale of those questions. This applies to the historian, the philosopher, and the practising scientist as well as to the Palaeolithic researchers introduced in this book. Change the question, look closer, and general patterns may disappear, to be replaced by a diversity of smaller, competing stories. It then becomes more difficult to identify the context of knowledge production with particular

---

[10] Gieryn 1983; Star and Griesemer 1989.   [11] Whitaker 1889 (i), 334–335.

interest groups or moments of controversy, or to find broad scientific boundaries beneath the variety of rhetoric and collaboration.

The aim of this book has not been to build general models of scientific knowledge or to judge researchers right or wrong; although these pages might, perhaps, offer case-study material for the theories of philosophers or sociologists of science, or contain information of interest to today's archaeologists or geologists. Instead, these chapters have looked closer at some of the turbulent undercurrents of research, reviving the roles and lives of forgotten researchers through snatches of biography, exploring what they thought about the age and character of ancient stone tools, why they came to those conclusions, and why some were more successful in winning support than others. On this scale of analysis, general historical patterns become scattered through the specific personal, social, and local contexts of individual researchers—just as the Palaeolithic patterns of Breuil or de Mortillet splintered when scrutinized more closely, or the histories written by practising scientists in terms of present research draw the past into another of Kuhn's mopping-up operations.

This journey through the shady realms of past Palaeolithic research has followed some awkward paths, but it will have been worthwhile if the traveller looks back and, instead of a few flickering references to a few famous figures, now sees a more vivid and interesting landscape. Dawkins, in 1870, wrote to an erstwhile colleague on the Geological Survey to share a joke: 'What is the difference between temptation and geological time? The one is a wile of the devil and the other is a devil of a while'.[12] In the century following the acceptance of human antiquity, the ancient stone tools of Britain were cast back 'a devil of a while', but there were also a surprising number of temptations along the way to draw researchers in different directions as they tried to trace patterns in the past.

[12] Dawkins to Hughes, 17 March 1870: SMC: TMH.

Glossary

anthropology   Used by the author in the restricted British sense: to refer to the study, physical or cultural, of contemporary humans (rather than extending the term to cover the archaeological study of their products and the palaeoanthropological study of ancient human remains). In historical context, however, the term has a slightly different meaning. During the nineteenth and early twentieth centuries, this term encompassed prehistoric as well as contemporary times: anthropological societies and journals welcomed archaeological papers. The historical component was lost in Europe during the early twentieth century, but was retained in America.

artefact   An object modified or used by humans, distinct from objects formed naturally and not associated with human activity.

assemblage   The entire group of artefacts found together in a single area at a given site, deposited over a restricted period of time.

core, core-tool   A piece of stone from which flakes have been deliberately removed. The core could serve as a tool (a 'core-tool'), or could be discarded as a by-product of flake production. Distinct from deep-sea cores: tubes of sediment from the sea-bed that preserve records of past climatic change.

culture   The varied and shifting meanings of this term stimulated numerous arguments about the British Palaeolithic sequence. Strictly, a 'culture' refers to the customs and material reflection of a particular group of people who share a similar outlook on the world. In Palaeolithic research, the term encompasses racial, aesthetic, geographical, and also chronological aspects; each might be given different emphasis. In earlier decades, the term 'culture-stage' described the products of peoples thought to be in the same *stage of progress* (even if there was no evidence of contact between them) and was often used interchangeably with 'industry'. Between the 1930s and 1950s, 'culture' became applied more broadly to industries of similar character: hand-axe cultures and flake cultures ran in parallel streams through time, and 'culture-contact' between their makers became an explanation for industrial diversity. This usage was criticised for the implication of a close connection between tool-making *peoples* who might, in fact, be distant in time or space.

drift   A nineteenth-century term used both for geological sediments, and for the period of time in which these were deposited (capitalised here, when used to denote time). Drift deposits were sometimes identified more specifically with the agencies that produced them: as 'river-drift' or 'glacial drift.' They were loosely equivalent to the deposits and the periods known as 'Recent', 'Superficial', or 'Quaternary'. The geological division was used in parallel with archaeological divisions: 'Palaeolithic' stone tools of River-Drift times were distinguished from 'Neolithic' tools of Alluvial times, and an intermediate 'Cave' (or 'Reindeer') period was often added.

eolith   From the Greek words *eos* (dawn) and *lithos* (stone): a name given to crude stones that some regarded as the earliest human artefacts, and assigned to an Eolithic period. The East Anglian eoliths became known as pre-palaeoliths, and were later described as a Pre-Chellean industry.

ethnography   The description and interpretation of social behaviour displayed by a group of people—their customs and culture—usually based on first-hand observation by anthropologists in the field.

flake, flake-tool   A flake is the sliver struck by a blow from a piece of stone (or 'core'). Here, 'flake-tool' refers to a flake that has been modified into a useful tool-form by further fine chipping. Unmodified or 'simple' flakes could be discarded as by-products of core-tool production, but could also serve as tools.

implement   A flexible term, sometimes restricted to core-tools with retouch such as the hand-axe (which excluded flake-tools from the definition), but also used interchangeably with the general term 'tool'. Here, 'implement' refers to a stone artefact modified by chipping (retouch): an artefact thus distinguished from a simple un-worked flake or from the waste product of tool-manufacture.

industry   Another flexible term: sometimes employed as an equivalent to the more specific 'assemblage'; but also used interchangeably with the broader 'culture.' Here, it is used to refer to assemblages of similar character, often from various different sites, left in the same broad period of time, and classed together under such a label as 'Chellean', 'Acheulian', 'Clactonian', and so on.

industrial phase   A finer division of an industry, popular in the 1930s and 1940s. The 'Acheulian II,' for example, was a phase of the 'Acheulian' industry, with its own, specific character, date, and associations with other industries.

Mammoth period   A division of Quaternary time popular in the nineteenth century and strongly associated with the work of Lartet in France and Dupont in Belgium. The Mammoth period was a loose equivalent to the 'River-Drift' or 'Palaeolithic' period, and was distinguished from the later Reindeer period. Mammoth seemed to have dominated Europe before reindeer increased in abundance.

Neolithic   From the Greek *neos* (new) and *lithos* (stone): a recent period of the Stone Age characterised by polished stone tools.

Palaeoanthropology   The study of ancient hominid remains.

Palaeolithic   From the Greek *palaios* (ancient) and *lithos* (stone): an early division of the Stone Age. In the nineteenth century, 'Palaeolithic' was often used in a restricted sense to describe the chipped stone tools from the river-drifts (up to, and often including, the Mousterian industry), and encompassed only the Lower and Middle Palaeolithic periods of today's researchers. These 'Palaeolithic' tools preceded the chipped stone tools of the Reindeer period (today's Upper Palaeolithic), and the later polished stone tools of the Neolithic period.

Palaeontology   The study of fossils.

Quaternary   Used here to refer to a geological period that began with the deposition of the Cromer Forest Bed, continued through the Glacial epoch, and ended with the Neolithic period. This period was described in various different terms in the past: as Newer Tertiary, Pliocene, Post-Tertiary, Pleistocene, Post-Pliocene and Post-Glacial. It was also sub-divided in various ways: on the basis of palaeontology, the glacial epoch, and the appearance of the Palaeolithic in Britain. Some of these confusions are avoided in this book by using the single term: 'Quaternary.'

race   A division of human groups that refers to physical rather than cultural differences.

Reindeer period   A division of the Stone Age popular in the nineteenth century, associated with the cave-dwelling tool-makers of southern France. It has a loose equivalent in the Upper Palaeolithic period of today's researchers. The Reindeer period succeeded the 'Mammoth' or 'Palaeolithic' period, which was associated with river-drift deposits.

retouch   A tool-working term: the chipping, trimming or flaking of a stone to alter its form: transforming a simple flake, or core, into an implement.

river-drift—see **drift**

river-terrace   Spreads of sediment deposited by an ancient river as it cut down through its valley floor, left in horizontal bands or steps on the valley sides above the current river.

stratigraphy   The order and relation between different geological layers, or strata. Here, the term 'stratigraphical geologist' is used occasionally to identify geologists who worked on the correlation of different deposits in time and space.

terrace—see **river-terrace**

Tertiary   A geological period, once the most recent division of the earth's rocks. In the nineteenth century, late Tertiary deposits were separated from the rest of this period and assigned various different names. Some researchers continued to describe these later deposits as 'late-Tertiary' or even 'Tertiary.' See **Quaternary** and Berry (1987, 77–78, 105–114).

tool   Used here as a general term for any stone artefact, from simple flakes to elaborately chipped implements, to avoid the often specific meaning of the term 'implement.' In practice, Palaeolithic researchers often used 'tool' interchangeably with 'implement'.

tradition   Also 'techno-complex': a large group of similar industries, which may have little proximity in time, space, or culture (in its restricted sense).

zone-fossil   A particular species of fossil creature, or a stone tool of distinct character, used to date a geological deposit to a specific period of time.

Bibliography

Manuscript Sources

AAM	University of Cambridge Museum of Archaeology & Anthropology
AMO	Ashmolean Museum, University of Oxford
BGS	Library of the Geological Survey of Great Britain, Keyworth
BLL	British Library, London (Add Ms)
BLO	Bodleian Library, University of Oxford
	CH C. Hawkes papers (un-catalogued at time of research; the temporary box numbers given here will change)
BMD	Buxton Museum and Art Gallery, Derbyshire (DERSB)
	WBD W. B. Dawkins papers
	JWJ J. W. Jackson papers
BM(F)	British Museum (Franks' House), London
FMF	Falconer Museum, Forres, Grampian
	HF H. Falconer papers. Numbers refer to Patrick Boylan's catalogue (Boylan 1977)
GSL	Archives of the Geological Society of London (LDGSL)
IAL	Institute of Archaeology Library, University College London
ICL	Imperial College London (College Archives)
NHM	Natural History Museum Archives, London
	DF140 series: Anthropology sub-department correspondence
NLW	National Library of Wales, Aberystwyth
PRM	Pitt Rivers Museum, University of Oxford
	EBT E. B. Tylor papers
RSL	Royal Society of London archives
SMC	Sedgwick Museum of Earth Sciences, University of Cambridge
	TGB T. G. Bonney papers (letters un-catalogued at time of research)
	TMH T. McKenny Hughes papers (letters un-catalogued at time of research)
SSW	Salisbury and South Wiltshire Museum
	P-R Pitt-Rivers papers
UCL	University College London, Special Collections (Institute of Archaeology Archives)
ULC	University Library, Cambridge (Add Ms)
ULE	Edinburgh University Library, Special Collections Department
ULL	University Library, Liverpool, Special Collections and Archives
UMO	Oxford University Museum of Natural History
	DB-P D. Baden-Powell papers

JP   J. Parker papers
WJA   W. J. Arkell papers
WJS   W. J. Sollas papers

Printed Sources

Abbott, W. J. L. (1897). Worked Flints from the Cromer Forest Bed. *Natural Science* **10**, 89–96.

—— (1911). On the Classification of the British Stone Age Industries, and some New, and Little Known, Well-marked Horizons and Cultures. *Journal of the Royal Anthropological Institute* **41**, 458–481.

Adams, D. (1993). *Dirk Gently's Holistic Detective Agency*. London: Pan Books Ltd.

Agassiz, E. C. (1885). *Louis Agassiz, his Life and Correspondence* (2 vols.). London: Macmillan and Company.

Agassiz, L. (1840). On Glaciers, and the Evidence of their having once existed in Scotland, Ireland, and England. *Proceedings of the Geological Society of London* **3**, 327–332.

Anon.(1859a). Society of Antiquaries. *The Athenaeum* **1650**, 781–782.

—— (1859b). On the Occurrence of Flint-implements, associated with the Remains of Animals of Extinct Species in Beds of a late Geological Period. *Annals and Magazine of Natural History* **4** (third series), 230–237.

—— (1867a). Obituary. Hugh Falconer. *Proceedings of the Royal Society* **15**, xiv–xx.

—— (1867b). Geological Society. On the Age of the Lower Brick-earths of the Thames Valley. *The London, Edinburgh, and Dublin Philosophical Magazine and Journal of Science* **33** (fourth series), 233–234.

—— (1867c). Lines on a Scratched Boulder. *The Geological Magazine* **4**, 94.

—— (1876). A Geological Ramble. The Antiquity of Man. *The Eastern Daily Press*, 11th October.

—— (1877). The Antiquity of Man. *Nature* **16**, 97–98.

—— (1891). Abstract and discussion of Prestwich, J., 'On the Age, Formation, and Successive Drift Stages of the Valley of the Darent; with Remarks on the Palaeolithic Implements of the District, and on the Origin of its Chalk Escarpment'. *Abstracts of the Proceedings of the Geological Society of London* **Session 1890–91** (566), 33–38.

—— (1900a). Exhibition of Flint Implements from the Thames and Lea Valleys, including examples of Eolithic type and Palaeolithic forms by S. H. Warren. *Proceedings of the Geologists' Association* **16**, 60.

—— (1900b). Note of paper by Archibald Geikie on 'Our Older Sea-margins'. *Proceedings of the Geologists' Association* **16**, 535.

—— (1909). Exhibits at the meeting of the Geological Society, March 24th 1909. *Quarterly Journal of the Geological Society of London* **65**, cxxv.

—— (1911). Exhibition of the Palaeolithic Wooden Spear from Clacton. *Quarterly Journal of the Geological Society of London* **67**, xcix.

—— (1913). Exhibits of Geological Interest presented at the meeting on November 1st, 1912. *Proceedings of the Geologists' Association* **24**, 301–302.

—— (1914). Summary of a meeting at the Geological Society, 1914. *Quarterly Journal of the Geological Society of London* **70**, ii–xii.

—— (1923a). Abstract and discussion of Warren, S. H., 'The Late Glacial Stage of the Lea Valley (Third Report)' and 'The *Elephas-antiquus* Bed of Clacton-on-Sea (Essex) and its Flora and Fauna'. *Abstracts of the Proceedings of the Geological Society of London* Session 1922–23 (1099), 53–57.

—— (1923b). Exhibition of Chellean Hand-axe from Sidestrand by J. E. Sainty. *Proceedings of the Prehistoric Society of East Anglia* **iv**, 126–127.

—— (1933). Early Man in East Africa (report of committees). *Nature* **131**, 477–478.

—— [Chambers, R.] (1844). *Vestiges of the Natural History of Creation*. London: John Churchill.

Argyll, Duke of (1874). Obituary. Louis Jean Rodolphe Agassiz. *Proceedings of the Geological Society of London* **30**, xxxvii–xliv.

Arkell, W. J. (1943). The Pleistocene Rocks at Trebetherick Point, North Cornwall: Their Interpretation and Correlation. *Proceedings of the Geologists' Association* **54**, 141–170.

—— (1951). Thames Terraces and Alpine Glaciations. Some Recent Correlations. *The Archaeological News Letter* **4**, 17–19.

Baden-Powell, D. (1948). The Chalky Boulder Clays of Norfolk and Suffolk. *The Geological Magazine* **85**, 279–296.

—— (1968). Obituary. James Edward Sainty. *Proceedings of the Geologists' Association* **79**, 267–269.

Bailey, E. (1952). *Geological Survey of Great Britain*. London: Thomas Murby & Co.

Barnes, A. S., and Moir, J. R. (1926). A Misleading Exhibit. *Man* **26**, 78–79.

Barr, J. (1985). Why the World was Created in 4004 B.C.: Archbishop Ussher and Biblical Chronology. *Bulletin of the John Rylands University Library of Manchester* **67**, 575–608.

Berry, W. B. N. (1987). *Growth of a Prehistoric Time Scale Based on Organic Evolution*. Oxford: Blackwell.

Blanckaert, C. (1988). On the Origins of French Ethnology: William Edwards and the Doctrine of Race. In G. W. Stocking (Ed.), *Bones, Bodies, Behavior. Essays on Biological Anthropology* (pp. 18–55). Wisconsin: University of Wisconsin Press.

Bonney, T. G. (1885). Obituary. Searles Valentine Wood. *Proceedings of the Geological Society of London* **41**, 40–41.

Bordes, F. (1956). Some Observations on the Pleistocene Succession in the Somme Valley. *Proceedings of the Prehistoric Society* **22**, 1–5.

—— and Bourgon, M. (1951). Le Complexe moustérien: Moustériens, levalloisien et tayacien. *L'Anthropologie* **55**, 1–23.

Boswell, P. G. H. (1914). On the Occurrence of the North Sea Drift (Lower Glacial) and Certain other Brick-Earths, in Suffolk. *Proceedings of the Geologists' Association* **25**, 121–153.

—— (1922). The Geology of the Country around Felixstowe and Ipswich. *Proceedings of the Geologists' Association* **33**, 285–305.

—— (1930). Early Man and the Correlation of Glacial Deposits. *Report of the British Association for the Advancement of Science* **100** (Bristol), 379–381.

Boswell, P. G. H. (1931). The Stratigraphy of the Glacial deposits of East Anglia in Relation to Early Man. *Proceedings of the Geologists' Association* 42, 87–111.

—— (1932a). The Contacts of Geology: The Ice Age and Early Man in Britain. *British Association for the Advancement of Science* 102 (York), 57–88.

—— (1932b). The Oldoway Human Skeleton. *Nature* 130, 237–238.

—— (1935). Human Remains from Kanam and Kanjera, Kenya Colony. *Nature* 135, 371.

—— (1936). Problems on the Borderland of Archaeology and Geology in Britain. *Proceedings of the Prehistoric Society* 2, 149–160.

—— (1940). Climates of the Past: A Review of the Geological Evidence. *Quarterly Journal of the Royal Meteorological Society* 66, 249–274.

—— (1945). James Reid Moir (1879–1944). *Proceedings of the Prehistoric Society* 11, 66–68.

—— and Moir, J. R. (1923). The Pleistocene Deposits and their Contained Palaeolithic Flint Implements at Foxhall Road, Ipswich. *Journal of the Royal Anthropological Institute* 53, 229–262.

Boule, M. (1888). Essai de paléontologie stratigraphique de l'homme. *Revue d'anthropologie* (1888), 129–144, 272–297, 385–411, 647–680.

—— (1895). Review of Smith, W. G., 'Man the Primeval Savage.' *L'Anthropologie* 6, 319–320.

—— (1905). L'Origine des eolithes. *L'Anthropologie* 16, 257–267.

—— (1908a). L'Homme fossile de la Chapelle-aux-Saints (Corrèze). *L'Anthropologie* 19, 519–525.

—— (1908b). Observations sur un silex taillé du Jura et sur la chronologie de M. Penck. *L'Anthropologie* 19, 1–13.

—— (1912a). Review of 'Proceedings of the Prehistoric Society of East Anglia. Vol. 1. Part 1.' *L'Anthropologie* 23, 426–429.

—— (1912b). L'Institut de paléontologie humaine. *Congrès international d'anthropologie et d'archéologie préhistoriques. Compte rendu de la 14 session, Genève, 1912* (pp. 493–496). Geneva: Albert Kündig.

—— (1915). La Paléontologie humaine en Angleterre. *L'Anthropologie* 26, 1–67.

—— (1918–19). Victor Commont. *L'Anthropologie* 29, 162–164.

Bourgeois, L. (1868). Étude sur des silex travaillés trouvés dans les dépôts tertiaires de la commune de Thenay près Pontlevoy (Loir-et-Cher), *Congrès international d'anthropologie et d'archéologie préhistoriques. Compte rendu de la 2 session, Paris, 1867* (pp. 67–75). Paris: C. Reinwald.

—— (1873). Sur les silex considérés comme portant les marques d'un travail humain et découverts dans le terrain miocène de Thenay, *Congrès international d'anthropologie et d'archéologie préhistoriques. Compte rendu de la 6 session, Bruxelles, 1872* (pp. 81–94). Brussels: C. Muquardt.

Bowden, M. (1993). *Pitt Rivers: the Life and Archaeological Work of Lieutenant-General Augustus Henry Lane Fox Pitt Rivers, DCL, FRS, FSA*. Cambridge: Cambridge University Press.

Bowler, P. J. (1986). *Theories of Human Evolution: A Century of Debate, 1844–1944*. Baltimore: John Hopkins University Press.

—— (1992). From 'savage' to 'primitive': Victorian Evolutionism and the Interpretation of Marginalized Peoples. *Antiquity* 66, 721–729.

—— (2004). Lankester, Sir (Edwin) Ray (1847–1929). In H. C. G. Matthew, & B. Harrison (Eds.), *Oxford Dictionary of National Biography* (vol. 32, pp. 535–538). Oxford: Oxford University Press.

Boylan, P. J. (1967). Dean William Buckland, 1784–1856. A Pioneer in Cave Science. *Studies in Speleology* 1, 237–253.

—— (1977). *The Falconer Papers, Forres*. Leicester: Leicestershire Museums, Art Galleries and Records Service.

—— (1981). The Role of William Buckland (1784–1856) in the Recognition of Glaciation in Great Britain. In J. Neale, and J. Flenley (Eds.), *The Quaternary in Britain. Essays, Reviews and Original Work on the Quaternary published in Honour of Lewis Penny on his Retirement* (pp. 1–8). Oxford: Pergamon Press.

Boyle, M. (1963). Recollections of the Abbé Breuil. *Antiquity* 37, 12–18.

Breuil, H. (1908). Les Divisions du quaternaire ancien. *Revue archéologique* 11, 415–417.

—— (1910). Sur la présence d'éoliths a la base de l'éocène Parisien. *L'Anthropologie* 21, 385–408.

—— (1913a). Comments on Bayer, J., 'Chronologie des temps quaternaires,' *Congrès international d'anthropologie et d'archéologie préhistoriques. Compte rendu de la 14 session, Genève, 1912* (pp. 162–164). Geneva: Albert Kündig.

—— (1913b). Les Subdivisions du paléolithique superieur et leur signification, *Congrès international d'anthropologie et d'archéologie préhistoriqes. Compte rendu de la 14 session, Genève, 1912* (pp. 165–238). Geneva: Albert Kündig.

—— (1921a). G. d'Ault du Mesnil. *L'Anthropologie* 31, 161–162.

—— (1921b). La question des industries primitives (Ipswich, Piltdown). *Revue anthropologique* 31, 356–359.

—— (1922). Les Industries pliocènes de la région d'Ipswich. *Revue anthropologique* 32, 226–229.

—— (1926). Palaeolithic Industries from the Beginning of the Rissian to the Beginning of the Wurmian Glaciation. *Man* 26, 176–179.

—— (1929). La Préhistoire. *Revue des cours et conférences (College de France)* 31, 97–113.

—— (1930). Le Clactonien et sa place dans la chronologie. *Bulletin de la société préhistorique Française* 27, 221–227.

—— (1931). The Pleistocene Sequence in the Thames Valley. *South-Eastern Naturalist and Antiquary* 36, 95–98.

—— (1932a). Le Paléolithique ancien en Europe occidentale et sa chronologie. *Bulletin de la société préhistorique Française* 29, 570–578.

—— (1932b). Les Industries a éclats du paléolithique ancien, 1.—Le Clactonien. *Préhistoire* 1, 125–190.

—— (1936). Hoxne et le clactonien. *L'Anthropologie* 46, 208–209.

—— (1937a). Discours de M. l'Abbé H. Breuil, président sortant. *Bulletin de la société préhistorique Française* 34, 52–67.

Breuil, H. (1937b). Reply on being presented with the Gold Medal of the Society of Antiquaries. *The Antiquaries Journal* **17**, 257–260.

—— (1939). The Pleistocene Succession in the Somme Valley. *Proceedings of the Prehistoric Society* **5**, 33–38.

—— (1947). Age of the Baker's Hole Coombe Rock, Northfleet, Kent. *Nature* **160**, 831.

—— (1948). Reply to the Geological Society upon receiving the Prestwich Medal. *Abstracts of the Proceedings of the Geological Society of London* (1440), 64–66.

—— (1949). *Beyond the Bounds of History: Scenes from the Old Stone Age.* London: P. R. Gawthorn Ltd.

—— (1950). Hugo Obermaier (1877–1946). *Revue archéologique* **35**, 105–110.

—— and Koslowski, L. (1931). Études de stratigraphie paléolithique dans le nord de la France, la Belgique et l'Angleterre. *L'Anthropologie* **41**, 449–488.

—— and —— (1932). Études de stratigraphie paléolithique dans le nord de la France, la Belgique et l'Angleterre. *L'Anthropologie* **42**, 27–47, 291–314.

—— and —— (1934). Études de stratigraphie paléolithique dans le nord de la France, la Belgique et l'Angleterre. *L'Anthropologie* **44**, 249–290.

—— and Lantier, R. (1965). *The Men of the Old Stone Age.* London: George G. Harrap & Co. Ltd.

Bridgland, D. (1994). *Quaternary of the Thames.* London: Chapman and Hall.

Brodrick, A. H. (1963). *The Abbé Henri Breuil, Prehistorian.* London: Hutchinson & Co.

Brown, E. E. S. (1958a). General Meetings. In G. S. Sweeting (Ed.), *The Geologists' Association 1858–1958. A History of the First Hundred Years* (pp. 27–37). Colchester: Benham and Company Limited.

—— (1958b). Obituary of Arthur Leonard Leach. *Proceedings of the Geologists' Association* **69**, 67–70.

Brown, J. A. (1886). The Thames-valley Surface-deposits of the Ealing District and their Associated Palaeolithic Floors. *Quarterly Journal of the Geological Society of London* **42**, 192–200.

—— (1887a). *Palaeolithic Man in N. W. Middlesex. The Evidence of his Existence and the Physical Conditions under which he lived in Ealing and its Neighbourhood, illustrated by the Condition and Culture presented by certain Existing Savages.* London: Macmillan & Co.

—— (1887b). On a Palaeolithic Workshop Floor discovered near Ealing. *Proceedings of the Society of Antiquaries of London* **11** (second series), 211–215.

—— (1888). On the Discovery of Elephas primigenius, associated with Flint Implements, at Southall. *Proceedings of the Geologists' Association* **10**, 361–372.

—— (1889). Working Sites and Inhabited Land Surfaces of the Palaeolithic Period in the Thames Valley. *Transactions of the County of Middlesex Natural History and Science Society* **Session 1889–89**, 40–73.

—— (1893). On the Continuity of the Palaeolithic and Neolithic Periods. *Journal of the Royal Anthropological Institute* **22**, 66–98.

—— (1896). Notes on the High-level River Drift between Hanwell and Iver. *Proceedings of the Geologists' Association* **14**, 153–173.

Buckland, W. (1823). *Reliquiae Diluvianae, or Observations on the Organic Remains contained in Caves, Fissures, and Diluvial Gravel, and on other Geological Phenomena attesting the action of an Universal Deluge.* London: John Murray.

—— (1826). Observations on the Bones of Hyaenas and other Animals in the Cavern of Lunel near Montpelier, and in the adjacent strata of Marine Formation. *Proceedings of the Geological Society of London* 1, 3–6.

—— (1836). *Geology and Mineralogy considered with reference to Natural Theology* (2 vols.). London: William Pickering.

—— (1840a). Address to the Geological Society. *Proceedings of the Geological Society of London* 3, 210–248.

—— (1840b). On the Evidences of Glaciers in Scotland and the North of England. *Proceedings of the Geological Society of London* 3, 332–337, 345–348.

—— (1841). On the Glacia-Diluvial Phaenomena in Snowdonia and the adjacent parts of North Wales. *Proceedings of the Geological Society of London* 3, 579–584.

Bull, A. J. (1942). Pleistocene Chronology. *Proceedings of the Geologists' Association* 53, 1–45.

Bulman, G. W. (1891). On the Sands and Gravels Intercalated in the Boulder-clay. *The Geological Magazine* Decade III, Vol. VIII, 337–348, 402–410.

Burchell, J. P. T. (1932). Early Neanthropic Man and his Relation to the Ice Age. *Proceedings of the Prehistoric Society of East Anglia* vi, 253–303.

—— (1934). The Middle Mousterian Culture and its Relation to the Coombe Rock of Post-Early Mousterian Times. *The Antiquaries Journal* 14, 33–39.

Burkitt, M. C. (1921). Congress at Liege. *Proceedings of the Prehistoric Society of East Anglia* iii, 453–457.

—— (1925). *Prehistory. A Study of Early Cultures in Europe and the Mediterranean Basin* (2nd ed.). Cambridge: Cambridge University Press.

—— (1933). *The Old Stone Age.* Cambridge: Cambridge University Press.

—— (1936a). Nomenclature of Palaeolithic Finds from Fresh Regions. *Man* 36, 103–104.

—— (1936b). Nomenclature of Palaeolithic Finds. *Man* 36, 215–216.

—— (1944). Mr. J. Reid Moir, F.R.S. *Nature* 153, 368–369.

Bury, H. (1916). The Palaeoliths of Farnham. *Proceedings of the Geologists' Association* 27, 151–192.

—— (1923). Some Aspects of the Hampshire Plateau Gravels. *Proceedings of the Prehistoric Society of East Anglia* iv, 15–41.

—— (1935). The Farnham Terraces and their Sequence. *Proceedings of the Prehistoric Society* 1, 60–69.

Butzer, K. W. (1972). *Environment and Archeology. An Ecological Approach to Prehistory.* London: Methuen & Co. Ltd.

Bynum, W. F. (1984). Charles Lyell's *Antiquity of Man* and its Critics. *Journal of the History of Biology* 17, 153–187.

Capitan, L. (1922). Cartailhac. *Revue anthropologique* 32, 1–6.

—— (1923). Les silex d'Ipswich—Rapport du Dr. Capitan. *Revue anthropologique* 33, 58–67.

Carneiro, A. (2005). Outside Government Science, 'Not a Single Tiny Bone to Cheer Us Up!' The Geological Survey of Portugal (1857–1908), The Involvement of Common Men, and the Reaction of Civil Society to Geological Research. *Annals of Science* **62**, 141–204.

Carreck, J. N. (1973). Obituary. Alvan Theophilus Marston. *Proceedings of the Geologists' Association* **84**, 118–120.

Cartailhac, E. (1898). Gabriel de Mortillet. *L'Anthropologie* **9**, 601–612.

Caton-Thompson, G. (1946a). The Aterian Industry: Its Place and Significance in the Palaeolithic World. *Journal of the Royal Anthropological Institute* **76**, 87–130.

—— (1946b). The Levalloisian Industries of Egypt. *Proceedings of the Prehistoric Society* **12**, 57–120.

Chambers, R. (1853). On Glacial Phenomena in Scotland and Parts of England. *Edinburgh New Philosophical Journal* **54**, 229–281.

—— (1855). Further Observations on Glacial Phenomena in Scotland and the North of England. *Edinburgh New Philosophical Journal* **1** (new series), 97–103.

Chambers, W., and Chambers, R. (1853). Note on 'A Visit to Banwell Caverns.' *Chambers's Edinburgh Journal* **19**, 76–77.

Chandler, R. H. (1914). The Pleistocene Deposits of Crayford, together with the Report of an Excursion to Slades Green and Barnhurst. *Proceedings of the Geologists' Association* **25**, 61–71.

—— (1916). The Implements and Cores of Crayford. *Proceedings of the Prehistoric Society of East Anglia* **ii**, 240–248.

—— (1930). On the Clactonian Industry at Swanscombe. *Proceedings of the Prehistoric Society of East Anglia* **vi**, 79–116.

—— (1931a). The Clactonian Industry and Report of Field Meeting at Swanscombe. *Proceedings of the Geologists' Association* **42**, 175–177.

—— (1931b). Report of Excursion to Swanscombe, Kent, August 30th, 1930. *Proceedings of the Prehistoric Society of East Anglia* **vi**, 250.

—— (1932). The Clactonian Industry and Report of Field Meeting at Swanscombe (II). *Proceedings of the Geologists' Association* **43**, 70–72.

—— and Leach, A. L. (1911). Excursion to Dartford Heath. *Proceedings of the Geologists' Association* **22**, 171–175.

—— and —— (1912). On the Dartford Heath Gravel and on a Palaeolithic Implement Factory. *Proceedings of the Geologists' Association* **23**, 102–111.

Childe, V. G. (1935). Changing Methods and Aims of Prehistory. *Proceedings of the Prehistoric Society* **1**, 1–15.

—— (1944). The Future of Archaeology. *Man* **43**, 18–19.

—— (1951). Review of Breuil, H., and R. Lantier (1951), 'Les Hommes de la Pierre Ancienne (Paléolithique et Mésolithique).' *Proceedings of the Prehistoric Society* **17**, 234–235.

Christy, H. (1865). On the Prehistoric Cave-Dwellers of Southern France. *Transactions of the Ethnological Society of London* **3** (new series), 362–372.

Clark, E. W. Le G., and Morant, G. M. (1938). The Swanscombe Fossil. *Report of the British Association for the Advancement of Science* **108** (Cambridge), 469.

Clark Howell, F. (1965). Review of Oakley, K. (1964), 'Frameworks for Dating Fossil Man.' *American Anthropologist* **67**, 1043–1048.

Clark, J. G. D. (1937). The London University Institute of Archaeology. *Proceedings of the Prehistoric Society* **3**, 166.

—— (1938). Anglo-French Co-operation in Archaeology. *Proceedings of the Prehistoric Society* **4**, 340–341.

—— (1941). Review of Childe, V.G. (1939), 'The Dawn of European Civilization.' *Proceedings of the Prehistoric Society* **7**, 147–148.

—— (1989). *Prehistory at Cambridge and Beyond*. Cambridge: Cambridge University Press.

Clarke, W. G. (1919). In Memoriam. W. Allen Sturge, M.V.O., M.D., F.R.C.P. *Proceedings of the Prehistoric Society of East Anglia* **iii**, 12–13.

Cohen, C. (1994). *The Fate of the Mammoth. Fossils, Myth, and History*. Chicago: Chicago University Press.

—— (1999). Abbé Breuil 1877–1961. In T. Murray (Ed.), *Encyclopaedia of Archaeology. The Great Archaeologists* (pp. 301–312). Oxford: Abc-Clio.

—— and Hublin, J. (1989). *Boucher de Perthes 1788–1868. Les origins romantiques de la préhistoire*. Paris: Belin.

Coles, J. M., and Higgs, E. S. (1969). The Archaeology of Early Man. London: Faber and Faber.

Collins, D. (1969). Culture Traditions and Environment of Early Man. *Current Anthropology* **10**, 267–316.

Collins, H. M. (1981). The Place of the 'Core Set' in Modern Science. *History of Science* **19**, 6–19.

Commont, V. (1908). Les Industries de l'ancien Saint-Acheul. *L'Anthropologie* **19**, 527–572.

—— (1909a). Saint-Acheul et Montières. Notes de géologie, de paléontologie et de préhistoire. *Memoires de la société géologique du Nord* **6**, 1–68.

—— (1909b). L'industrie moustérienne dans la région du nord de la France. *Congrès préhistorique de France. Compte rendu de la 5 session, Beauvais, 1909*, 115–157.

—— (1910). Note préliminaire sur les terrasses fluviatiles de la vallée de la Somme. *Annales de la société géologique du Nord* **39**, 185–210.

—— (1911). Evolution de l'industrie chelléenne dans les alluvions fluviatiles de la vallée de la Somme. *Revue préhistorique* **6**, 65–80.

—— (1912a). Chronologie et stratigraphie des industries protohistoriques, néolithiques et paléolithiques dans les dépôts holocènes et pleistocènes du nord de la France et en particulier de la vallée de la Somme. Remarques et comparisons relatives aux loess et aux glaciations. *Congrès international d'anthropologie et d'archéologie préhistoriques. Compte rendu de la 14 session, Genève, 1912* (pp. 239–254). Geneva: Albert Kündig.

—— (1912b). Moustérien a faune chaude dans la vallée de la Somme a Montières-les-Amiens. *Congrès international d'anthropologie et d'archéologie préhistoriques. Compte rendu de la 14 session, Genève, 1912* (pp. 291–300). Geneva: Albert Kündig.

Commont, V. (1912c). La Chronologie et la stratigraphie des dépôts quaternaires dans la vallée de la Somme. *Annales de la société géologique de Belgique* **39**, 156–178.

Conway, B. (1996). An Historical Perspective on Geological Research at Barnfield Pit, Swanscombe. In B. Conway, J. McNabb, and N. Ashton (Eds.), *Excavations at Barnfield Pit, Swanscombe 1968–1972* (pp. 9–30). London: British Museum Occasional Paper 94.

Cornwall, I. W. (1964). Obituary. Frederick Zeuner. *Proceedings of the Geologists' Association* **75**, 117–120.

Coye, N. (2000). Boucher de Perthes: un médiateur en préhistoire (1837–1864). *Praehistoria* **1**, 9–18.

—— (2006). Sur les chemins de la préhistoire. L'abbé Breuil du Périgord à l'Afrique du Sud. Paris: Somogy éditions d'art.

Cranshaw, S. (1983). *Handaxes and Cleavers. Selected English Acheulian Industries*. Oxford: British Archaeological Reports British Series, No. 113.

Cremo, M. A., and Thompson, R. L. (1998). *Forbidden Archeology. The Hidden History of the Human Race* (1st ed., revised). Los Angeles: Bhaktivedanta Book Publishing, Inc.

Croll, J. (1864). On the Physical Cause of the Change of Climate during Geological Epochs. *The London, Edinburgh, and Dublin Philosophical Magazine and Journal of Science* **28** (fourth series), 121–137.

—— (1868). On Geological Time, and the probable Date of the Glacial and the Upper Miocene Period. (Part III). *The London, Edinburgh, and Dublin Philosophical Magazine and Journal of Science* **36** (fourth series), 362–386.

—— (1875). *Climate and Time in their Geological Relations. A Theory of Secular Changes of the Earth's Crust*. London: Daldy, Isbister, & Co.

Cuvier, G. (1821). *Recherches sur les ossemens fossiles, tome premier*. Paris: Chez. G. Dufour et E. D'Ocagne.

D'Acy, E. (1875). Quelques observations sur la succession chronologique des types appelés généralement type de Saint-Acheul et type du Moustier. *Matériaux pour l'histoire primitive et naturelle de l'homme* **10**, 281–287.

—— (1887). De l'emmanchement des silex taillés, du type généralement connu sous le nom de type de Saint-Acheul ou de Chelles. *Bulletins de la société d'anthropologie de Paris* **10** (third series), 158–182, 219–237.

Dalrymple, G. B., and Lanphere, M. A. (1969). *Potassium-Argon Dating. Principles, Techniques and Applications to Geochronology*. San Francisco: W.H. Freeman and Company.

Daniel, G. (1975). *A Hundred and Fifty Years of Archaeology* (2nd ed.). London: Duckworth.

Darwin, C. (1871). *The Descent of Man and Selection in Relation to Sex* (2 vols.). London: John Murray.

Davis, J. B. (1864). *The Neanderthal Skull: Its Peculiar Conformation Explained Anatomically*. London: Taylor and Francis.

Dawkins, W. B. (1862). On a Hyaena-den at Wookey-Hole, near Wells. *Quarterly Journal of the Geological Society of London* **18**, 115–125.

—— (1863). On a Hyaena-den at Wookey-Hole, near Wells. No.II. *Quarterly Journal of the Geological Society of London* 19, 260–274.

—— (1866a). Esquimaux in the South of France. *The Saturday Review* 22, 712–713.

—— (1866b). On the Habits and Condition of the Two Earliest Known Races of Men. *Quarterly Journal of Science* 3, 333–346.

—— (1867). On the Age of the Lower Brick-earths of the Thames Valley. *Quarterly Journal of the Geological Society of London* 23, 92–109.

—— (1868). The Former Range of the Reindeer in Europe. *Popular Science Review* 7, 34–45.

—— (1869). On the Distribution of the British Postglacial Mammals. *Quarterly Journal of the Geological Society of London* 25, 192–217.

—— (1871a). The Relation of the Quaternary Mammalia to the Glacial Period. *Report of the British Association for the Advancement of Science* 41 (Edinburgh), 95–96.

—— (1871b). On Pleistocene Climate and the Relation of the Pleistocene Mammalia to the Glacial Period. *Popular Science Review* 10, 388–397.

—— (1872). The Classification of the Pleistocene Strata of Britain and the Continent by means of the Mammalia. *Quarterly Journal of the Geological Society of London* 28, 410–446.

—— (1874). *Cave Hunting. Researches on the Evidence of Caves respecting the Early Inhabitants of Europe.* London: Macmillan and Co.

—— (1877). On the Mammal-fauna of the Caves of Creswell Crags. *Quarterly Journal of the Geological Society of London* 33, 589–612.

—— (1878). On the Evidence afforded by the Caves of Great Britain as to the Antiquity of Man. *Journal of the Royal Anthropological Institute* 7, 151–162.

—— (1880). *Early Man in Britain and his Place in the Tertiary Period.* London: Macmillan and Co.

—— (1881a). Review of Geikie, J. (1881), 'Prehistoric Europe.' *Nature* 23, 309–310.

—— (1881b). Prehistoric Europe. *Nature* 23, 482.

—— (1882). On the Present Phase of the Antiquity of Man. *Report of the British Association for the Advancement of Science* 52 (Southampton), 597–604.

—— (1910). The Arrival of Man in Britain in the Pleistocene Age. *Journal of the Royal Anthropological Institute* 40, 233–263.

Dawson, C., and Woodward, A. S. (1913). On the Discovery of a Palaeolithic Human Skull and Mandible in a Flint-bearing Gravel overlying the Wealden (Hastings Beds) at Piltdown, Fletching (Sussex). *Quarterly Journal of the Geological Society of London* 69, 117–151.

De Bont, R. (2003). The Creation of Prehistoric Man: Aimé Rutot and the Eolith Controversy, 1900–1920. *Isis* 94, 604–630.

De Mortillet, G. (1869). Classification chronologique des cavernes de l'époque de la pierre simplement eclatée, et observations sur le diluvium a cailloux brises. *Bulletin de la société géologique de France* 2, 583–587.

—— (1871). Nécrologie. Édouard Lartet. *Revue scientifique* 1 (second series), 307–308.

—— (1872). Classification de l'âge de la pierre. *Matériaux pour l'histoire primitive et naturelle de l'homme* 7 (second series), 464–465.

De Mortillet, G. (1873). Classification des diverses périodes de l'âge de la pierre, *Congrès international d'anthropologie et d'archéologie préhistoriques. Compte rendu de la 6 session, Bruxelles, 1872* (pp. 432–444). Bruxelles: Weissenbruch.

—— (1883). *Le Préhistorique. Antiquité de l'homme.* Paris: C. Reinwald.

—— and de Mortillet, A. (1881). *Musée préhistorique.* Paris: C. Reinwald.

—— and —— (1900). *Le Préhistorique; origine et antiquité de l'homme* (3rd ed.). Paris: C. Reinwald.

De Quatrefages, A. (1881). *The Human Species.* International Scientific Series (2nd ed.). London: C. Kegan Paul & Co.

—— (1884). *Hommes fossiles et hommes sauvages.* Paris: Librairie J.-B. Baillière et Fils.

Déchelette, J. (1908). Review of Obermaier, H., 'Die Steingeräte des französischen Altpaläolithikums. Eine kritische Studie über ihre Stratigraphie und Evolution.' *L'Anthropologie* **19**, 460–463.

Dennell, R. (1990). Progressive Gradualism, Imperialism and Academic Fashion: Lower Palaeolithic Archaeology in the Twentieth Century. *Antiquity* **64**, 549–558.

Desnoyers, J. (1863). Note sur des indices materiels de la coexistence de l'homme avec l'*Elephas meridionalis* dans un terrain des environs de Chartres, plus anciens que les terrains de transport quaternaires des vallées de la Somme et de la Seine. *Comptes rendus de l'académie des sciences* **56**, 1073–1083.

Dewey, H. (1913). The Raised Beach of North Devon: its Relation to Others and to Palaeolithic Man. *The Geological Magazine* Decade V, Vol. X, 154–163.

—— (1915). Surface Changes since the Palaeolithic Period in Kent and Surrey. *Proceedings of the Prehistoric Society of East Anglia* ii, 107–116.

—— (1919). On some Palaeolithic Flake-implements from the High Level Terraces of the Thames Valley. *Geological Magazine* Decade VI, Vol. VI, 49–57.

—— (1926). Obituary of William Whitaker. *Proceedings of the Geologists' Association* **37**, 231–235.

—— (1931). Palaeolithic Thames Deposits. *Proceedings of the Prehistoric Society of East Anglia* vi, 147–155.

—— (1932). The Palaeolithic Deposits of the Lower Thames Valley. *Quarterly Journal of the Geological Society of London* **88**, 35–56.

—— (1944). Obituary. Llewellyn Treacher. *Proceedings of the Geologists' Association* **55**, 42–44.

—— and Bromehead, C. E. N. (1921). *The Geology of South London. Explanation of Sheet 270.* Geological Survey Memoir. London: H.M.S.O.

—— and Smith, R. (1914). The Palaeolithic Sequence at Swanscombe, Kent. *Proceedings of the Geologists' Association* **25**, 90–97.

—— and Smith, R. (1924). Flints from the Sturry gravels, Kent. *Archaeologia* **74**, 117–136.

Dines, H. G. (1929). The Flint Industries of Bapchild. *Proceedings of the Prehistoric Society of East Anglia* vi, 12–26.

Dubois, E. (1894). *Pithcanthropus erectus. Eine menschenaehnliche Uebergangsform aus Java.* Batavia: Landesdruckerei.

Duckworth, W. L. H. (1912). *Prehistoric Man.* Cambridge: Cambridge University Press.

Dupont, E. (1872). *Le temps préhistoriques en Belgique. L'Homme pendant les âges de la pierre dans les environs de Dinant-sur-Meuse* (2nd ed.). Brussels: C. Muquardt.

—— (1873). Classement des âges de la pierre en Belgique, *Congrès international d'anthropologie et d'archéologie préhistoriques. Compte rendu de la 6 session, Bruxelles, 1872* (pp. 459–479). Brussels: Weissenbruch.

Dury, G. H. (1986). *The Face of the Earth* (5th ed.). London: Allen & Unwin.

Dyer, J. (1959). 'Middling for Wrecks': Extracts from the story of Worthington and Henrietta Smith. *Journal of the South Bedfordshire Archaeological Society* 2, 1–15.

—— (1967–68). 'W.G.S.' and the Potato Blight Mystery. *Bedfordshire Magazine* 11, 91–96.

—— (1978). Worthington George Smith. *Bedfordshire Historical Record Society* 57, 141–179.

East, G. E. (1888). Spurious Flint Implements. *The Geological Magazine* Decade III, Vol. V, 239.

Ebbatson, L. (1994). Context and Discourse: Royal Archaeological Institute Membership 1845–1942. In B. Vyner (Ed.), *Building on the Past. Papers celebrating 150 years of the Royal Archaeological Institute* (pp. 22–74). London: The Royal Archaeological Institute.

Edwards, K., Clark, G., Mitchell, F., Walker, D., Willis, E., West, R., Smith, A., and Dickson, J. and C. (1986). Professor Sir Harry Godwin, F.R.S., 1901–1985—A Tribute. *Journal of Archaeological Science* 13, 299–306.

Emiliani, C. (1955). Pleistocene Temperatures. *Journal of Geology* 63, 538–578.

Evans, J. (1860). On the Occurrence of Flint Implements in Undisturbed Beds of Gravel, Sand, and Clay. *Archaeologia* 38, 280–307.

—— (1863). Account of some Further Discoveries of Flint Implements in the Drift on the Continent and in England. *Archaeologia* 39, 57–84.

—— (1864a). On some Recent Discoveries of Flint Implements in Drift-deposits in Hants and Wilts. *Quarterly Journal of the Geological Society of London* 20, 188–194.

—— (1864b). On some Bone- and Cave-deposits of the Reindeer-period in the South of France. *Quarterly Journal of the Geological Society of London* 20, 444.

—— (1869). On the Manufacture of Stone Implements in Prehistoric Times, *International Congress of Prehistoric Archaeology: Transactions of the 3rd Session, London, 1868* (pp. 191–193). London: Longmans, Green, and Co.

—— (1872). *The Ancient Stone Implements, Weapons, and Ornaments, of Great Britain.* London: Longmans, Green, Reader, and Dyer.

—— (1875). Presidential Address. *Quarterly Journal of the Geological Society of London* 31, xxxvii–lxxvi.

—— (1878). Presidential Address to 'The Meeting on the Present State of the Question of the Antiquity of Man' and subsequent discussion. *Journal of the Royal Anthropological Institute* 7, 149–151, 174–185.

—— (1891). The Progress of Archaeology. Opening Address of the Antiquaries Section at the Edinburgh Meeting. *The Archaeological Journal* 48, 251–262.

Evans, J. (1897a). Obituary. Joseph Prestwich. *Proceedings of the Royal Society* **60**, xii–xvi.

—— (1897b). *The Ancient Stone Implements, Weapons, and Ornaments, of Great Britain* (2nd ed.). London: Longmans, Green and Co.

Evans, J. (1943). *Time and Chance. The Story of Arthur Evans and his Forebears.* London: Longmans, Green and Co.

—— (1956). *A History of the Society of Antiquaries.* Oxford: Oxford University Press.

Fagan, B. (2001). *Grahame Clark. An Intellectual Biography of an Archaeologist.* Cambridge: Westview Press.

Falconer, H. (1857). On the Species of Mastodon and Elephant occurring in the Fossil State in Great Britain. Part I. Mastodon. *Quarterly Journal of the Geological Society of London* **13**, 307–360.

—— (1858). On the Species of Mastodon and Elephant occurring in the Fossil State in England. Part II. Elephas. *Quarterly Journal of the Geological Society of London* **14**, 81–84.

—— (1860a). On the Ossiferous Grotta di Maccagnone, near Palermo. *Quarterly Journal of the Geological Society of London* **16**, 99–106.

—— (1860b). On the Ossiferous Caves of the Peninsula of Gower, in Glamorganshire, South Wales. *Quarterly Journal of the Geological Society of London* **16**, 487–491.

—— (1863). Primeval Man—What Led to the Question? *The Athenaeum* 1849, 459–460.

—— and Cautley, P. T. (1845). *Fauna Antiqua Sivalensis, being the Fossil Zoology of the Sewalik Hills, in the North of India. Illustrations. Proboscidea.* London: Smith, Elder and Co.

—— and —— (1846). *Fauna Antiqua Sivalensis, being the Fossil Zoology of the Sewalik Hills, in the North of India. Proboscidea.* London: Smith, Elder and Co.

Finnegan, D. A. (2004). The Work of Ice: Glacial Theory and Scientific Culture in Early Victorian Edinburgh. *British Journal for the History of Science* **37**, 29–52.

Fischer, P. (1872). The Scientific Labors of Edward Lartet. *Annual Report of the Board of Regents of the Smithsonian Institution for 1872*, 172–184.

Fisher, O. (1866). On the Warp (of Mr. Trimmer)—its Age and probable Connexion with the Last Geological Events. *Quarterly Journal of the Geological Society of London* **22**, 553–565.

—— (1872). On a Worked Flint from the Brick-earth of Crayford, Kent. *The Geological Magazine* **9**, 268–269.

—— (1879). On Implement-bearing Loams in Suffolk. *Proceedings of the Cambridge Philosophical Society* **3**, 285–289.

Flett, J. S. (1937). *The First Hundred Years of the Geological Survey of Great Britain.* London: H.M.S.O.

Fleure, H. J. (2004). Peake, Harold John Edward (1867–1946), revised by M. Pottle. In H. C. G. Matthew, & B. Harrison (Eds.), *Oxford Dictionary of National Biography* (vol. 43, pp. 268–269). Oxford: Oxford University Press.

Flower, J. W. (1867). On some Flint Implements lately found in the Valley of the Little Ouse River, at Thetford, Norfolk. *Quarterly Journal of the Geological Society of London* **23**, 45–56.

—— (1872). On the Relative Ages of the Stone Implement Periods in England. *Journal of the Royal Anthropological Institute* 1, 274–295.

Foote, Y. (2004). Evans, Sir John (1823–1908). In H. C. G. Matthew, & B. Harrison (Eds.), *Oxford Dictionary of National Biography* (vol. 18, pp. 719–722). Oxford: Oxford University Press.

Forbes, E. (1846). On the Connexion between the Distribution of the existing Fauna and Flora of the British Isles, and the Geological Changes which have Affected their Area, especially during the Epoch of the Northern Drift. In H. De la Beche (Ed.), *Memoirs of the Geological Survey of Great Britain and of the Museum of Economic Geology in London, Vol. 1* (pp. 336–432). London: Longman, Brown, Green and Longmans.

—— (1854). Anniversary Address of the President. *Proceedings of the Geological Society of London* 10, xxviii–lxxxi.

Fraipont, J., & Lohest, M. (1887). La race humaine de Néanderthal ou de Canstadt en Belgique. *Archives de Biologie* 7, 587–757.

Frere, J. (1800). Account of Flint Weapons discovered at Hoxne in Suffolk. *Archaeologia* 13, 204–205.

Fry, P. H. (1999). Samuel Taylor Coleridge. The Rime of the Ancient Mariner, *Case Studies in Contemporary Criticism*. Boston and New York: Bedford / St. Martin's.

Fuller, J. G. C. M. (2001). Before the Hills in Order Stood: The Beginning of the Geology of Time in England. In C. L. E. Lewis, and S. J. Knell (Eds.), *The Age of the Earth: From 4004 BC to AD 2002* (pp. 15–23). London: Geological Society, Special Publication No. 190.

Garrod, D. (1928). Nova et Vetera: A Plea for a New Method in Palaeolithic Archaeology. *Proceedings of the Prehistoric Society of East Anglia* v, 260–267.

—— (1938a). The Upper Palaeolithic in the Light of Recent Discovery. *Proceedings of the Prehistoric Society* 4, 1–26.

—— (1938b). The Swanscombe Find. *Report of the British Association for the Advancement of Science* 108 (Cambridge), 469.

—— (1946). *Environment, Tools and Man. An Inaugural Lecture.* Cambridge: Cambridge University Press.

—— (1961). Obituary (Henri Breuil). *Man* 61, 205–207.

Garwood, E. J. (1932). Award of the Lyell Medal. *Quarterly Journal of the Geological Society of London* 88, lv–lvi.

Geikie, A. (1863). On the Phenomena of the Glacial Drift of Scotland. *Transactions of the Geological Society of Glasgow* 1, 1–190.

—— (1895). *Memoir of Sir Andrew Crombie Ramsay.* London: Macmillan and Co.

—— (1898). Summary of Progress of the Geological Survey of the United Kingdom for 1897. London: H.M.S.O.

—— (1924). *A Long Life's Work. An Autobiography.* London: Macmillan and Co. Ltd.

Geikie, J. (1871). On Changes of Climate during the Glacial Epoch. *The Geological Magazine* 8, 545–553.

—— (1872). On Changes of Climate during the Glacial Epoch. *The Geological Magazine* 9, 23–31, 61–69, 105–111, 164–170, 215–222, 254–265.

Geikie, J. (1874). *The Great Ice Age and its Relation to the Antiquity of Man.* London: W. Isbister & Co.

—— (1877a). *The Great Ice Age and its Relation to the Antiquity of Man* (2nd ed.). London: W. Isbister & Co.

—— (1877b). The Antiquity of Man. *Nature* 16, 141–142.

—— (1881a). *Prehistoric Europe. A Geological Sketch.* London: Edward Stanford.

—— (1881b). Prehistoric Europe. *Nature* 23, 336.

—— (1881c). Prehistoric Europe. *Nature* 23, 433–434.

—— (1893). *Fragments of Earth Lore. Sketches & Addresses Geological and Geographical.* Edinburgh: John Bartholomew & Co.

—— (1894). *The Great Ice Age and its Relation to the Antiquity of Man* (3rd ed.). London: Edward Stanford.

—— (1914). *The Antiquity of Man in Europe.* Edinburgh: Oliver and Boyd.

George, W. H. (2004). William Whitaker (1836–1925)—Geologist, Bibliographer and a Pioneer of British Hydrogeology. In J. D. Mather (Ed.), *200 Years of British Hydrogeology* (pp. 51–65). London: Geological Society, Special Publication No. 225.

Gibbard, P. L. (1994). *Pleistocene History of the Lower Thames Valley.* Cambridge: Cambridge University Press.

Gieryn, T. F. (1983). Boundary-Work and the Demarcation of Science from Non-Science: Strains and Interests in Professional Ideologies of Scientists. *American Sociological Review* 48, 781–895.

Godwin, H. (1941). Pollen-Analysis and Quaternary Geology. *Proceedings of the Geologists' Association* 52, 328–361.

Godwin-Austen, R. A. C. (1850). On the Valley of the English Channel. *Quarterly Journal of the Geological Society of London* 6, 69–97.

—— (1851). On the Superficial Accumulations of the Coasts of the English Channel and the Changes they Indicate. *Quarterly Journal of the Geological Society of London* 7, 118–136.

—— (1855). On Land-surfaces beneath the Drift-gravel. *Quarterly Journal of the Geological Society of London* 11, 112–119.

—— (1863). On the Alluvial Accumulation in the Valley of the Somme and Ouse. *Report of the British Association for the Advancement of Science* 33 (Newcastle-on-Tyne), 68.

Goodrich, E. S. (1930). Edwin Ray Lankester. 1847–1929. *Proceedings of the Royal Society of London* 106, x–xv.

Goodwin, A. J. H. (1946). The Terminology of Prehistory. *South African Archaeological Bulletin* 1, 91–100.

—— and Lowe, C. van R. (1929). *The Stone Age Cultures of South Africa.* Annals of the South African Museum. Edinburgh: Neill and Co, Ltd.

Gras, S. (1862). Antiquity of the Human Race. *The Parthenon*, 19th July, 12.

Grayson, D. K. (1983). *The Establishment of Human Antiquity.* New York: Academic Press.

—— (1985). The First Three Editions of Charles Lyell's *The Geological Evidences of the Antiquity of Man. Archives of Natural History* 13, 105–121.

——— (1986). Eoliths, Archaeological Ambiguity, and the Generation of 'Middle-Range' Research. In D. J. Meltzer, D. D. Fowler, and J. A. Sabloff (Eds.), *American Archaeology: Past and Future. A Celebration of the Society for American Archaeology* (pp. 77–133). Washington D. C.: Smithsonian Institution Press.

Green, J. F. N., and Wooldridge, S. W. (1933). Obituary. George Barrow. *Proceedings of the Geologists' Association* 44, 111–112.

Greenwell, W. (1863). Address to the Members of the Tyneside Naturalists' Field Club, read by the President, the Rev. William Greenwell, M.A. *Transactions of the Tyneside Naturalists' Field Club* 6, 1–30.

——— (1870). On the Opening of Grime's Graves in Norfolk. *Journal of the Ethnological Society* 2 (second series), 419–439.

Gregory, J. W. (1930). The Correlation of Pluvial Periods and the German Drifts. *Report of the British Association for the Advancement of Science* 100 (Bristol), 376–379.

Gregory, W. K. (1938). Biographical Memoir of Henry Fairfield Osborn 1857–1935. *National Academy of Sciences Biographical Memoirs* 19, 52–119.

Gruber, J. W. (1965). Brixham Cave and the Antiquity of Man. In M. E. Spiro (Ed.), *Context and Meaning in Cultural Anthropology* (pp. 373–402). New York: The Free Press.

Gustafsson, A. (1998). The History of Archaeology: Good Archaeology as Bad History? In A-C. Andersson, Å. Gillberg, O. W. Jensen, H. Karlsson, and M. V. Rolö (Eds.), *The Kaleidoscopic Past* (pp. 285–293). Göteborg: Göteborg University.

Halls, H. H., and Sainty, J. E. (1926). In Memoriam. W. G. C. *Proceedings of the Prehistoric Society of East Anglia* v, 89–90.

Hamal-Nandrin, J., and Fraipont, C. (1923). Les silex d'Ipswich—Rapport de MM. Hamal-Nandrin et Charles Fraipont. *Revue anthropologique* 23, 57–58.

Hamlin, C. (1982). James Geikie, James Croll, and the Eventful Ice Age. *Annals of Science* 39, 565–583.

Hammond, M. (1980). Anthropology as a Weapon of Social Combat in Late-Nineteenth-Century France. *Journal of the History of the Behavioural Sciences* 16, 118–132.

——— (1982). The Expulsion of the Neanderthals from Human Ancestry: Marcellin Boule and the Social Context of Scientific Research. *Social Studies of Science* 12, 1–36.

Hare, F. K. (1947). The Geomorphology of a part of the Middle Thames. *Proceedings of the Geologists' Association* 58, 294–339.

Harmer, F. W. (1895a). The Southern Character of the Molluscan Fauna of the Coralline Crag Tested by an Analysis of its Characteristic and Abundant Species. *Report of the British Association for the Advancement of Science* 65 (Ipswich), 675–676.

——— (1895b). On the Derivative Shells of the Red Crag. *Report of the British Association for the Advancement of Science* 65 (Ipswich), 676–677.

——— (1899). On a Proposed New Classification of the Pliocene Deposits of the East of England. *Report of the British Association for the Advancement of Science* 69 (Dover), 751–753.

Harmer, F. W. (1902). A Sketch of the Later Tertiary History of East Anglia. *Proceedings of the Geologists' Association* **17**, 416–479.

—— (1904). The Great Eastern Glacier. *Report of the British Association for the Advancement of Science* **74** (Cambridge), 542–543.

—— (1910a). *The Glacial Geology of Norfolk and Suffolk* (reprint ed.). London: Jarrold and Sons.

—— (1910b). The Pleistocene Period in the Eastern Counties of England. In H. W. Monckton, and R. S. Herries (Eds.), *Geology in the Field: The Jubilee Volume of the Geologists' Association (1858–1908)* (pp. 103–123). London: Edward Stanford.

Harrison, B. (1892). On certain Rude Implements from the North Downs. *The Journal of the Royal Anthropological Institute* **21**, 263–267.

—— (1895). High-level Flint-drift of the Chalk. Report of the Committee, consisting of Sir John Evans (Chairman), Mr. B. Harrison (Secretary), Professor J. Prestwich, and Professor H.G. Seeley. *Report of the British Association for the Advancement of Science* **65** (Ipswich), 349–351.

—— (1904). *An Outline of the History of the Eolithic Flint Implements:* 'To be obtained from the Author. Price 6d nett; post free.'

Harrison, E. R. (1928). *Harrison of Ightham.* London: Oxford University Press.

Harrison, W. J. (2004). Christy, Henry (1810–1865), revised by A. Bowdoin Van Riper. In H. C. G. Matthew, & B. Harrison (Eds.), *Oxford Dictionary of National Biography* (vol. 11, pp. 562–563). Oxford: Oxford University Press.

Haward, F. N. (1912). The Chipping of Flints by Natural Agencies. *Proceedings of the Prehistoric Society of East Anglia* **i**, 185–195.

—— (1914). The Problem of the Eoliths. *Proceedings of the Prehistoric Society of East Anglia* **i**, 347–360.

—— (1919). The Origin of the 'Rostro-carinate Implements' and Other Chipped Flints from the Basement Beds of East Anglia. *Proceedings of the Prehistoric Society of East Anglia* **iii**, 118–146.

—— (1921). The Fracture of Flint: A Reply to the Criticisms of Prof. A. S. Barnes. *Proceedings of the Prehistoric Society of East Anglia* **iii**, 448–452.

Hawkes, C. F. C. (1938). The Archaeological Associations of the Swanscombe Skull. *Report of the British Association for the Advancement of Science* **108** (Cambridge), 468.

—— (1951). British Prehistory Half-way through the Century. *Proceedings of the Prehistoric Society* **17**, 1–15.

Hearne, T. (1770). *Joannis Lelandi Antiquarii de Rebus Britannicis Collectanea.* London: Richardson.

Heekeren, H. R. van (1948). Prehistoric Discoveries in Siam, 1943–44. *Proceedings of the Prehistoric Society* **14**, 24–32.

Herries Davies, G. L. (1969). *The Earth in Decay: A History of British Geomorphology, 1578–1878.* London: MacDonald Technical and Scientific.

—— (2004). Geikie, James Murdoch (1839–1915). In H. C. G. Matthew, & B. Harrison (Eds.), *Oxford Dictionary of National Biography* (vol. 21, pp. 723–724). Oxford: Oxford University Press.

Hicks, H. (1885). On the Fynnon Beuno and Cae Gwyn Bone-Caves, North Wales. *Report of the British Association for the Advancement of Science* 55 (Aberdeen), 1021–1023.

—— (1886a). On Some Recent Researches in Bone-caves in Wales. *Proceedings of the Geologists' Association* 9, 1–20.

—— (1886b). Results of Recent Researches in some Bone-caves in North Wales (Ffynnon Beuno and Cae Gwyn), with a Note on the Animal Remains, by W. Davies. *Quarterly Journal of the Geological Society of London* 42, 3–19.

—— (1888). On the Migrations of Pre-Glacial Man. *The Geological Magazine* Decade III, Vol. V, 29–30.

—— (1897). Obituary. Sir Joseph Prestwich. *Proceedings of the Geological Society of London* 53, xlix–lii.

Himus, G. W. (1954). The Geologists' Association and its Field Meetings. *Proceedings of the Geologists' Association* 65, 1–10.

Hinton, M. A. C. (1900). The Pleistocene Deposits of the Ilford and Wanstead District. *Proceedings of the Geologists' Association* 16, 271–281.

—— (1910). A Preliminary Account of the British Fossil Voles and Lemmings; with some Remarks on the Pleistocene Climate and Geography. *Proceedings of the Geologists' Association* 21, 489–507.

—— (1926a). The Pleistocene Mammalia of the British Isles and their Bearing upon the Date of the Glacial Period. *Proceedings of the Yorkshire Geological Society* 20, 325–348.

—— (1926b). *Monograph of the Voles and Lemmings (Microtinae) Living and Extinct.* London: British Museum (Natural History).

—— and Kennard, A. S. (1905). The Relative Ages of the Stone Age Implements of the Thames Valley. *Proceedings of the Geologists' Association* 19, 76–100.

—— Oakley, K. P., Dines, H. G., *et al.* (1938). Report on the Swanscombe Skull. *Journal of the Royal Anthropological Institute* 68, 17–98.

Hollingworth, S. E. (1947). Review of Zeuner, F.E. (1945), 'The Pleistocene Period, its Climate, Chronology and Faunal Successions.' *Proceedings of the Prehistoric Society* 13, 187–188.

—— Allison, J., & Godwin, H. (1950). Interglacial Deposits from the Histon Road, Cambridge. *Quarterly Journal of the Geological Society of London* 105, 495–509.

Holmes, T. V. (1892). The New Railway from Grays Thurrock to Romford: Sections between Upminster and Romford. *Quarterly Journal of the Geological Society of London* 48, 365–372.

Hopwood, A. T. (1935). Fossil Elephants and Man. *Proceedings of the Geologists' Association* 46, 46–60.

Horner, L. (1861). Annivarsary Address of the President. *Proceedings of the Geological Society of London* 17, xxxi–lxxii.

Howorth, H. (1894). Mr Harker and Mr Deeley on the Scandinavian Ice-Sheet. *The Geological Magazine* Decade IV, Vol I, 496–499.

—— (1896). The Chalky and other Post-Tertiary Clays of Eastern England. *The Geological Magazine* Decade IV, Vol. III, 449–463.

Hughes, T. M. (1878). On the Evidence afforded by the Gravels and Brick-earth. *Journal of the Royal Anthropological Institute* 7, 162–165.

—— (1887). On the Drifts of the Vale of Clwyd and their Relation to the Caves and Cave-deposits. *Quarterly Journal of the Geological Society of London* 43, 73–120.

—— (1913). Flints. *Proceedings of the Cambridge Antiquarian Society with Communications made to the Society* 18, 26–66.

Hutchinson, H. G. (1914). *Life of Sir John Lubbock, Lord Avebury.* London: Macmillan and Co.

Hutchinson, H. N. (1896). *Prehistoric Man and Beast.* London: Smith, Elder, & Co.

Huxley, T. H. (1863). *Evidence as to Man's Place in Nature.* London: Williams and Norgate.

—— (1869). Obituary. A. Morlot. *Proceedings of the Geological Society of London* 25, xxxi.

—— (1890). The Aryan Question and Pre-Historic Man. *The Nineteenth Century* 28, 750–777.

Imbrie, J., and Imbrie, K. P. (1979). *Ice Ages. Solving the Mystery.* London: The Macmillan Press Ltd.

Irons, J. C. (1896). *Autobiographical Sketch of James Croll with Memoir of his Life and Work.* London: Edward Stanford.

Isaac, G. L. (1975). Sorting out the Muddle in the Middle: An Anthropologist's Post-Conference Appraisal. In K. Butzer, and G. Issac (Eds.), *After the Australopithecines. Stratigraphy, Ecology, and Culture Change in the Middle Pleistocene* (pp. 875–887). Paris: Mouton Publishers.

Jewitt, L. (1867). Flint Jack; A Memoir and an Appeal. *The Reliquary* 8, 65–76.

Jones, T. R. (1877). *Lecture on the Antiquity of Man; Illustrated by the Contents of Caves and Relics of the Cave-folk.* London: John Van Voorst.

Keith, A. (1912). Description of the Ipswich Skeleton. *Proceedings of the Prehistoric Society of East Anglia* i, 203–209.

—— (1938). The Swanscombe Fossil. *Report of the British Association for the Advancement of Science* 108 (Cambridge), 468.

—— (1944). James Reid Moir, 1879–1944. *Royal Society Obituary Notices* 4, 733–745.

Kelley, H. (1937). Acheulian Flake Tools. *Proceedings of the Prehistoric Society* 3, 15–28.

Kendall, G. H. O. (1915). Some Palaeolithic Pits and Periods in Hertfordshire, etc. *Proceedings of the Prehistoric Society of East Anglia* ii, 135–139.

Kendall, P. F. (1923). Mr. F. W. Harmer. *Nature* 111, 779–780.

Kendrick, T. D. (1971). In the 1920s. *The British Museum Quarterly* 35, 2–8.

Kennard, A. S. (1916). The Pleistocene Succession in England. *Proceedings of the Prehistoric Society of East Anglia* ii, 249–267.

—— (1931). Obituary: Bernard Barham Woodward. *Proceedings of the Geologists' Association* 42, 72–73.

—— (1944). The Crayford Brickearths. *Proceedings of the Geologists' Association* 55, 121–169.

—— (1947). Fifty and One Years of the Geologists' Association. *Proceedings of the Geologists' Association* 58, 271–293.

—— and Woodward, B. B. (1897). The Post-Pliocene Non-Marine Mollusca of Essex. *Essex Naturalist* 10, 87–109.

Kenyon, F. (1937). The presentation of the Gold Medal of the Society of Antiquaries to Henri Breuil. *The Antiquaries Journal* 17, 255–257.

King, W. B. R. (1955). The Pleistocene Epoch in England. *Quarterly Journal of the Geological Society of London* 111, 187–208.

—— and Oakley, K. P. (1936). The Pleistocene Succession in the Lower Part of the Thames Valley. *Proceedings of the Prehistoric Society* 2, 52–76.

Kitchin, F. L. (1927). Obituary. George William Lamplugh. *The Geological Magazine* 64, 91–92.

Knell, S. J. (2000). *The Culture of English Geology, 1815–1851. A Science Revealed Through its Collecting.* Aldershot: Ashgate Publishing Limited.

Kuhn, T. S. (1996). *The Structure of Scientific Revolutions* (3rd ed.). Chicago: University of Chicago Press.

Lacaille, A. D. (1936). The Palaeolithic Sequence at Ivor, Bucks. *The Antiquaries Journal* 16, 420–443.

—— (1964). Review: Frameworks for Dating Fossil Man by Kenneth Oakley. *The Antiquaries Journal*, 109–110.

Laming-Emperaire, A. (1964). *Origines de L'archéologie préhistorique en France: des superstitions medievales a la découverte de l'homme fossile.* Paris: A. et J. Picard.

Lamplugh, G. W. (1906). On British Drifts and the Interglacial Problem. *Report of the British Association for the Advancement of Science* 76 (York), 532–558.

—— (1917). Clement Reid. Obituary. *Quarterly Journal of the Geological Society of London* 73, lxi–lxiv.

Lane Fox, A. H. (1867). Primitive Warfare: Illustrated by Specimens from the Museum of the Institution. *Journal of the Royal United Service Institution* 11, 612–645.

—— (1868). Primitive Warfare, Section II. On the Resemblance of the Weapons of Early Races; their Variations, Continuity, and Development of Form. *Journal of the Royal United Service Institution* 12, 399–439.

—— (1869). On the Discovery of Flint Implements of Palaeolithic Type in the Gravel of the Thames Valley at Acton and Ealing. *Report of the British Association for the Advancement of Science* 39 (Exeter), 130–132.

—— (1872). On the Discovery of Palaeolithic Implements in Association with *Elephas primigenius* in the Gravels of the Thames Valley at Acton. *Quarterly Journal of the Geological Society of London* 28, 449–465.

—— (1875). On the Principles of Classification Adopted in the Arrangement of his Anthropological Collection, now Exhibited at the Bethnal Green Museum. *Journal of the Royal Anthropological Institute* 4, 293–308.

Lankester, E. R. (1865). On the Crags of Suffolk and Antwerp. *The Geological Magazine* Vol. II, No. III, 103–106.

—— (1870). The Bone-bed of Suffolk and Stone-bed of Norfolk. *Quarterly Journal of the Geological Society of London* 26, 493–514.

—— (1906). President's Address. *Report of the British Association for the Advancement of Science* 76 (York), 3–42.

Lankester, E. R. (1912a). The Investigation of Flint. *Nature* **90**, 331–333.

—— (1912b). On the Discovery of a Novel Type of Flint Implement Below the Base of the Red Crag of Suffolk, Proving the Existence of Skilled Workers of Flint in the Pliocene Age. *Philosophical Transactions of the Royal Society of London* **202**, 283–336.

—— (1912c). The Sub-Crag Flint Implements. *Nature* **90**, 249–250.

—— (1912d). Past and Present Man. *The Saturday Review* **113**, 332–334.

—— (1913). Science from an Easy Chair. Facts and Theories about Primeval Man, *The Daily Telegraph*, 6th June.

—— (1914). Description of the Test Specimen of the Rostro-Carinate Industry found beneath the Norwich Crag. *Occasional Papers of the Royal Anthropological Institute* No. 4.

—— (1921). Letter to the Editor (Flint Implements from the Cromer Forest Bed). *Nature* **106**, 757.

Lantier, R. (1945). Marcellin Boule (1861–1942). *Revue archéologique* **23** (sixth series), 131–133.

—— (1961). L'Abbé Henri Breuil (1877–1961). *Bulletin de la societe préhistorique Française* **58**, 650–653.

Lartet, E. (1861). Nouvelles recherches sur la coexistence de l'homme et des grands mammifères fossiles réputes charactéristiques de la dernière période géologique. *Annales des sciences naturelles, quatrième série, zoologie* **15**, 177–253.

—— (1862). New Researches respecting the Co-existence of Man with the Great Fossil Mammals, regarded as Characteristic of the Latest Geological Period. *Natural History Review* **2**, 53–71.

—— and Christy, H. (1864a). Cavernes du Périgord. Objets gravés et sculptés des temps pré-historiques dans l'Europe occidentale. *Revue archéologique* **9**, 233–265.

—— and —— (1864b). On the Périgord Caves, and on the Engraved and Carved Objects of Prehistoric Date found in Western Europe. *Quarterly Journal of the Geological Society of London* **20** (part 2), 19–20.

—— and —— (1864c). Note sur de nouvelles observations relatives a l'existence de l'homme dans le centre de la France a une époque ou cette contrée était habitée par le renne et d'autres animaux qui n'y vivent pas de nos jours. *Annales des sciences naturelles, cinquième série, zoologie et paléontologie* **1**, 232–240.

—— and —— (1864d). New Observations on the Existence of Man in Central France at a Period when that Country was inhabited by the Reindeer and other Animals which are now Extinct there. *Annals and Magazine of Natural History* **13** (third series), 323–329.

—— and —— (1865–75). *Reliquiae Aquitanicae: Being Contributions to the Archaeology and Palaeontology of Perigord and the Adjoining Provinces of Southern France.* London: Rupert Jones.

Layard, N. (1903). A Recent Discovery of Palaeolithic Implements in Ipswich. *Journal of the Royal Anthropological Institute* **33**, 41–43.

Leakey, L. S. B. (1931). *The Stone Age Cultures of Kenya Colony.* Cambridge: Cambridge University Press.

—— (1932). The Oldoway Human Skeleton. *Nature* **129**, 721–722.

—— (1933). The Oldoway Human Skeleton. *Nature* **131**, 397–398.

—— (1934). *Adam's Ancestors*. London: Methuen.

—— (1935). *The Stone Age Races of Kenya*. London: Oxford University Press.

—— (1951). *Olduvai Gorge. A Report on the Evolution of the Hand-axe Culture in Beds I–IV*. Cambridge: Cambridge University Press.

—— (1974). *By the Evidence. Memoirs, 1932–1951*. London: Harcourt Brace Jovanovich.

Leakey, M. (1984). *Disclosing the Past*. London: Weidenfeld and Nicolson.

Lester, J. (1995). *E. Ray Lankester and the Making of Modern British Biology, Edited and with Additional Material by Peter J. Bowler*. Oxford: British Society for the History of Science, Monograph No. 9.

Lewis, C. L. E. (2000). *The Dating Game. One Man's Search for the Age of the Earth*. Cambridge: Cambridge University Press.

—— and Knell, S. J. (2001). The Age of the Earth: From 4004 BC to AD 2002. London: Geological Society, Special Publication No. 190.

Lohest, M., and Fourmarier, P. (1923). Les silex d'Ipswich—Rapport de Max Lohest et Paul Fourmarier. *Revue anthropologique* **23**, 54–57.

Lowe, C. van R. (1936). The Nomenclature of Palaeolithic Finds. *Man* **36**, 199–200.

—— and Breuil, H. (1945). The Evolution of the Levallois Technique in South Africa. *Man* **45**, 49–59.

Lubbock, J. (1862). On the Evidence of the Antiquity of Man, afforded by the Physical Structure of the Somme Valley. *Natural History Review* **2**, 244–269.

—— (1864). Cave Men. *Natural History Review* **4**, 407–428.

—— (1865). *Pre-Historic Times as illustrated by Ancient Remains and the Manners and Customs of Modern Savages*. London: Williams and Norgate.

—— (1866). Address delivered to the Section of 'Primaeval Antiquities' at the London Meeting of the Archaeological Institute, 1866. *The Archaeological Journal* **23**, 190–211.

—— (1869). *Prehistoric Times as illustrated by Ancient Remains and the Manners and Customs of Modern Savages* (2nd ed.). London: Williams and Norgate.

—— (1878). *Prehistoric Times as illustrated by Ancient Remains and the Manners and Customs of Modern Savages* (4th ed.). London: Williams and Norgate.

—— (1913). *Prehistoric Times as illustrated by Ancient Remains and the Manners and Customs of Modern Savages* (7th ed.). London: Williams and Norgate.

Lyell, C. (1830–33). *Principles of Geology, being an Attempt to Explain the Former Changes of the Earth's Surface, by Reference to Causes Now in Operation* (3 vols.). London: John Murray.

—— (1840a). On the Boulder Formation or Drift, and Associated Freshwater Deposits composing the Mud Cliffs of Eastern Norfolk. *Proceedings of the Geological Society of London* **3**, 171–179.

—— (1840b). On the Geological Evidence of the former existence of Glaciers in Forfarshire. *Proceedings of the Geological Society of London* **3**, 337–345.

—— (1841). *Elements of Geology* (2 vols; 2nd ed.). London: John Murray.

Lyell, C. (1853). *Principles of Geology, or the Modern Changes of the Earth and its Inhabitants Considered as Illustrative of Geology* (9th ed.). London: John Murray.

—— (1855). *A Manual of Elementary Geology: the Ancient Changes of the Earth and its Inhabitants as Illustrated by Geological Monuments* (5th ed.). London: John Murray.

—— (1859). On the Occurrence of Works of Human Art in Post-Pliocene Deposits. *Report of the British Association for the Advancement of Science* 29 (Aberdeen), 93–95.

—— (1863a). *The Geological Evidences of the Antiquity of Man* (3rd ed.). London: John Murray.

—— (1863b). The Antiquity of Man. *The Athenaeum* 1851, 523–525.

—— (1867). *Principles of Geology, or the Modern Changes of the Earth and its Inhabitants Considered as Illustrative of Geology* (10th ed.). London: John Murray.

Lyon, J. (1970). The Search for Fossil Man: Cinq personnages à la recherche du temps perdu. *Isis* 61, 68–84.

MacCurdy, G. (1905). The Eolithic Problem—Evidences of a Rude Industry Antedating the Paleolithic. *American Anthropologist* 7, 425–479.

—— (1907). Some Phases of Prehistoric Archaeology. *Proceedings of the American Association for the Advancement of Science* 56, 543–567.

—— (1910). Recent Discoveries Bearing on the Antiquity of Man in Europe. *Annual Report of the Board of Regents of the Smithsonian Institute for 1909*, 531–583.

Mackie, S. J. (1865a). Notice to Subscribers, Contributors, & Advertisers. *The Geological and Natural History Repertory; and Journal of Pre-Historic Archaeology and Ethnology* 1.

—— (1865b). The Editor's Introduction of his Work to the World. *The Geological and Natural History Repertory; and Journal of Pre-Historic Archaeology and Ethnology* 1, 5–8.

Mantell, G. (1850–51). On the Remains of Man, and Works of Art imbedded in Rocks and Strata, as Illustrative of the Connexion between Archaeology and Geology. *The Edinburgh New Philosophical Journal* 50, 235–254.

Marr, J. E. (1920). Man and the Ice Age. *Proceedings of the Prehistoric Society of East Anglia* iii, 177–191.

—— Excavations at High Lodge, Mildenhall, in 1920, A.D. Report on the Geology. *Proceedings of the Prehistoric Society of East Anglia* iii, 353–367.

Marston, A. T. (1936). Preliminary Note on a New Fossil Human Skull from Swanscombe, Kent. *Nature* 138, 200–201.

—— (1937). The Swanscombe Skull. *Journal of the Royal Anthropological Institute* 67, 339–406.

Mason, R. J. (1965). The Abbé H. Breuil in South Africa (1929–1950). In E. Ripoll Perelló (Ed.), *Miscelánea en homenaje al abate Henri Breuil* (pp. 141–147). Barcelona: Instituto de Prehistoria y Arqueologia.

—— (1967). Questions of Terminology in Regard to the Study of Earlier Stone Age Cultures in South Africa. In W. W. Bishop, and J. D. Clark (Eds.), *Background to Evolution in Africa* (pp. 765–769). Chicago: University of Chicago Press.

McBurney, C. B. M. (1950). The Geographical Study of the Older Palaeolithic Stages in Europe. *Proceedings of the Prehistoric Society* 16, 163–183.

—— (1953). Review of Leakey, L.S.B. (1953), 'Adam's Ancestors.' *Proceedings of the Prehistoric Society* **19**, 127.

McNabb, J. (1996). Through the Looking Glass: An Historical Perspective on Archaeological Research at Barnfield Pit, Swanscombe, ca.1900–1964. In B. Conway, J. McNabb, and N. Ashton (Eds.), *Excavations at Barnfield Pit, Swanscombe, 1968–72* (pp. 31–51). London: British Museum Press.

—— (1998). The History of Investigations at East Farm Pit, Barnham. In N. Ashton, G. Lewis, and S. Parfitt (Eds.), *Excavations at the Lower Palaeolithic site at East Farm, Barnham, Suffolk 1989–1994* (pp. 5–12). London: British Museum Occasional Paper 125.

Miller, S. H., and Skertchly, S. B. J. (1878). *The Fenland Past and Present*. London: Longmans, Green, and Co.

Milne-Edwards, H. (1864). Sur de nouvelles observations de MM. Lartet et Christy, relatives à l'existence de l'homme dans le centre de la France à une époque ou cette contrée était habitée par le renne et d'autres animaux qui n'y vivent pas de nos jours. *Comptes rendues hebdomadaires des séances de l'académie des sciences* **58**, 401–408.

Mitchell, G. H. (1961). Percy George Hamnall Boswell 1886–1960. *Biographical Memoirs of Fellows of the Royal Society* **7**, 17–30.

Moir, J. R. (1910). The Discovery of the Flint Implements of Pre-Crag Man, *The Times*, 17th October, 8.

—— (1911). The Flint Implements of Sub-Crag Man. *Proceedings of the Prehistoric Society of East Anglia* **i**, 17–24.

—— (1912a). The Natural Fracture of Flint and its Bearing upon Rudimentary Flint Implements. *Proceedings of the Prehistoric Society of East Anglia* **i**, 171–185.

—— (1912b). The Natural Fracture of Flint. *Nature* **90**, 461–463.

—— (1912c). The Making of a Rostro-carinate Flint Implement. *Nature* **90**, 334.

—— (1912d). The Occurrence of a Human Skeleton in a Glacial Deposit at Ipswich. *Proceedings of the Prehistoric Society of East Anglia* **i**, 194–202.

—— (1913a). Flint Implements of Man from the Middle Glacial Gravel and the Chalky Boulder Clay of Suffolk. *Proceedings of the Prehistoric Society of East Anglia* **i**, 307–319.

—— (1913b). A Defence of the 'Humanity' of the Pre-River Valley Implements of the Ipswich District. *Proceedings of the Prehistoric Society of East Anglia* **i**, 368–374.

—— (1913c). The Sub-Crag Flints. *The Geological Magazine* **10**, 553–555.

—— (1914). The Fractured Flints of the Eocene 'Bull-Head' Bed at Coe's Pit, Bramford near Ipswich. *Proceedings of the Prehistoric Society of East Anglia* **i**, 397–404.

—— (1915a). On the Further Discoveries of Flint Implements of Man beneath the Base of the Red Crag of Suffolk. *Proceedings of the Prehistoric Society of East Anglia* **ii**, 12–31.

—— (1915b). Pre-Palaeolithic Flint Implements. *The Geological Magazine* **Decade VI, Vol. II**, 46–47.

Moir, J. R. (1915c). Flints from the Suffolk Bone Bed. *The Geological Magazine* **Decade VI, Vol. II**, 191.

Moir, J. R. (1915d). Human Palaeontology in England. *The Geological Magazine* Decade VI, Vol. II, 476–478.

—— (1916). On the Evolution of the Earliest Palaeoliths from the Rostro-carinate Implements. *Journal of the Royal Anthropological Institute* 46, 197–220.

—— (1918). The Ancestry of the Mousterian Palaeolithic Flint Implements. *Proceedings of the Prehistoric Society of East Anglia* ii, 508–519.

—— (1919). *Pre-Palaeolithic Man.* Ipswich: W.E. Harrison.

—— (1920a). The Transition from Rostro-carinate Flint Implements to the Tongue-shaped Implements of River-terrace Gravels. *Philosophical Transactions of the Royal Society of London* 209, 329–350.

—— (1920b). The Geological Age of the Earliest Palaeolithic Flint Implements. *The Geological Magazine* 57, 221–224.

—— (1921a). Further Discoveries of Humanly-fashioned Flints in and beneath the Red Crag of Suffolk. *Proceedings of the Prehistoric Society of East Anglia* iii, 389–430.

—— (1921b). On an Early Chellian-Palaeolithic Workshop-Site in the Pliocene 'Forest Bed' of Cromer, Norfolk. *Journal of the Royal Anthropological Institute* 51, 385–418.

—— (1921c). Flint Implements from the Cromer Forest Bed. *Nature* 106, 756–757.

—— (1921d). II. A Description of the Humanly-fashioned Flints found during the Excavations at High Lodge, Mildenhall. *Proceedings of the Prehistoric Society of East Anglia* iii, 367–372.

—— (1922). The Red Crag Flints of Foxhall. *Man* 22, 104–105.

—— (1923). An Early Palaeolith from the Glacial Till at Sidestrand, Norfolk. *The Antiquaries Journal* 3, 135–137.

—— (1924a). Tertiary Man in England. *Natural History* 24, 636–654.

—— (1924b). Some Archaeological Problems. *Proceedings of the Prehistoric Society of East Anglia* iv, 234–240.

—— (1927a). The Silted-up Lake of Hoxne and its Contained Flint Implements. *Proceedings of the Prehistoric Society of East Anglia* v, 137–165.

—— (1927b). *The Antiquity of Man in East Anglia.* Cambridge: Cambridge University Press.

—— (1930). Obituary. Sir Edwin Ray Lankester, K.C.B., F.R.S., D.Sc., LL.D., 1847–1929. *Proceedings of the Prehistoric Society of East Anglia* vi, 140–142.

—— (1935). *Prehistoric Archaeology and Sir Ray Lankester.* Ipswich: Norman Adlard & Co. Ltd.

—— (1939). *The Earliest Men.* Huxley Memorial Lecture. London: Macmillan and Co., Ltd.

—— and Barnes, A. S. (1923). A Criticism of Mr. S. H. Warren's Views on Subsoil Pressure-flaking of Flints. *The Geological Magazine* 60, 526–528.

—— and Burchell, J. P. T. (1930). Flint Implements of Upper Palaeolithic facies from beneath the uppermost Boulder Clay of Norfolk and Yorkshire. *The Antiquaries Journal* 10, 359–383.

—— and —— (1935). Diminutive Flint Implements of Pliocene and Pleistocene Age. *The Antiquaries Journal* 15, 119–133.

Molleson, T. I. (1983). Obituary. Kenneth Page Oakley. *Proceedings of the Geologists' Association* **94**, 95.

Moore, D. T. (2004). Falconer, Hugh (1808–1865). In H. C. G. Matthew, & B. Harrison (Eds.), *Oxford Dictionary of National Biography* (vol. 18, pp. 967–968). Oxford: Oxford University Press.

Morlot, A. (1855). On the Post-tertiary and Quaternary Formations of Switzerland. *Edinburgh New Philosophical Journal* **2** (new series), 14–29.

Morrell, J., and Thackray, A. (1981). Gentlemen of Science: Early Years of the British Association for the Advancement of Science. Oxford: Clarendon Press.

Morris, J. (1838). On the Deposits containing Carnivora and other Mammalia in the Thames Valley. *The Magazine of Natural History* **2** (new series), 539–546.

Moser, S. (1998). *Ancestral Images: The Iconography of Human Origins*. Stroud: Sutton Publishing Limited.

Movius, H. (1948). The Lower Palaeolithic Cultures of Southern and Eastern Asia. *Transactions of the American Philosophical Society* **38**, 329–420.

—— (1949). Old World Palaeolithic Archaeology. *American Geological Society Bulletin* **50**, 1443–1456.

—— (1953). Old World Prehistory: Paleolithic. In A. L. Kroeber (Ed.), *Anthropology Today: An Encyclopedic Inventory* (pp. 163–192). Chicago: University of Chicago Press.

Murchison, C. (1868). *Palaeontological Memoirs and Notes of the late Hugh Falconer* (2 vols.). London: Robert Hardwicke.

Murphy, P. J., and Lord, T. (2003). Victoria Cave, Yorkshire, UK: New Thoughts on an Old Site. *Cave and Karst Science* **30**, 83–88.

Newbigin, M. I., and Flett, J. S. (1917). *James Geikie. The Man and the Geologist*. Edinburgh: Oliver and Boyd.

Newton, E. T. (1895). On a Human Skull and Limb-bones found in the Palaeolithic Terrace-gravel at Galley Hill, Kent. *Quarterly Journal of the Geological Society of London* **51**, 505–527.

—— (1898). The Evidence for the Existence of Man in the Tertiary Period. *Proceedings of the Geologists' Association* **15**, 63–82.

Nicholas, T. C. (1963). Obituary. William Bernard Robinson King. *Proceedings of the Geological Society of London*, 150–153.

Oakley, K. P. (1937a). Field Meeting at Taplow, Burnham and Iver, Bucks. *Proceedings of the Geologists' Association* **48**, 276–279.

—— (1937b). Review of a translation by R. A. S. Macalister of Schmidt, R. R. (1936), 'The Dawn of the Human Mind.' *Proceedings of the Prehistoric Society* **3**, 187–188.

—— (1939). Field Meeting at Swanscombe, Kent. *Proceedings of the Geologists' Association* **50**, 357–361.

—— (1943). The Future of Quaternary Research in Britain. *South-Eastern Naturalist and Antiquary* **48**, 25–32.

Oakley, K. P. assisted by Lacaille, A. D., in conjunction with Hawkes, C. F. C. (1948a). Survey of the Present Position in British Prehistoric and Early Historic Archaeology.

1: The Palaeolithic Age, *A Survey and Policy of Field Research in the Archaeology of Great Britain. I. The Prehistoric and Early Historic Ages to the Seventh Century A.D.* (pp. 13–25). London: Council for British Archaeology.

—— assisted by Lacaille, A. D., in conjunction with Hawkes, C. F. C. (1948b). Outstanding Problems in British Prehistoric and Early Historic Archaeology: Policy and Recommendations for Field Research. 1: The Palaeolithic Age, *A Survey and Policy of Field Research in the Archaeology of Great Britain. I. The Prehistoric and Early Historic Ages to the Seventh Century A.D.* (pp. 79–82). London: Council for British Archaeology.

—— (1952). Swanscombe Man. *Proceedings of the Geologists' Association* 63, 271–300.

—— (1959). The Life and Work of Samuel Hazzledine Warren. *Essex Naturalist* 30, 143–161.

—— (1964a). The Problem of Man's Antiquity. An Historical Survey. *Bulletin of the British Museum (Natural History) Geology Series* 9, 83–155.

—— (1964b). *Frameworks for Dating Fossil Man.* London: Weidenfeld and Nicolson.

—— (1969). *Frameworks for Dating Fossil Man* (3rd ed.). London: Weidenfeld and Nicolson.

—— and King, W. B. R. (1945). Age of the Baker's Hole Coombe Rock, Northfleet, Kent. *Nature* 155, 51–52.

—— and Leakey, L. S. B. (1935). Fossil Human Occipital Bone from Thames Gravels. *Nature* 136, 916–917.

—— and Leakey, M. (1937). Report on Excavations at Jaywick Sands, Essex (1934), with some Observations on the Clactonian Industry, and on the Fauna and Geological Significance of the Clacton Channel. *Proceedings of the Prehistoric Society* 3, 217–260.

Obermaier, H. (1904). Le Quaternaire des Alpes, et la nouvelle classification du Professeur Albrecht Penck. *L'Anthropologie* 15, 25–36.

—— (1906a). Quaternary Human Remains in Central Europe. *Annual Report of the Board of Regents of the Smithsonian Institution for 1906*, 373–397.

—— (1906b). Beiträge zur Kenntnis des Quartärs in den Pyrenäen. *Archiv fur Anthropologie* 32, 299–310.

—— (1908). Die Steingerate des franzosischen Altpalaolithikums. Eine kritische Studie uber ihre Stratigraphie und Evolution. *Mitteilungen der Prähistorischen Kommission der Akademie der Wissenschaften* 2, 41–125.

—— (1909). Les Formations glaciaires des Alpes et l'homme paléolithique. *L'Anthropologie* 20, 497–522.

—— (1919). Das Palaolithikum und Epipalaolithikum Spaniens. *Anthropos* 14, 143–179.

O'Connor, A. (2005). The Competition for the Woodwardian Chair of Geology, Cambridge 1873. *British Journal for the History of Science* 38, 437–461.

O'Connor, R. (2007) forthcoming. *The Earth on Show: Fossils and the Poetics of Popular Science, 1802–1856.* Chicago: Chicago University Press.

O'Connor, J. G., and Meadows, A. J. (1976). Specialization and Professionalization in British Geology. *Social Studies of Science* 6, 77–89.

Oldroyd, D. (1990). *The Highlands Controversy. Constructing Geological Knowledge through Fieldwork in Nineteenth-Century Britain.* Chicago: University of Chicago Press.

—— (1996). *Thinking About the Earth: A History of Ideas in Geology.* London: Athlone Press.

—— (1999). Early Ideas about Glaciation in the English Lake District: The Problem of Making Sense of Glaciation in a Glaciated Region. *Annals of Science* **56**, 175–203.

—— (2004a). Geikie, Sir Archibald (1835–1924). In H. C. G. Matthew, & B. Harrison (Eds.), *Oxford Dictionary of National Biography* (vol. 21, pp. 721–723). Oxford: Oxford University Press.

—— (2004b). Marr, John Edward (1857–1933). In H. C. G. Matthew, & B. Harrison (Eds.), *Oxford Dictionary of National Biography* (vol. 36, pp. 753–754). Oxford: Oxford University Press.

—— (2004c). Barrow, George (1853–1932). In H. C. G. Matthew, & B. Harrison (Eds.), *Oxford Dictionary of National Biography* (vol. 4, pp. 94–95). Oxford: Oxford University Press.

—— (2006). *Earth Cycles: A Historical Perspective.* Westport: Greenwood Press.

—— and McKenna, G. (2005). Conditions of Employment and Work Practices in the Early Years of the Geological Survey of Great Britain. *Earth Sciences History* **24**, 197–223.

Osborn, H. F. (1915). Review of the Pleistocene of Europe, Asia and Northern Africa. *Annals of the New York Academy of Sciences* **26**, 215–315.

—— (1916). *Men of the Old Stone Age, their Environment, Life and Art.* London: G. Bell and Sons, Ltd.

—— (1922). Pliocene (Tertiary) and Early Pleistocene (Quaternary) Mammalia of East Anglia, Great Britain, in Relation to the Appearance of Man. *The Geological Magazine* **59**, 434–441.

—— and Reeds, C. A. (1922). Old and New Standards of Pleistocene Division in Relation to the Prehistory of Man in Europe. *Bulletin of the Geological Society of America* **33**, 411–490.

Owen, R. (1843). Report on the British Fossil Mammalia. Part II. Ungulata. *Report of the British Association for the Advancement of Science* **13** (Cork), 208–241.

—— (1846). *A History of British Fossil Mammals, and Birds.* London: John Van Voorst.

Paterson, T. T. (1937). Studies on the Palaeolithic Succession in England. No. I. The Barnham Sequence. *Proceedings of the Prehistoric Society* **3**, 87–135.

—— (1940a). The Swanscombe Skull: A Defence. *Proceedings of the Prehistoric Society* **6**, 166–169.

—— (1940b). Geology and Early Man. *Nature* **146**, 12–15, 49–52.

—— (1941). On a World Correlation of the Pleistocene. *Transactions of the Royal Society of Edinburgh* **60**, 373–425.

Paterson, T. T. (1942). *Lower Palaeolithic Man in the Cambridge District.* Ph.D. thesis: Department of Archaeology, University of Cambridge.

Paterson, T. T. (1945). Core, Culture and Complex in the Old Stone Age. *Proceedings of the Prehistoric Society* 11, 1–19.

—— and Fagg, B. E. B. (1940). Studies on the Palaeolithic Succession in England. No. II. The Upper Brecklandian (Elveden). *Proceedings of the Prehistoric Society* 6, 1–29.

Peake, H. (1922). Archaeological Finds in the Kennet Gravels near Newbury. *The Antiquaries Journal* 2, 125–130.

—— (1930). Reply to Boswell, P. G. H. (1930) 'Early Man and the Correlation of Glacial Deposits.' *Report of the British Association for the Advancement of Science* 100 (Bristol), 382–383.

—— and Fleure, H. J. (1927). *Apes and Men*. Oxford: Clarendon Press.

Penck, A. (1897). On the Glacial Formations of the Alps. *Report of the British Association for the Advancement of Science* 67 (Toronto), 647.

—— and Brückner, E. (1901–09). *Die Alpen im Eiszeitalter*. Leipzig: Tauchnitz.

Pengelly, H. (1897). *A Memoir of William Pengelly, of Torquay, F.R.S., Geologist, with a Selection from his Correspondence*. London: John Murray.

Pengelly, W. (1873). The Cavern discovered in 1858 in Windmill Hill, Brixham, South Devon. *Transactions of the Devonshire Association for the Advancement of Science, Literature, and Art* 6, 775–856.

—— (1874). On the Flint and Chert Implements found in Kent's Cavern, near Torquay, Devonshire. *The Quarterly Journal of Science* 4, **new series**, 141–155.

—— (1877). Address to Section C. Geology. *Report of the British Association for the Advancement of Science* 47 (Plymouth), 54–66.

—— (1883). Address by the Vice President of the Anthropology Section. *Report of the British Association for the Advancement of Science* 53 (Southport), 549–561.

Peyrony, D. (1930). Le Moustier. Ses gisements, ses industries, ses couches géologiques. *Revue anthropologique* 40, 155–176.

—— (1933). La Micoque et ses diverses industries. *Congrès international d'anthropologie et d'archéologie préhistorique. Compte rendu de la 15 session, Paris, 1931*, 435–441. Paris: Librairie E. Nourry.

Phillips, J. (1855). *Manual of Geology: Practical and Theoretical*. London and Glasgow: Richard Griffin and Company.

Pike, K., and Godwin, H. (1952). The Interglacial at Clacton-on-Sea, Essex. *Quarterly Journal of the Geological Society of London* 108, 261–272.

Plunkett, S. (1999). Nina Frances Layard, Prehistorian (1853–1935). In W. Davies, and R. Charles (Eds.), *Dorothy Garrod and the Progress of the Palaeolithic. Studies in the Prehistoric Archaeology of the Near East and Europe* (pp. 242–262). Oxford: Oxbow Books.

Pocock, T. I. (1903). On the Drifts of the Thames Valley near London. In J. Teall (Ed.), *Summary of Progress of the Geological Survey of the United Kingdom and Museum of Practical Geology for 1902* (pp. 199–207). London: H.M.S.O.

Porter, R. (1978). Gentlemen and Geology: The Emergence of a Scientific Career, 1660–1920. *The Historical Journal* 21, 809–836.

—— (1982). The Natural Sciences Tripos and the 'Cambridge School of Geology,' 1850–1914. *History of Universities* 2, 193–216.

Portlock, J. E. (1857). Obituary. William Buckland. *Proceedings of the Geological Society of London* 13, xxvi–xlv.

—— (1858). Obituary. Joshua Trimmer. *Proceedings of the Geological Society of London* 14, xxxii–xxxvi.

Preller, C. S. Du R. (1894). On the Three Glaciations in Switzerland. *The Geological Magazine* Decade IV, Vol I, 27–35.

Prestwich, G. (1899). *Life and Letters of Sir Joseph Prestwich.* London: William Blackwood and Sons.

Prestwich, J. (1856). Note on the Gravel near Maidenhead, in which the Skull of the Musk Buffalo was found. *Quarterly Journal of the Geological Society of London* 12, 131–133.

—— (1859). Sur la découverte d'instruments en silex associés à des restes de mammifères d'espèces perdues dans des couches non remaniées d'une formation géologique recent. Lettre de M. Prestwich à M. Élie de Beaumont. *Comptes rendues hebdomadaires des séances de l'académie des sciences* 49, 634–636.

—— (1860a). On the Occurrence of Flint-implements, associated with the Remains of Animals of Extinct Species in Beds of a Late Geological Period, in France at Amiens and Abbeville, and in England at Hoxne. *Philosophical Transactions of the Royal Society of London* 150, 277–318.

—— (1860b). On the Occurrence of Flint-implements, associated with the Remains of Extinct Mammalia, in Undisturbed Beds of a Late Geological Period. *Proceedings of the Royal Society of London* 10, 50–59.

—— (1861a). Notes on some Further Discoveries of Flint Implements in Beds of Post-Pliocene Gravel and Clay; with a few Suggestions for Search Elsewhere. *Quarterly Journal of the Geological Society of London* 17, 362–368.

—— (1861b). On the Occurrence of the *Cyrena fluminalis,* together with Marine Shells of Recent Species, in Beds of Sand and Gravel over Beds of Boulder-clay near Hull; with an Account of some Borings and Well-sections in the same District. *Quarterly Journal of the Geological Society of London* 17, 446–456.

—— (1863). The Antiquity of Man. *The Athenaeum* 1852, 555.

—— (1864). Theoretical Considerations on the Conditions under which the (Drift) Deposits containing the Remains of Extinct Mammalia and Flint Implements were accumulated, and on their Geological Age. *Philosophical Transactions of the Royal Society* 154, 247–310.

—— (1872). Obituary. Edouard Lartet. *Quarterly Journal of the Geological Society of London* 28, xlv–xlix.

—— (1873). Report on the Exploration of Brixham Cave. *Philosophical Transactions of the Royal Society of London* 163, 471–572.

—— (1887). Consideration on the Date, Duration, and Conditions of the Glacial Period, with reference to the Antiquity of Man. *Quarterly Journal of the Geological Society of London* 43, 393–408.

Prestwich, J. (1888). *Geology. Chemical, Physical and Stratigraphical* (2 vols.). Oxford: Clarendon Press.

Prestwich, J. (1889). On the Occurrence of Palaeolithic Flint Implements in the Neighbourhood of Ightham, Kent; their Distribution and Probable Age. *Quarterly Journal of the Geological Society of London* **45**, 270–297.

—— (1891). On the Age, Formation, and Successive Drift Stages of the Valley of the Darent; with Remarks on the Palaeolithic Implements of the District, and on the Origin of its Chalk Escarpment. *Quarterly Journal of the Geological Society of London* **47**, 126–163.

—— (1892). On the Primitive Characters of the Flint Implements of the Chalk Plateau of Kent, with reference to the Question of their Glacial or Pre-Glacial Age. *Journal of the Royal Anthropological Institute* **21**, 246–262.

Rainger, R. (1991). *An Agenda for Antiquity: Henry Fairfield Osborn and Vertebrate Paleontology at the American Museum of Natural History, 1890–1935*. Tuscaloosa: University of Alabama Press.

Rames, J. (1884). Géologie du Puy Courny. Éclats de silex tortoniens du basin d'Aurillac (Cantal). *Matériaux pour l'histoire primitive et naturelle de l'homme* **1** (third series), 385–406.

Ramsay, A. C. (1852). On the Superficial Accumulations and Surface-markings of North Wales. *Quarterly Journal of the Geological Society of London* **8**, 371–376.

—— (1862). On the Glacial Origin of certain Lakes in Switzerland, the Black Forest, Great Britain, Sweden, North America, and Elsewhere. *Quarterly Journal of the Geological Society of London* **18**, 185–205.

—— (1878). *The Physical Geology and Geography of Great Britain* (5th ed.). London: Edward Stanford.

—— (1894). *The Physical Geology and Geography of Great Britain* (6th ed.). London: Edward Stanford.

Read, C. H. (1902). *A Guide to the Antiquities of the Stone Age in the Department of British and Mediaeval Antiquities*. London: William Clowes and Sons, Limited.

—— (1911). *A Guide to the Antiquities of the Stone Age in the Department of British and Mediaeval Antiquities* (2nd ed.). London: William Clowes and Sons, Limited.

Reboux, J. (1869). Note on Flint Implements found in the Paris Beds, *International Congress of Prehistoric Archaeology: Transactions of the 3rd Session, London, 1868* (pp. 222–223). London: Longmans, Green, and Co.

—— (1873). Recherches paléontologiques dans le bassin de Paris, *Congrès international d'anthropologie et d'archéologie préhistoriques. Compte rendu de la 5 session, Bologne, 1871* (pp. 98–102). Bologne: Fava et Garagnani.

—— (1876). Résumé des recherches faites dans le quaternaire de Paris, *Congrès international d'anthropologie et d'archéologie préhistoriques. Compte rendu de la 7 session, Stockholm, 1874* (pp. 64–68). Stockholm: P.A. Norstedt & Söner.

Regal, B. (2002). *Henry Fairfield Osborn. Race, and the Search for the Origins of Man*. Aldershot: Ashgate Publishing Limited.

—— (2004). *Human Evolution: A Guide to the Debates*. Oxford: Abc-Clio.

Reid, C. (1896). The Relation of Palaeolithic Man to the Glacial Epoch. Report of the Committee [...] appointed to ascertain by excavation at Hoxne the relation of the Palaeolithic Deposits to the Boulder Clay, and to the deposits with Arctic and

Temperate Plants. *Report of the British Association for the Advancement of Science* **66** (Liverpool), 400–416.

—— (1897). The Palaeolithic Deposits at Hitchin and their Relation to the Glacial Epoch. *Proceedings of the Royal Society of London* **61**, 40–49.

Reinach, S. (1889). *Antiquités nationales. Description raisonnée du musée de Saint-Germain-en-Laye. I. Époque des alluvions et des cavernes.* Paris: Librairie de Firmin-Didot.

—— (1897). La station de Taubach, près de Weimar. *L'Anthropologie* **8**, 53–60.

—— (1899). Gabriel de Mortillet. *Revue historique* **69**, 67–95.

—— (1908). Review of Obermaier, H. (1908), 'Die Steingerate des franzosischen Altpalaolithikums.' *Revue archéologique* **1**, 305.

—— (1919). Victor Commont. *Revue archéologique* **9** (fifth series), 197–198.

—— (1929). Les silex dits rostro-carénés. *Revue archéologique* **30** (fifth series), 327.

Ribeiro, C. (1873). Sur les silex taillés, découverts dans les terrains miocène et pliocène du Portugal, *Congrès international d'anthropologie et d'archéologie préhistoriques. Compte rendu de la 6 session, Bruxelles, 1872* (pp. 95–100). Brussels: C. Muquardt.

Richard, N. (1999a). Gabriel de Mortillet 1821–1898. In T. Murray (Ed.), *Encyclopaedia of Archaeology. The Great Archaeologists* (pp. 93–107). Oxford: Abc-Clio.

—— (1999b). Marcellin Boule 1861–1942. In T. Murray (Ed.), *Encyclopaedia of Archaeology. The Great Archaeologists* (pp. 263–273). Oxford: Abc-Clio, Inc.

—— (2002). Archaeological Arguments in National Debates in Late 19th-Century France: Gabriel de Mortillet's *La Formation de la nation français* (1897). *Antiquity* **76**, 177–184.

Riper, A. B. Van (1990). *Discovering Prehistory: Geology, Archaeology and the Human Antiquity Problem in Mid-Victorian Britain.* Ph.D. thesis: University of Wisconsin, Madison.

—— (1993). *Men among the Mammoths. Victorian Science and the Discovery of Prehistory.* Chicago: University of Chicago Press.

Robinson, W. (1874). On the Antiquity of Man. *The Cambridge Independent Press*, 2nd May.

Roe, D. (1981). Amateurs and Archaeologists: Some Early Contributions to British Palaeolithic Studies. In J. D. Evans, B. Cunliffe, and C. Renfrew (Eds.), *Antiquity and Man: Essays in honour of Glyn Daniel* (pp. 214–220). London: Thames and Hudson.

—— (2004a). Oakley, Kenneth Page (1911–1981). In H. C. G. Matthew, & B. Harrison (Eds.), *Oxford Dictionary of National Biography* (vol. 41, pp. 325–326). Oxford: Oxford University Press.

—— (2004b). McBurney, Charles Brian Montagu (1914–1979). In H. C. G. Matthew, and B. Harrison (Eds.), *Oxford Dictionary of National Biography.* Oxford: Oxford University Press.

Rowley-Conwy, P. (1996). Why didn't Westropp's 'Mesolithic' catch on in 1872? *Antiquity* **70**, 940–944.

Rudwick, M. J. S. (1985). *The Great Devonian Controversy. The Shaping of Scientific Knowledge among Gentlemanly Specialists.* Chicago: University of Chicago Press.

Rudwick, M. J. S. (1986). The Shape and Meaning of Earth History. In D. C. Lindberg, and R. L. Numbers (Eds.), *God and Nature: Historical Essays on the Encounter between Christianity and Science* (pp. 296–321). Berkeley: University of California Press.

—— (1992). *Scenes from Deep Time: Early Pictorial Representations of the Prehistoric World.* Chicago: University of Chicago Press.

—— (1997). *Georges Cuvier, Fossil Bones, and Geological Catastrophes.* Chicago: University of Chicago Press.

—— (2005). *Bursting the Limits of Time: The Reconstruction of Geohistory in the Age of Revolution.* Chicago: University of Chicago Press.

Rupke, N. (1983). *The Great Chain of History. William Buckland and the English School of Geology (1814–1849).* Oxford: Clarendon Press.

Rutot, A. (1898). Sur les silex mesviniens de la Flandre occidentale. *Bulletin de la société d'anthropologie de Bruxelles* 17, 382–383.

—— (1900). Sur la distribution des industries paléolithiques dans les couches quaternaires de la Belgique. *L'Anthropologie* 11, 707–746.

—— (1903). L'État actuel de la question de l'antiquité de l'homme. *Bulletin de la société Belge de géologie de paléontologie et d'hydrologie* 17, 425–438.

—— (1921). The Discoveries at Spiennes. *The Antiquaries Journal* 1, 54–55.

Sackett, J. R. (1981). From de Mortillet to Bordes: A Century of French Palaeolithic Research. In G. Daniel (Ed.), *Towards a History of Archaeology* (pp. 85–99). London: Thames and Hudson.

—— (1991). Straight Archaeology French Style: The Phylogenetic Paradigm in Historic Perspective. In G. Clark (Ed.), *Perspectives on the Past: Theoretical Biases in Mediterranean Hunter-gatherer Research* (pp. 109–139). Philadelphia: University of Pennsylvania Press.

Sainty, J. E. (1927). An Acheulean Palaeolithic Workshop Site at Whitlingham, near Norwich. *Proceedings of the Prehistoric Society of East Anglia* v, 176–213.

Sandford, K. S. (1930). Contribution to the discussion on 'The Relation between past Pluvial and Glacial Periods.' *Report of the British Association for the Advancement of Science* 100 (Bristol), 379.

—— (1932). Some Recent Contributions to the Pleistocene Succession in England. *The Geological Magazine* 69, 1–18.

Savage, R. J. G. (1963). Martin Alister Campbell Hinton, 1883–1961. *Biographical Memoirs of Fellows of the Royal Society* 9, 155–170.

Schoetensack, O. (1908). *Der Unterkiefer des Homo heidelbergensis aus den Sanden von Mauer bei Heidelberg.* Leipzig: Verlag von Wilhelm Engelmann.

Schwalbe, G. (1906). *Studien zur Vorgescgichte des Menschen.* Stuttgart: E. Schweizerbartsche Verlagsbuschhandlung.

Secord, A. (1994a). Science in the Pub: Artisan Botanists in Early Nineteenth-Century Lancashire. *History of Science* 32, 269–315.

—— (1994b). Corresponding Interests: Artisans and Gentlemen in Nineteenth-Century Natural History. *British Journal for the History of Science* 27, 383–408.

Secord, J. A. (2000). *Victorian Sensation. The Extraordinary Publication, Reception, and Secret Authorship of Vestiges of the Natural History of Creation.* Chicago: University of Chicago Press.

Sedgwick, A. (1831). Address to the Geological Society on Retiring from the President's Chair. *Quarterly Journal of the Geological Society of London* **1**, 281–316.

Shaaffhausen, H. (1861). On the Crania of the Most Ancient Races of Man. *Natural History Review* **1**, 155–175.

Shackleton, N. J. (1967). Oxygen Isotope Analyses and Pleistocene Temperatures Re-assessed. *Nature* **215**, 15–17.

—— (1975). The Stratigraphic Record of Deep-Sea Cores and Its Implications for the Assessment of Glacials, Interglacials, Stadials, and Interstadials in the Mid-Pleistocene. In K. W. Butzer, and G. L. Isaac (Eds.), *After the Australopithecines. Stratigraphy, Ecology, and Culture Change in the Middle Pleistocene* (pp. 1–24). Paris: Mouton Publishers.

Sheets-Pyenson, S. (1982). Geological Communication in the Nineteenth Century: the Ellen S. Woodward Autograph Collection at McGill University. *Bulletin of the British Museum (Natural History), Historical Series* **10**, 179–226.

Sherlock, R. L. (1935). *British Regional Geology: London and Thames Valley.* Geological Survey publication. London: H.M.S.O.

—— with additions by Casey, R., Holmes, S. C. A., & Wilson, V (1960). *British Regional Geology: London and Thames Valley.* Geological Survey publication (3rd ed.). London: H.M.S.O.

Shipman, P. (2001). *The Man Who Found the Missing Link. The Extraordinary Life of Eugene Dubois.* London: Weidenfeld & Nicolson.

Shotton, F. W. (1963). William Bernard Robinson King, 1889–1963. *Biographical Memoirs of Fellows of the Royal Society* **9**, 171–182.

Skertchly, S. B. J. (1876a). On the Discovery of Palaeolithic Implements of Inter-glacial Age. *Nature* **14**, 448–449.

—— (1876b). Inter-Glacial Man. *The Geological Magazine* **Decade II, Vol. III**, 476.

—— (1877). The Antiquity of Man. *Nature* **16**, 142.

—— (1879). Evidence of the Existence of Palaeolithic Man during the Glacial Period in East Anglia. *Report of the British Association for the Advancement of Science* **49** (Sheffield), 379–380.

Slater, G. (1911). Excursion to Ipswich. *Proceedings of the Geologists' Association* **22**, 11–16.

Smith, P. E. L. (1962). The Abbé Henri Breuil and Prehistoric Archaeology. *Anthropologica* **4**, 199–208.

Smith, P. J. (2004). *A Splendid Idiosyncracy: Prehistory at Cambridge 1915–1950.* Ph.D. thesis: Department of Archaeology, University of Cambridge.

Smith, R. (1912). On the Classification of Palaeolithic Stone Implements. *Proceedings of the Geologists' Association* **23**, 137–147.

—— (1915a). Prehistoric Problems in Geology. *Proceedings of the Geologists' Association* **26**, 1–20.

Smith, R. (1915b). Researches at Rickmansworth: Report on Excavations made in 1914 on Behalf of the British Museum. *Archaeologia* 66, 195–224.

—— (1922). Flint Implements of Special Interest. *Archaeologia* 72, 25–40.

—— (1935). Notes on Foreign Periodicals. *Proceedings of the Prehistoric Society of East Anglia* vii, 417–418.

—— and Dewey, H. (1913a). Stratification at Swanscombe: Report on Excavations made on Behalf of the British Museum and H.M. Geological Survey. *Archaeologia* 64, 177–204.

—— and —— (1913b). Appendix 1. On some Palaeolithic Gravels near Swanscombe, Kent. In J. Teall (Ed.), *Summary of Progress of the Geological Survey of the United Kingdom and Museum of Practical Geology for 1912* (pp. 82–85). London: H.M.S.O.

—— and —— (1914). The High Terrace of the Thames: Report on Excavations made on Behalf of the British Museum and H.M. Geological Survey in 1913. *Archaeologia* 65, 187–212.

Smith, T., Hislop, W., and Wakefield, J. E. (1858). The Geologists' Association. Leaflet.

Smith, W. (1816–1819). *Strata Identified by Organized Fossils.* London: W. Arding.

—— (1817). *Stratigraphical System of Organized Fossils.* London: E. Williams.

Smith, W. G. (1864). Effects of Eating a Poisonous Fungus. *The Journal of Botany, British and Foreign* 2, 215–218.

—— (1879). On Palaeolithic Implements from the Valley of the Lea. *Journal of the Royal Anthropological Institute* 8, 275–279.

—— (1880). Palaeolithic Implements from the Valley of the Brent. *Journal of the Royal Anthropological Institute* 9, 316–320.

—— (1882). Palaeolithic Floors. *Nature* 25, 460.

—— (1883a). Palaeolithic Implements of North-east London. *Nature* 27, 270–274.

—— (1883b). Primeval Man in the Valley of the Lea. *Transactions of the Essex Field Club* 3, 102–147.

—— (1884). On a Palaeolithic Floor at North East London. *Journal of the Royal Anthropological Institute* 13, 357–384.

—— (1887). Primaeval Man in the Valley of the Lea. *Essex Naturalist* 1, 36–38, 83–91, 125–137.

—— (1888). Lepores Palaeolithici: Or, The Humorous Side of Flint Implement Hunting. *Essex Naturalist* 2, 7–12.

—— (1894). *Man the Primeval Savage.* London: Edward Stanford.

—— (1908). Eoliths. *Man* 8, 49–53.

—— (1916). Notes on the Palaeolithic Floor near Caddington. *Archaeologia* 67, 49–74.

Sollas, W. (1909). Obituary. Sir John Evans. *Proceedings of the Geological Society of London* 65, lviii–lxi.

—— (1910). Anniversary Address of the President. *Quarterly Journal of the Geological Society of London* 66, xlviii–lxxxviii.

—— (1911). *Ancient Hunters and their Modern Representatives.* London: Macmillan and Co.

—— (1913). The Formation of 'Rostro-carinate' Flints. *Report of the British Association for the Advancement of Science* 83 (Birmingham), 788–790.

—— (1915). *Ancient Hunters and their Modern Representatives* (2nd ed.). London: Macmillan and Co.

—— (1920). A Flaked Flint from the Red Crag. *Proceedings of the Prehistoric Society of East Anglia* **iii**, 261–267.

—— (1923). Man and the Ice Age. *Nature* **111**, 332–334.

—— (1924). *Ancient Hunters and their Modern Representatives* (3rd ed.). London: Macmillan and Co.

Solomon, J. D. (1930). Contribution to the discussion on 'The Relation between past Pluvial and Glacial Periods.' *Report of the British Association for the Advancement of Science* **100** (Bristol), 381–382.

—— (1932). The Glacial Succession on the North Norfolk Coast. *Proceedings of the Geologists' Association* **43**, 241–271.

Sommer, M. (2004a). 'An amusing account of a cave in Wales': William Buckland (1784–1856) and the Red Lady of Paviland. *The British Journal for the History of Science* **37**, 53–74.

—— (2004b). Eoliths as Evidence for Human Origins? The British Context. *History and Philosophy of Life Science* **26**, 209–241.

—— (2005). *Ancient Hunters and Their Modern Representatives*: William Sollas's (1849–1936) Anthropology from Disappointed Bridge to Trunkless Tree and the Instrumentalisation of Racial Conflict. *Journal of the History of Biology* **38**, 327–365.

—— (2006). Mirror, Mirror on the Wall: Neanderthal as Image and 'Distortion' in Early 20th-Century French Science and Press. *Social Studies of Science* **36**, 207–240.

Sorby, H. C. (1880). Award of the Lyell Medal to John Evans. *Proceedings of the Geological Society of London* **36**, 30.

Spaulding, A. C. (1953). Statistical Techniques for the Discovery of Artefact Types. *American Antiquity* **18**, 305–313.

Spencer, F. (1984). The Neanderthals and Their Evolutionary Significance: A Brief Historical Survey. In F. H. Smith, & F. Spencer (Eds.), *The Origins of Modern Humans: A World Survey of the Evidence* (pp. 1–49). New York: Alan R. Liss, Inc.

—— (1988). Prologue to a Scientific Forgery: The British Eolithic Movement from Abbeville to Piltdown. In G. W. Stocking (Ed.), *Bones, Bodies, Behavior. Essays on Biological Anthropology* (pp. 84–116). Wisconsin: University of Wisconsin Press.

—— (1990). *Piltdown: A Scientific Forgery.* London: Natural History Museum Publications.

Spurrell, F. C. J. (1880a). On the Discovery of the Place where Palaeolithic Implements were made at Crayford. *Quarterly Journal of the Geological Society of London* **36**, 544–548.

—— (1880b). On Implements and Chips from the Floor of a Palaeolithic Workshop. *The Archaeological Journal* **37**, 294–299.

—— (1883). Palaeolithic Implements found in West Kent. *Archaeologia Cantiana* **15**, 89–103.

Spurrell, F. C. J. (1884). On some Palaeolithic Knapping Tools and Modes of Using Them. *Journal of the Royal Anthropological Institute* **13**, 109–118.

Star, S. L., & Griesemer, J. R. (1989). Institutional Ecology, 'Translations' and Boundary Objects: Amateurs and Professionals in Berkeley's Museum of Vertebrate Zoology, 1907–39. *Social Studies of Science* 19, 387–420.

Stevens, E. T. (1870). *Flint Chips. A Guide to Pre-Historic Archaeology as illustrated by the Collection in the Blackmore Museum, Salisbury.* London: Bell and Daldy.

Stocking, G. W. (1987). *Victorian Anthropology.* Oxford: Macmillan.

Stockmans, F. (1966). Notice sur Aime Rutot. *Annuaire de academie royale de Belgique* 132, 3–123.

Strahan, A. (1914). Summary of Progress of the Geological Survey of the United Kingdom and Museum of Practical Geology for 1913. London: H.M.S.O.

—— (1925). William Whitaker. 1836–1925. *Proceedings of the Royal Society of London* 97, ix–xii.

Stubblefield, C. J. (1967). Henry Dewey. *Proceedings of the Geological Society of London* Session 1965–**66** (1636), 189–190.

Sturge, W. A. (1908). Presidential Address. *Proceedings of the Prehistoric Society of East Anglia* i, 9–16.

—— (1911). The Chronology of the Stone Age. *Proceedings of the Prehistoric Society of East Anglia* i, 43–105.

—— (1912). Implements of the Later Palaeolithic 'Cave' Periods in East Anglia. *Proceedings of the Prehistoric Society of East Anglia* i, 210–232.

Sutcliffe, A. J. (1998). Frederic Everard Zeuner—A Student's Appreciation and Appraisal. *Geologia Sudetica* 31, 129–132.

Sweeting, G. S. (1958). Origin and Growth. In G. S. Sweeting (Ed.), *The Geologists' Association 1858–1958. A History of the First Hundred Years* (pp. 1–26). Colchester: Benham and Company Limited.

Teall, J. (1911). Summary of Progress of the Geological Survey of the United Kingdom and Museum of Practical Geology for 1910. London: H.M.S.O.

—— (1912). Summary of Progress of the Geological Survey of the United Kingdom and Museum of Practical Geology for 1911. London: H.M.S.O.

—— (1913). Summary of Progress of the Geological Survey of the United Kingdom and Museum of Practical Geology for 1912. London: H.M.S.O.

Thackray, J. C. (2003). To See the Fellows Fight. Eye Witness Accounts of Meetings at the Geological Society of London and its Club, 1822–1868. Oxford: The British Society for the History of Science.

—— (2004). Prestwich, Sir Joseph (1812–1896). In H. C. G. Matthew, & B. Harrison (Eds.), *Oxford Dictionary of National Biography* (vol. 45, pp. 276–278). Oxford: Oxford University Press.

Thomas, H. D. (1954). Obituary of Frederick James Nairn Haward. *Proceedings of the Geologists' Association* 65, 88–89.

Thomas, H. H. (1929). Obituary. Aubrey Strahan. *Proceedings of the Geological Society of London* 85, lviii–lix.

Tiddeman, R. H. (1873). The Relation of Man to the Ice-sheet in the North of England. *Nature* 9, 14–15.

—— (1874). Second Report of the Committee [...] Appointed for the Purpose of Assisting in the Exploration of the Settle Caves (Victoria Cave). *Report of the British Association for the Advancement of Science* **44** (Belfast), 133–138.

—— (1876). The Age of Palaeolithic Man. *Nature* **14**, 505–506.

—— (1878). On the Age of the Hyaena-Bed at the Victoria Cave, Settle, and its Bearing on the Antiquity of Man. *Journal of the Royal Anthropological Institute* **7**, 165–173.

Tonnochy, A. B. (1953). Four Keepers of the Department of British and Medieval Antiquities. *The British Museum Quarterly* **18**, 83–88.

Topley, W., and Jones, T. R. (1865). Obituary Notice of Henry Christy. *The Geological Magazine* **2**, 286–288.

Treacher, L. (1904). On the Occurrence of Stone Implements in the Thames Valley between Reading and Maidenhead. *Man* **4**, 17–19.

—— (1910). Excursion to Maidenhead. *Proceedings of the Geologists' Association* **21**, 198–201.

Treacher, M. S., Arkell, W. J., and Oakley, K. P. (1948). On the Ancient Channel between Caversham and Henley, Oxfordshire, and its Contained Flint Implements. *Proceedings of the Prehistoric Society* **14**, 126–154.

Trigger, B. G. (1989). *A History of Archaeological Thought.* Cambridge: Cambridge University Press.

Trimmer, J. (1831). On the Diluvial Deposits of Caernarvonshire, between the Snowdon Chain of Hills and the Menai Strait, and on the Discovery of Marine Shells in Sand and Gravel on the Summit of Moel Tryfane, near Caernarvon, 1000ft. above the level of the Sea. *Proceedings of the Geological Society of London* **1**, 331–332.

—— (1847). On the Geology of Norfolk as illustrating the Laws of the Distribution of Soils. *Journal of the Royal Agricultural Society of England* **7**, 444–485.

—— (1851a). Generalizations respecting the Erratic Tertiaries or Northern Drift, founded on the Mapping of the Superficial Deposits of a Large Portion of Norfolk. With a Description of the Freshwater Deposits of the Gaytonthorpe Valley; and a Note on the Contorted Strata of Cromer Cliffs. *Quarterly Journal of the Geological Society of London* **7**, 19–31.

—— (1851b). On the Origin of the Soils which cover the Chalk of Kent. *Quarterly Journal of the Geological Society of London* **7**, 31–38.

Trinkaus, E., and Shipman, P. (1994). *The Neandertals. Changing the Image of Mankind.* London: Pimlico.

Turner, C. (1970). The Middle Pleistocene Deposits at Marks Tey, Essex. *Philosophical Transactions of the Royal Society of London, Series B* **257**, 373–440.

—— and Gibbard, P. (1996). Richard West—An Appreciation. *Quaternary Science Reviews* **15**, 375–389.

Tweedale, G. (1991). Geology and Industrial Consultancy: Sir William Boyd Dawkins (1837–1929) and the Kent Coalfield. *British Journal for the History of Science* **24**, 435–451.

Tweedale, G. (2004). Dawkins, Sir William Boyd (1837–1929). In H. C. G. Matthew, & B. Harrison (Eds.), *Oxford Dictionary of National Biography* (vol. 15, pp. 540–542). Oxford: Oxford University Press.

Tylor, E. B. (1865). *Researches into the Early History of Mankind and the Development of Civilization*. London: John Murray.

—— (1869). The Condition of Prehistoric Races, as Inferred from Observation of Modern Tribes, *International Congress of Prehistoric Archaeology: Transactions of the 3rd Session, London, 1868* (pp. 11–26). London: Longmans, Green, and Co.

—— (1893). On the Tasmanians as Representatives of Palaeolithic Man. *Journal of the Royal Anthropological Institute* **23**, 141–152.

—— (1898). On the Survival of Palaeolithic Conditions in Tasmania and Australia, with Especial Reference to the Modern Use of Unground Stone Implements in West Australia. *Report of the British Association for the Advancement of Science* **68** (Bristol), 1014–1015.

Underwood, W. C. (1912). Recent Discoveries in Palethnology and the Works of Early Man. *Proceedings of the Prehistoric Society of East Anglia* **i**, 135–139.

—— Sturge, W. A., Clarke, W. G., Layard, N., and Corner, F. (1911). Report of the Special Committee. *Proceedings of the Prehistoric Society of East Anglia* **i**, 24–41.

Urey, H. C. (1947). The Thermodynamic Properties of Isotopic Substances. *Journal of the Chemical Society* **1947**, 562–581.

Vallois, H. V. (1941–46). Marcellin Boule. *L'Anthropologie* **50**, 203–210.

Vaufrey, R. (1962). L'Abbé Henri Breuil. *L'Anthropologie* **66**, 158–165.

Vauville, M. O. (1891). Instruments variés provenant des gisements quaternaires de Mont-Notre-Dame, Lime et Ciry (Aisne). *Bulletins de la société d'anthropologie de Paris* **2**, (fourth series), 343–363.

Virchow, R. (1872). Untersuchung des Neanderthal-Schädels. *Berliner Gesellschaft für Anthropologie, Ethnologie und Urgeschichte* **1871–1872** (27 April 1872), 4–12.

Wade, A. G., and Smith, R. (1935). A Palaeolithic Succession at Farnham, Surrey. *Proceedings of the Prehistoric Society* **vii**, 348–353.

Wallace, A. R. (1870). *Contributions to the Theory of Natural Selection*. London: Macmillan and Co.

Warren, S. H. (1900). Palaeolithic Flint Implements from the Chalk Downs of the Isle of Wight and the Valleys of the Rivers Western Yar and Stour. *The Geological Magazine* **7**, 406–412.

—— (1902). On the Value of Mineral Condition in Determining the Relative Age of Stone Implements. *The Geological Magazine* **9**, 97–105.

—— (1905). On the Origin of 'Eolithic' Flints by Natural Causes, Especially by the Foundering of Drifts. *Journal of the Royal Anthropological Institute* **35**, 337–364.

—— (1910). Arctic Plants from the Valley Gravels of the River Lea. *Nature* **85**, 206.

—— (1911). Excursion to Ponder's End and Chingford. *Proceedings of the Geologists' Association* **22**, 166–171.

—— (1912a). On a Late Glacial Stage in the Valley of the River Lea, Subsequent to the Epoch of River-Drift Man. *Quarterly Journal of the Geological Society of London* **68**, 213–226.

—— (1912b). Palaeolithic Remains from Clacton-on-Sea. *Essex Naturalist* **17**, 15.

—— (1913). Problems of Flint Fracture. *Man* **13**, 37–38.

—— (1914a). The Experimental Investigation of Flint Fracture and its Application to Problems of Human Implements. *Journal of the Royal Anthropological Institute* **44**, 412–450.

—— (1914b). Some Points in the Eolithic Controversy. *The Geological Magazine* Decade VI, Vol. I, 546–552.

—— (1916). Archaeology: Palaeolithic Art. A review of Sollas, W. J. (1915), 'Ancient Hunters and their Modern Representatives,' 2nd edition, London: Macmillan & Co. *Man* **16**, 79–80.

—— (1920). A Natural 'Eolith' Factory beneath the Thanet Sand. *Quarterly Journal of the Geological Society of London* **76**, 238–253.

—— (1922a). The Red Crag Flints of Foxhall. *Man* **22**, 87–89.

—— (1922b). The Mesvinian Industry of Clacton-on-Sea, Essex. *Proceedings of the Prehistoric Society of East Anglia* **iii**, 597–602.

—— (1923a). Sub-Soil Pressure-Flaking. *Proceedings of the Geologists' Association* **34**, 153–175.

—— (1923b). The Eolithic Problem: A Reply. *Man* **23**, 82–83.

—— (1923c). The Sub-Soil Flint Flaking Sites at Grays. *Proceedings of the Geologists' Association* **34**, 38–42.

—— (1923d). The *Elephas antiquus* Bed of Clacton-on-Sea (Essex) and its Flora and Fauna. *Quarterly Journal of the Geological Society of London* **79**, 606–636.

—— (1924a). The Flint Flakings of the Weybourne Crag. *The Geological Magazine* **61**, 309–311.

—— (1924b). Pleistocene Classifications. *Proceedings of the Geologists' Association* **35**, 265–282.

—— (1924c). Palaeolithic and Neolithic Implements from the Thames Valley and Elsewhere. *Essex Naturalist* **21**, 67–77.

—— (1924d). Visit to Dordogne, France. June 4th to June 11th, 1923. *Proceedings of the Geologists' Association* **35**, 135–141.

—— (1924e). The Elephant-Bed of Clacton-on-Sea. *Essex Naturalist* **21**, 32–40.

—— (1925). The Study of Flint Flaking. *Proceedings of the University of Bristol Speleological Society* **3**, 302–308.

—— (1926). The Classification of the Lower Palaeolithic with Especial Reference to Essex. *Proceedings and Transactions of the South-Eastern Union of Scientific Societies*, 38–50.

—— (1928). The Study of Comparative Flaking in 1927. *Man* **28**, 6–8.

—— (1929). Beach Flaking Sites in Somerset. *Man* **29**, 33–34.

—— (1932a). The Palaeolithic Industries of the Clacton and Dovercourt Districts. *Essex Naturalist* **24**, 1–29.

—— (1932b). The Palaeolithic Industry of Clacton-on-Sea, Essex. *Reprinted from the Proceedings of the First International Congress of Prehistoric and Protohistoric Sciences, London, 1932*, 69.

—— (1940). Geological and Prehistoric Traps. *Essex Naturalist* **27**, 2–19.

Warren, S. H. (1942). The Drifts of South-Western Essex, Part II. *Essex Naturalist* **27**, 171–179.

—— (1949). Alfred Santer Kennard. *Quarterly Journal of the Geological Society of London* **104**, lvii–lviii.

—— (1951). The Clacton Flint Industry: A New Interpretation. *Proceedings of the Geologists' Association* **62**, 107–135.

—— (1958). The Clacton Flint Industry: A Supplementary Note. *Proceedings of the Geologists' Association* **69**, 123–129.

Watkins, C. F. (1864). The Antiquity of Man. *The Northampton Herald*, 12th March.

Watson, W. (1950). *Flint Implements. An Account of Stone Age Techniques and Cultures.* London: Trustees of the British Museum.

—— (1956). *Flint Implements. An Account of Stone Age Techniques and Cultures* (2nd ed., revised by W. Watson). London: Trustees of the British Museum.

—— (1968). *Flint Implements. An Account of Stone Age Techniques and Cultures* (3rd ed., revised by G. Sieveking). London: Trustees of the British Museum.

Wayland, E. J. (1923). Palaeolithic Types of Implements in Relation to the Pleistocene Deposits of Uganda. *Proceedings of the Prehistoric Society of East Anglia* **iv**, 96–112.

Webster, D. B. (1991). *Hawkeseye. The Early Life of Christopher Hawkes.* Stroud: Alan Sutton.

West, R. G. (1954). The Hoxne Interglacial Reconsidered. *Nature* **173**, 187–188.

—— (1956). The Quaternary Deposits at Hoxne, Suffolk. *Philosophical Transactions of the Royal Society of London, Series B* **239**, 265–356.

—— (1963). Problems of the British Quaternary. *Proceedings of the Geologists' Association* **74**, 147–186.

—— (1981). Palaeobotany and Pleistocene Stratigraphy in Britain. *The New Phytologist* **87**, 127–137.

—— (1988). Harry Godwin: 9 May 1901–12 August 1985. *Biographical Memoirs of Fellows of the Royal Society* **34**, 259–292.

—— and Donner, J. J. (1956). The Glaciations of East Anglia and the East Midlands: A Differentiation based on Stone-orientation Measurements of the Tills. *Quarterly Journal of the Geological Society of London* **112**, 69–91.

—— and McBurney, C. B. M. (1954). The Quaternary Deposits at Hoxne, Suffolk, and their Archaeology. *Proceedings of the Prehistoric Society* **20**, 131–154.

Whitaker, W. (1864). *The Geology of parts of Middlesex, Hertfordshire, Buckingham, Berkshire, and Surrey* (*Sheet 7 of the Map of the Geological Survey of Great Britain*). London: Longman, Green, Longman, Roberts and Green.

—— (1875). *Guide to the Geology of London and the Neighbourhood.* London: H.M.S.O.

—— (1887). What is the Use of the Essex Field Club? *Essex Naturalist* **1**, 180–181.

—— (1889). *The Geology of London and of part of the Thames Valley* (2 vols.). Geological Survey Memoir. London: H.M.S.O.

—— (1902). Twelve Years of London Geology. *Proceedings of the Geologists' Association* **17**, 81–109.

—— and Marr, J. E. (1911). Report on the Geological Position of Bolton and Laughlin's Brick-pit. *Proceedings of the Prehistoric Society of East Anglia* i, 43.

White, M. (1997). The Earlier Palaeolithic Occupation of the Chilterns (Southern England): Re-assessing the Sites of Worthington G. Smith. *Antiquity* 71, 912–931.

—— (2000). The Clactonian Question: On the Interpretation of Core-and-Flake Assemblages in the British Lower Palaeolithic. *Journal of World Prehistory* 14, 1–63.

—— and Plunkett, S. (2004). *Miss Layard Excavates: A Palaeolithic Site at Foxhall Road, Ipswich, 1903–1905*. Bristol: Western Academic and Specialist Press.

Whitley, N. (1865). *The 'Flint Implements' from Drift, Not Authentic. Being a Reply to the Geological Evidences of the Antiquity of Man*. London: Longman, Green, Longman, Roberts, and Green.

—— (1874). Brixham Cavern. *The Globe*, 31st March.

Williams, D. (2004). Boswell, Percy George Hamnall (1886–1960), revised by Robert M. Shackleton. In H. C. G. Matthew, & B. Harrison (Eds.), *Oxford Dictionary of National Biography* (vol. 6, pp. 746–747). Oxford: Oxford University Press.

Wilson, H. E. (1985). *Down to Earth. One Hundred and Fifty Years of the British Geological Survey*. Edinburgh: Scottish Academic Press.

Wilson, L. G. (1998). Lyell: The Man and his Times. In D. J. Blundell, and A. C. Scott (Eds.), *Lyell: The Past is the Key to the Present* (pp. 21–37). London: Geological Society, Special Publication No. 143.

Wilson, T. (1899). The Beginnings of the Science of Prehistoric Anthropology. *Proceedings of the American Association for the Advancement of Science* 48, 309–353.

Wiltshire, T. (1862). On the Ancient Flint Implements of Yorkshire. *Proceedings of the Geologists' Association* 1, 215–226.

Wiseman, J. D. H., and Ovey, C. D. (1950). Recent Investigations on the Deep-Sea Floor. *Proceedings of the Geologists' Association* 61, 28–84.

Woldstedt, P. (1950). Comparison of the East Anglian and Continental Pleistocene. *Nature* 165, 1002–1003.

Wood, S. V. (1866). On the Structure of the Thames Valley and of its Contained Deposits. *The Geological Magazine* 3, 57–63; 99–107.

—— (1867). On the Structure of the Postglacial Deposits of the South-East of England. *Quarterly Journal of the Geological Society of London* 23, 394–417.

—— (1868). Synchronous Age of the Grays and Erith Brickearths. *The Geological Magazine* 5, 534.

—— (1870a). On the Relation of the Boulder-clay, without Chalk, of the North of England, to the Great Chalky Boulder-clay of the South. *Quarterly Journal of the Geological Society of London* 26, 90–111.

—— (1870b). Observations on the Sequence of the Glacial Beds. *The Geological Magazine* 7, 17–22.

Wood, S. V. (1880). The Newer Pliocene Period in England. *Quarterly Journal of the Geological Society of London* 36, 457–528.

Wood, S. V. and Rome, J. L. (1868). On the Glacial and Postglacial Structure of Lincolnshire and South-east Yorkshire. *Quarterly Journal of the Geological Society of London* **24**, 146–184.

Woodward, B. B. (1904). Obituary. John Allen Brown. *Annual Report of the Ealing Natural Science and Microscopical Society* **27**, xiv–xvii.

Woodward, H. (1878). Eminent Living Geologists. No. 3. Professor John Morris. *The Geological Magazine* **Decade II, Vol. V**, 481–487.

—— (1885). Obituary. Searles V. Wood, the Younger. *The Geological Magazine* **Decade III, Vol. II**, 138–142.

—— (1913). Eminent Living Geologists: James Geikie. *The Geological Magazine* **Decade V, Vol. X**, 241–248.

—— (1916). Eminent Living Geologists. John Edward Marr. *The Geological Magazine* **Decade VI, Vol. III**, 289–295.

Woodward, H. B. (1878). Review of Skertchly, S. B. J. (1877), 'The Geology of the Fenland.' *The Geological Magazine* **Decade II, Vol. V**, 230–233.

—— (1885). Robert Alfred Cloyne Godwin-Austen. *The Geological Magazine* **Decade III, Vol. II**, 1–10.

—— (1886). The Glacial Drifts of Norfolk. *Proceedings of the Geologists' Association* **9**, 111–129.

—— (1907). *The History of the Geological Society of London*. London: Geological Society.

—— (1909). *The Geology of the London District*. Geological Survey Memoir. London: H.M.S.O.

—— (1922). *The Geology of the London District* (2nd ed., revised by C.E.N. Bromehead). Geological Survey Memoir. London: H.M.S.O.

—— Bennett, F. J., Skertchly, S. B. J., and Jukes-Browne, A. J. (1891). *The Geology of parts of Cambridgeshire and of Suffolk (Ely, Mildenhall, Thetford)*. Geological Survey Memoir. London: H.M.S.O.

Wooldridge, S. W. (1957). Some Aspects of the Physiography of the Thames Valley in Relation to the Ice Age and Early Man. *Proceedings of the Prehistoric Society* **23**, 1–19.

Worsaae, J. J. A. (1859). Works of Art in the Drift. *The Athenaeum* **1679**, 889–890.

Wright, T. (1859). Flint Implements in the Drift. *The Athenaeum* **1651**, 809.

Wright, W. B. (1937). *The Quaternary Ice Age* (2nd ed.). London: Macmillan and Co.

Wrigley, A. (1949–51). Alfred Santer Kennard, 1870–1948. *Proceedings of the Malacological Society of London* **28**, 4–5.

Wyatt, J. (1862a). On some Further Discoveries of Flint Implements in the Gravel. *Quarterly Journal of the Geological Society of London* **18**, 113–114.

—— (1862b). Flint Implements in the Drift in Bedfordshire. *Notes of the Bedfordshire Architectural and Archaeological Society* **10**, 145–156.

—— (1864). Further Discoveries of Flint Implements and Fossil Mammals in the Valley of the Ouse. *Quarterly Journal of the Geological Society of London* **20**, 183–188.

Zeuner, F. E. (1937). A Comparison of the Pleistocene of East Anglia with that of Germany. *Proceedings of the Prehistoric Society* 3, 136–157.

—— (1944). Review of the Chronology of the Palaeolithic Period, *Conference on the Problems and Prospects of European Archaeology* (pp. 14–19). London: University of London, Institute of Archaeology.

—— (1945). *The Pleistocene Period, its Climate, Chronology, and Faunal Successions.* London: Printed for the Ray Society.

—— (1946). *Dating the Past: An Introduction to Geochronology.* London: Methuen.

—— (1959). *The Pleistocene Period, its Climate, Chronology, and Faunal Successions* (2nd ed.). London: Hutchinson Scientific & Technical.

Index